Structure and Dynamics of Biopolymers

NATO ASI Series

Advanced Science Institutes Series

A Series presenting the results of activities sponsored by the NATO Science Committee, which aims at the dissemination of advanced scientific and technological knowledge, with a view to strengthening links between scientific communities.

The Series is published by an international board of publishers in conjunction with the NATO Scientific Affairs Division

A	**Life Sciences**	Plenum Publishing Corporation
B	**Physics**	London and New York
C	**Mathematical and Physical Sciences**	D. Reidel Publishing Company Dordrecht, Boston, Lancaster and Tokyo
D	**Behavioural and Social Sciences**	Martinus Nijhoff Publishers Dordrecht, Boston and Lancaster
E	**Applied Sciences**	
F	**Computer and Systems Sciences**	Springer-Verlag Berlin, Heidelberg, New York
G	**Ecological Sciences**	London, Paris and Tokyo
H	**Cell Biology**	

Series E: Applied Sciences – No. 133

Structure and Dynamics of Biopolymers

edited by:

C. Nicolini

Chair of Biophysics
School of Medicine
University of Genova
Genova
Italy

1987 **Martinus Nijhoff Publishers**
Dordrecht / Boston / Lancaster
Published in cooperation with NATO Scientific Affairs Division

Proceedings of the NATO Advanced Study Institute on "Biopolymers Structure and Dynamics", Erice, Italy, June 22-July 4, 1986

Library of Congress Cataloging in Publication Data

NATO Advanced Study Institute on "Biopolymer Structure and Dynamics" (1986 : Erice, Sicily)
Structure and dynamics of biopolymers.

(NATO ASI series. Series E, Applied sciences ; no. 133)
"Published in cooperation with NATO Scientific Affairs Division."
"Proceedings of the NATO Advanced Study Institute on 'Biopolymer Structure and Dynamics,' Erice, Italy, June 22-July 4, 1986"--T.p. verso.
Includes index.
1. Deoxyribonucleic acid--Structrue--Congresses.
2. Biopolymers--Structure--Congresses. 3. Molecular dynamics--Congresses. I. Nicolini, Claudio A.
II. NATO Scientific Affairs Division. III. Series.
QP624.N368 1986 574.87'3282 87-11291

ISBN-13:978-94-010-8120-7 e-ISBN-13:978-94-009-3619-5
DOI:10.1007/978-94-009-3619-5

Distributors for the United States and Canada: Kluwer Academic Publishers, P.O. Box 358, Accord-Station, Hingham, MA 02018-0358, USA

Distributors for the UK and Ireland: Kluwer Academic Publishers, MTP Press Ltd, Falcon House, Queen Square, Lancaster LA1 1RN, UK

Distributors for all other countries: Kluwer Academic Publishers Group, Distribution Center, P.O. Box 322, 3300 AH Dordrecht, The Netherlands

PREFACE

This NATO-ASI on BIOPOLYMERS STRUCTURE AND DYNAMICS held between 22nd June - 4th July 1986 at Erice (Italy) has brought together scientists from a broad variety of biophysical disciplines - polymer physics, biophysics and physical chemistry, structure and dynamics of polynucleotides, proteins, and polysaccharides - to present the current state of knowledge in their fields, both experimental and theoretical. This Advanced Study Institute was indeed a successfull attempt to enhance the possibility of intersection of a number of research lines that currently are progressing well but are still running largely in parallel with one another: protein folding, single-polymer phase transitions, DNA condensation into liquid crystalline-like arrays, packaging in viruses, and polysaccharide gel formation. Although each phenomenon is distinctive, an awareness of similarities may lead to new ideas.
The program has emphasized "condensed" forms of biopolymers. We are universally confronted in biology by chain polymers folded on themselves or interlinked in gel-like assemblies, whether we look at the native structure of proteins, the role of polysaccharides in connective tissue, or the genetic apparatus. A number of lectures have been devoted to condensed forms of DNA - closed circular supercoils, toruses, chromatin. The physical principles that govern protein folding received attention in the program. Polymer physicists have lectured on phase transitions and gels. The annihilation of polyelectrolyte repulsion as a factor promoting folding and elastic instability has been analyzed. In the case of DNA it has been proposed that folding and binding of small molecules may occur at the sites of transiently opened base pairs, so we have included in the program expert representation on various aspects of these subjects. Lectures on DNA toruses, polysaccharide-polynucleotide analogies, the higher order structure of polynucleosome array and dynamic characteristics of polynucleotide complexes have been finally included in order to ensure a wide participation of researchers in these areas.

Several prominent senior scientists have offered main lectures in the course and all have also given specific contributions to daily workshops and tutorials. I am grateful to all, but particularly to Professor Gerald Manning of Rutgers University for his active co-leadership in the conduction of this course at Erice and for the outstanding unique contribution given in planning and organizing such a complex Institute with participants coming from universities, research institutes and research laboratories of industries and organizations selected on the basis of geographic criteria and scientific background.

Claudio Nicolini

CONTENTS

Participants of the NATO ASI on "Biopolymers Structure and Dynamics"
Erice (Italy), June 22 - July 4, 1986

DYNAMICS OF POLYNUCLEOTIDES AS PROBED BY INTERACTING IONS AND MOLECULES

M. DAUNE

I.B.M.C. du CNRS et Université Louis Pasteur, 15 rue René Descartes, 67084 STRASBOURG Cedex, France.

INTRODUCTION

An increased number of in vitro studies have revealed the role played by local or extended conformational movements at the molecular level in the biological role of both proteins and nucleic acids. Specific recognition between enzyme and substrate, mechanisms of procaryotic transcription, antigenic efficiency of macromolecular domains were the first examples in which a definite role of the molecular dynamics can be assigned to at least a part of the biological process. The original concept of a unique relationship between structure and function, which was one of the building stone of molecular biology, is giving progressively the way to another one. If the fitting of two structures remains one essential prerequisite in any biological event at the molecular level, it has now to be completed by taking into considerations the respective movements of the partners. In other words specificity and efficiency which are the characteristics of any biological event have now to be understood not only in terms of structure but also in terms of kinetics.

Therefore, in any attempt to understand the role of molecular dynamics, a structural study has first to be made then followed by a precise evaluation of the kinetic processes. This is the way we will proceed in this lecture devoted to the use of molecules and ions as probes of DNA dynamics. Therefore we will consider successively :

I) The main features of the structure of B-DNA
II) The use of probes already attached to DNA
III) The study of reaction kinetics between some molecules and DNA
IV) A comparison between the data thus obtained and those gained from a direct study of dynamics of DNA.

I. STRUCTURE OF B-DNA

The primary structure of a nucleotidic chain is well known and made up of a successive array of 5'-nucleotide and 3'-nucleotides, i.e. of alternating sugar and phosphodiester. A monomeric unit is thus defined as comprised between a 5'-phosphate and a 3'-phosphate.

In fig. 1 the different torsional angles around the successive covalent bonds are named according to Sundaralingam together with another convention which is using the six first letters of the greek alphabet (α to ζ). In both systems χ represents the rotation angle about the C-N bond between the sugar and the base. The sugar is a puckered five-membered ring. Each atom of the ring can be displaced ("above" or "under" a reference plane) or two of them can be found alternatively up and down. Twenty possible conformations are finally obtained which can be placed at an angular distance of 18° on a circle (pseudo rotation). Each of the five endocyclic torsional angles τ can then be expressed in terms of the amplitude τ_m and the phase P of the pseudorotation and a continuous variation of τ could be imagined.

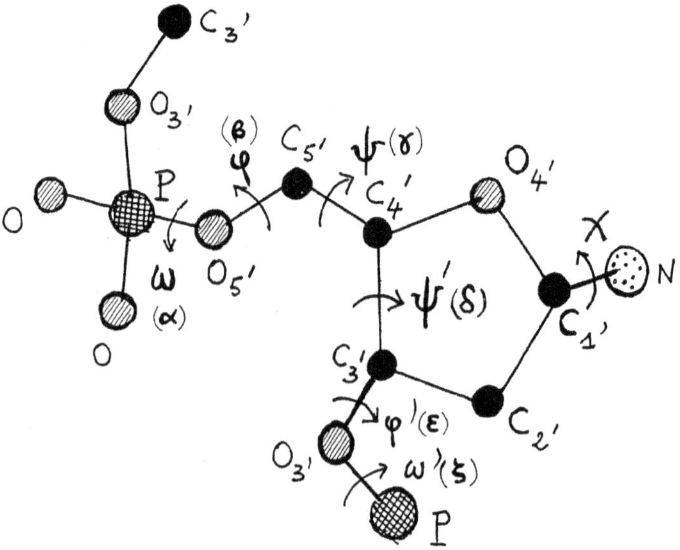

FIGURE 1. Torsional angles of the nucleotidic chain

Actually two major conformations occur which are named C2' endo around P = 180° and C3' endo around P = 0°, according to the displacement of the corresponding carbon atom toward the C5' out-of-plane carbon. In double stranded DNA, the A and B form are associated to C3' endo and C2' endo respectively. The normal pathway to go from one conformation to the other is via the O4' endo state (P = 90°) and through a potential barrier estimated to about 4 to 5 KCal/mole. In solution one thus expect to find the sugar in dynamic equilibrium between these two major conformations. We will see later this conformational fluctuation of the deoxyribose to be at the origin of internal movements of DNA molecule. To have a better understanding of DNA structure and dynamics we have to examine carefully, for each of the seven angles, the range of values which are usually found and their stereochemical significance.

- ψ'(δ) - The rotation about the C4' - C3' is determined by the conformation of the sugar in the restricted range 70° - 170° because of the ring closure. The variation of ψ' with the phase P of the pseudorotation is given in fig. 2. The C3' endo and C2' endo domains have ψ' values in the range 80-90 (g^+) and 140-150 (t) respectively.
- χ - It defines the orientation of the base relative to the sugar. In the range $-90° < \chi < 90°$ the base is said to be in anti, but syn when χ is close to 180°. There is a correlation between χ and the sugar puckering. In the anti domain, the range 0°-10° corresponds to a C3' endo conformation and the range 30°-80° to C2' endo ("low" χ and "high" χ).
- ψ(γ) - The torsion around the C4'-C5' bond defines the respective orientation of O5' and sugar. Three positions (g^+, t, g^-) can be adopted but g^+ is the most observed and in this case O5' can interact with the hydrogen atom bound to C6 of pyrimidines and C8 of purine, when the base is in anti.

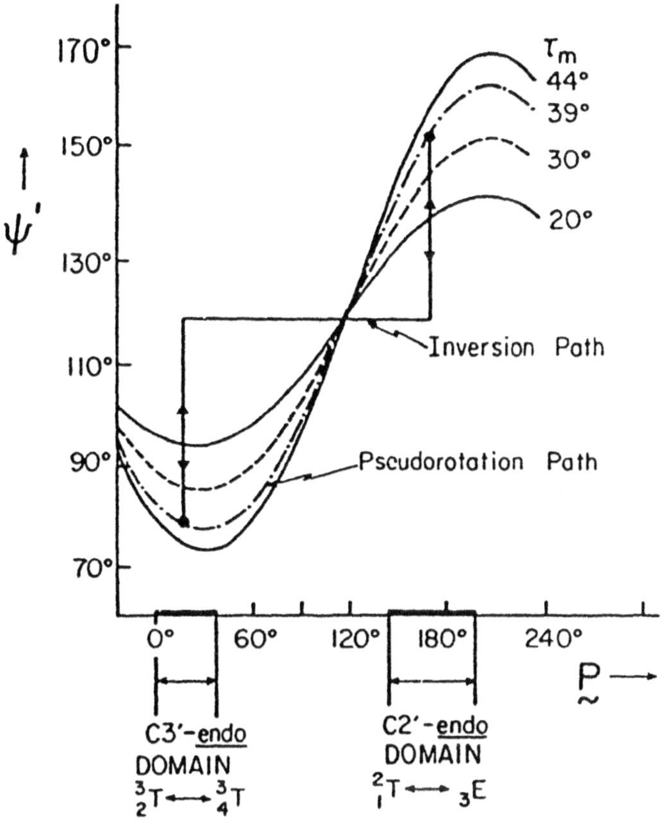

FIGURE 2. Variation of psi' with P (phase) and Tau_m (amplitude) of pseudorotation

- $\phi(\beta)$ and $\phi'(\varepsilon)$ - These torsion angles around C-O bonds (5' and 3' sides remain generally close to 180°, i.e. in trans position.
- $\omega(\alpha)$ and $\omega'(\zeta)$ - Variations of these two angles i.e. torsional movements around the P-O bonds are able to generate a great variety of topological forms of nucleic acids. A right handed helix will correspond to a $g^- g^-$ couple and a left handed to a $g^+ g^+$ one. In the latter case however it is impossible, for steric hindrance, to build a left handed double helix with all the ω and ω' angles in g^+ (see table I). Couples like g^-t, tg^+ and $g^- g^+$ create turns along the polynucleotidic chains, while tg^- or g^+t initiate local reorientation of the helix (kink). The PO 3' side (ω') is more flexible than the PO 5' side (ω) in the case of DNA. Actually and because PO 3' bond is almost parallel to the helix axis in B DNA, any change of ω' will correspond to a rotation of the base plane around the helix axis. Correlation between ω and ψ in one hand and ω' and ψ' in the other have been verified in recent cristallographic structure of short B helices and tRNA. Generally a decrease of ω is correlated with an increase of ψ and the same is true for the other (ω', ψ') couple (fig. 3). In view of this coupling between ω' and ψ', any variation of ψ' (i.e. of sugar puckering) will lead to a change of winding angle (the angle between two successive base pairs).

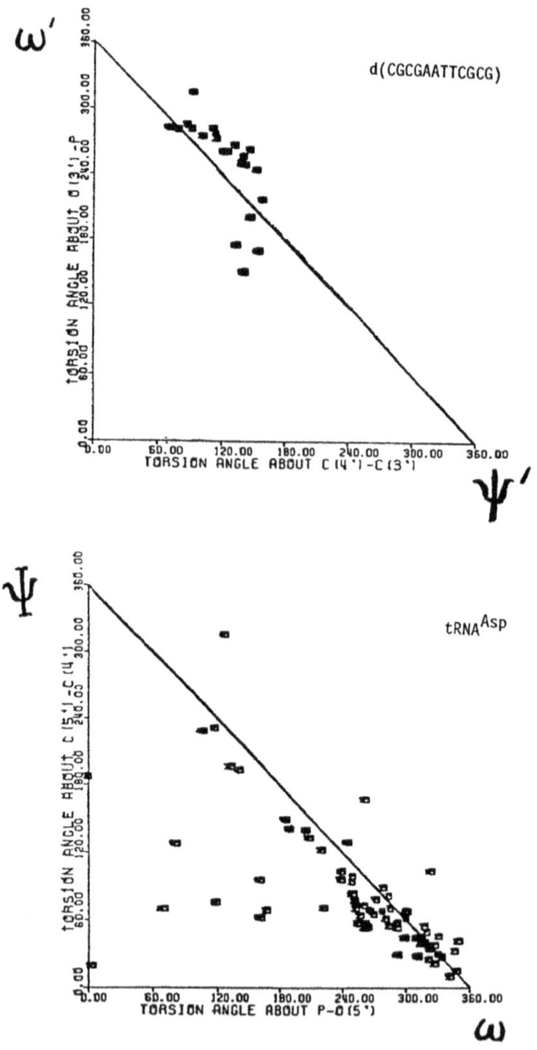

FIGURE 3. Correlation between torsional angles (each square corresponds to one phosphate)

According to Fig. 2, the range of ψ' variations is much larger inside the C2' endo domain than inside the C3' endo. A polymorphism in term of winding angle is thus expected to be higher in DNA than in RNA. These strong correlation (ω, ψ) and (ω', ψ') are easily explained by the trans orientation of ϕ and ϕ'.

From these stereochemical considerations two conclusions can be outlined :

1) According to conformational range for each of the seven angles of the

nucleotidic chain, a classification of DNA helices can be made (Fig. 4).
2) In terms of conformational dynamics, the B-DNA presents specific features :

a) due to the absence of a 2'-OH group in deoxyribose, the potential barrier between C3' endo and C2' endo is lower and the C2' endo sugar puckering is a preferred conformation. Because of the large domain of ψ' in the C2' endo conformation (Fig. 2) and the correlation between ψ' and ω', the B-DNA helix can undergo large variations of its winding angle.

b) Due to an increased probability of intermediate transient puckering states of the sugar (C1' exo, O4' endo, ...), the syn <—> anti transconformation of the base is facilitated.

For these two reasons, the dynamic flexibility of B-DNA is much higher than that of RNA which explains the interest of studying any interaction with DNA in terms of local dynamic specificity.

ANGLES	B-DNA	A-DNA	Z-DNA		IODO CpG WITH INTERCALATED EtBr		"G*"
			C	G	C	G	
ω	g^-	g^-	g^-(-129)	g^+(48)		g^-	t
ϕ	t - g^+	t	t *(-139)	t (179)		t	
ψ	g^+	g^+	g^+(56)	t (-170)	g^+	g^+	t
ψ'	t	g^+	t-g^+(138)	g^+ (100)	g^+	t-g^+	
ϕ'	t *	t	g^-(-94)	g^-(-108)	t		
ω'	t *	g^-	g^+ (80)	g^- (-70)	g^-		
χ	high anti	low anti	low anti	syn	low anti	high anti	

t * = in the range 200, 300 (-160, -60)

FIGURE 4. Types of helices and their corresponding structural ranges of torsional angles

II. USE OF PROBES ALREADY ATTACHED TO DNA

The idea is simply to introduce inside the helical structure ions or molecules which are physically or chemically linked to DNA molecule. These probes have of course to disturb as small as possible the original structure and dynamics of the double helix. They also must have physical properties which can be followed easily, independently of those of DNA. Monitoring their behaviour through a convenient physical device will be revealing the movements of the DNA molecule they are supposed to undergo at the same time.

1. Fluorescence polarisation anisotropy (FPA) and torsional rigidity

As a rough but illustrative manner to treat torsional internal movements of the double helix, one could consider a cylindrical model. Cylinders of length 2 L and radius a are randomly rotating in a medium of viscosity η_o. When the rotation about the helix axis is only taken into account, the rotational brownian diffusion coefficient D_1 is given by the classical expression :

$$D_1 = \frac{kT}{8 \Pi \eta_o a^2 L} = \frac{kT}{4 V \eta_o}$$

where V is the volume of the cylinder. The corresponding relaxation time $\tau_1{}^o = (4 D_1)^{-1}$ could represent as a first approximation the torsional movement of a fragment of DNA, the length L of which is supposed to be negligible as compared to the total length of the DNA molecule.
With the geometry of B-DNA (radius of 12.5 Å and 3.4 Å distance between adjacent base pairs) $\tau_1{}^o$ is a linear function of the number N of base pairs according to the approximate relation (at 20°C and in a medium with a viscosity $\eta_o = 10^{-3}$ MKSA), $\tau_1{}^o \sim 0{,}4$ N (ns).

One thus would expect to detect torsional movements of DNA segments in the nanosecond range which corresponds precisely to the life time of many fluorescent molecules. This was the basic idea of the pioneering work of Wahl, Paoletti and Le Pecq (1970) that is to use an intercalated dye as a probe of the internal motion of DNA. Indeed in their experiment the polarization coefficient p_o of the intercalated ethidium bromide was found to be equal to that measured in a rigid medium ($p_o \sim 0.410$). In a first approximation the dye molecule can therefore be viewed as embedded rigidly in the helix and able to follow its torsional movement with a relaxation time found equal to 28 ns.

Following this first and somewhat qualitative approach, a series of experimental and theoretical papers have appeared during the last ten years in order A) to measure more accurately the time-resolved fluorescence polarisation anisotropy and to obtain accurate relaxation times, B) to correlate experimental data with molecular parameters e.g. the torsional and the bending rigidity of DNA double-helix.

A. In a recent paper (Genest et al., 1985), a comparison was made between expected values of FPA calculated by assuming the ethidium bromide originally bound to a cylinder (or an ellipsoïd) modeling short DNA fragments. In this case the brownian rotation about a transverse axis (perpendicular to the helix axis) has also to be considered. The corresponding rotational diffusion coefficient D_2 is given for a cylinder by Broersma (1960).

$$D_2 = \frac{3 kT}{8 \Pi \eta_o L^3} \left[Ln \frac{2L}{a} - 1.57 + 7 \left(\frac{1}{2 Ln 2L/a} - 0.28 \right)^2 \right]$$

alid only when Ln 2L/a > 1.8 (i.e. when L/a > 3). With lower values of axial ratio, a spherical model could be used leading to

$$D_2 = \frac{kT}{6 \eta_o V} = \frac{kT}{8 \Pi \eta_o L^3}$$

Assuming the transition moment of the dye lying in a plane perpendicular to the helix axis, two relaxation times τ_1 and τ_2 will describe the process (see table I for a series of calculated values)

$$\tau_1{}^{-1} = 4 D_1 + 2 D_2 \qquad \tau_2{}^{-1} = 6 D_2$$

N MER	MONO	DI	HEXA	DODECA	x 20	x 100	x 200
L (Å)	3.4	6.8	20.4	40.8	68	340	680
D_1 (s^{-1})	6.06×10^8	3.03×10^8	1×10^8	5.05×10^7	3.03×10^7	6.06×10^6	3.03×10^6
D_2 (s^{-1})			SPHERE		CYLINDER (COMPLETE FORMULA)		
			6.7×10^7	3.3×10^7	1.54×10^6	1.81×10^5	1×10^5
τ_1	0.4 ns	0.8 ns	2.5 ns	5 ns	8 ns	41 ns	78 ns
			($1/6D_2$)				
τ_2					108 ns	0.9 μs	1.67 μs

TABLE 1. Rotational brownian movements of rod-shaped DNA fragments ($\eta_0 = 10^{-3}$ MKSA, T = 293 K)

Actually the expected variation of FPA values with increasing DNA length does not represent the experimental data. In Fig. 5 the discrepancy becomes apparent for about a 20 bp DNA fragment and increases with the length suggesting a fast elastic torsional movement.

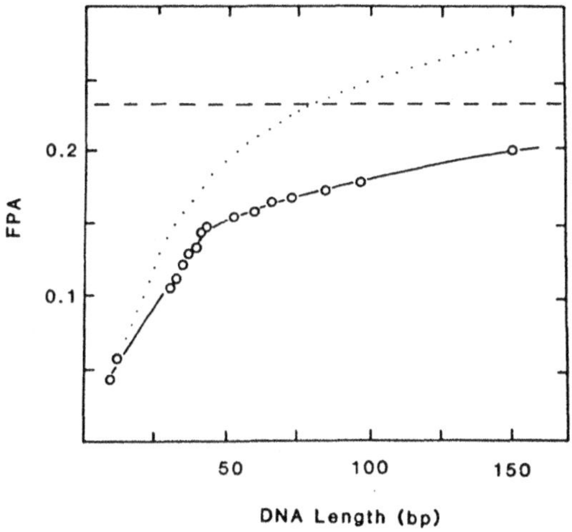

FIGURE 5. The effect of helix length on the <FPA> of EB-DNA complexes. The samples were obtained by DNase digestion and fractionation and were analyzed by gel electrophoresis. The solid line connects the experimental data, the dotted line shows the <FPA> expected if the depolarization were due only to helix tumbling, and the broken line shows the <FPA> observed for high molecular weight DNA at 28°C (Genest et al., 1985).

However the limiting value (0.37) of FPA when $n_0 \to \infty$ is lower than the theoretical one (0.41) indicating a rapid wobbling ($\sim$ 100 ps) of the dye inside its intercalation site. Finally three relaxation times are necessary to describe the complete motion of the dye. The rapid initial decay is attributed to the wobbling. A τ value very close to that found previously ($\sim$ 25 ns) corresponds to the relative torsional movement of base pairs. The largest τ (long time decay) would appear to characterize the bending movement of large enough DNA fragments.

Through the different experimental approaches, some features have to be pointed out :

a) Relaxation times cannot be attributed to exchange between bound and free probes but they characterize torsional movements of the DNA (Robinson et al., 1980).

b) These torsional movements do not occur primarily at sites of high enthalpy perturbed structure (open base pairs, strand breaks, kinks, ...) but likely reflect the movement of the whole helix structure (Thomas and Schurr, 1983).

c) In the limit of experimental errors no distinction could be made between GC or AT rich regions.

d) When a 146 pb DNA is wrapped around the core histone to define the chromatin subunit called a core particle, the same type of torsional movements is still present at least in some regions of the DNA (Genest et al., 1982). However this local motion disappears in condensed chromatin (Ashikawa et al., 1983).

e) In recent experiment (Genest et al., 1985) the FPA of ethidium probe molecules (about 1 dye per 200 base pairs) does not depend on a high level (up to 1 molecule per 2 pb) of another intercalated dye (proflavine).

One could conclude that either the dynamical characteristics of the DNA molecule are unaffected by high level of an intercalated molecule or the FPA is only correlated to highly localized motions of the base pairs.

B. In the theoretical approach the problem is to solve a diffusion equation for the relative twist angle α between two successive cross-sections of the rod-shaped DNA molecule.

The first difficulty is the choice of a convenient set of orthogonal coordinates in order to describe the process and to find the distribution function $F(\alpha,t)$. The second one is to take care properly of the bending and of the hydrodynamical and mechanical correlation between twisting and bending (Fig. 10).

In the first model (Barkley and Zimm, 1979) one dye per persistence length is assumed to be bound in average and situated in the middle of the rod-shaped segment. The bending displacements occuring in the fluorescence life-time are small and uncoupling of bending and twisting can be assumed. Finally component rotations are relative to a frame of reference attached to the helix.

In the last model (Schurr, 1984) the diffusion equation is obtained with Euler coordinates and corresponds therefore to a flexible macromolecule. In this case the rotation of rod-shaped subunits depends on both the overall tumbling of the DNA molecule and its flexibility. The effective rotational diffusion coefficient of a given rod is initially higher than in the case of a rigid macromolecule but decreases with time toward a limiting value. In other words, the rotational diffusion is a non-stationary Markov process.

The final result is the function $r(t)$ which represents the evolution with time of the FPA. Generally $r(t)$ appears as the product of twisting correlation functions (involving torsion and bending of the chain) and tumbling correlation functions. The analytical expression is too complicated to preclude any direct calculation but rather experimental behaviour of FPA is compared to the predicted one in different time ranges.

From this comparison the main parameter which is determined is the torsional rigidity of the DNA molecule. It is more often expressed in erg cm and related to a unit length. It is more convenient to use a torsional coefficient C in joules mole^{-1} (degree)$^{-2}$ or kcal mole^{-1} (degree)$^{-2}$ which let us possible to determine the torsional potential energy E_T between two successive base pairs with a variation α of their twist angle, according to

$$E_T \ = \ 1/2 \ C\alpha^2$$

It is easy to show that C is obtained (in KCal mole^{-1} (degree)$^{-2}$ by multiplying the torsional rigidity by 1.29 10^{17}. The following values are thus obtained from the literature :
0.022 or 0.053 (Barkley and Zimm, 1979), 0.017 (Thomas et al., 1980), 0.04 (Robinson et al., 1980), 0.0053 (Genest et al., 1982), 0.026 (Ashikawa et al., 1983), 0.023 or 0.038 (Allison and Schurr, 1979).

2. Electronic paramagnetic resonance of spin-labeled probe

Instead to follow the movement of the intercalated dye by measuring the anisotropy decay r(t) of the fluorescence it was also possible to covalently attach a spin-labeled probe to an intercalated molecule (Fig. 6) (Robinson et al., 1980) and then to analyse the EPR spectrum in terms of movements of the probe.

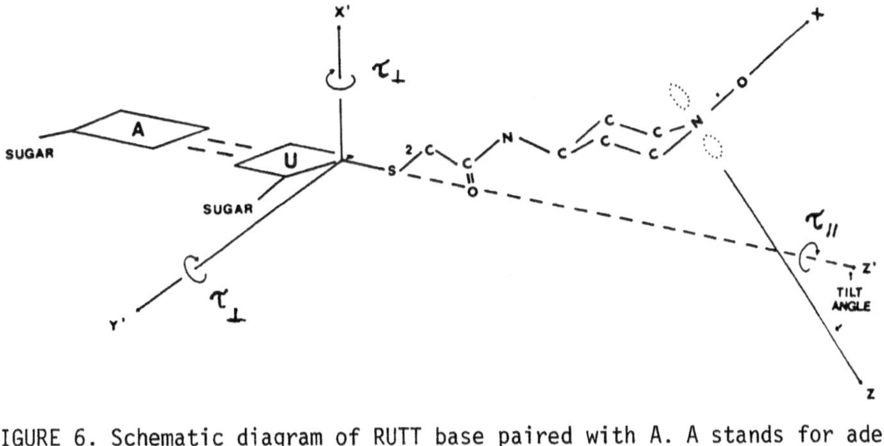

FIGURE 6. Schematic diagram of RUTT base paired with A. A stands for adenine and U for uridine. The principal rotational diffusion axis z' is tilted by an angle with respect to the molecular hyperfine principal z axis. τ is the correlation time for rotation about the z' axis, and τ is the correlation time for rotation about an axis (x' or y' axis) perpendicular to the z' axis. For simplicity the methyl groups of the nitroxide ring are not shown (Kao et al., 1985).

Experiments were made with DNA fibers or condensed DNA phase in presence of polyoxyethylene glycol. Analysis is again a best fitting between an experimental spectrum and a calculated one, by using three adjustable parameters : (i) the rotational correlation time around the helix axis, (ii) the tilt of the nitroxyl radical i.e. its orientation relative to the intercalator plane assumed to be perpendicular to the helix axis, (iii) the fraction of disordered helices into the fiber.

With acridine derivatives correlation times between 0.2 and 9 ns are found at room temperature. They have to be compared to a correlation time of 0.05 ns when the probes are free in solution. Higher value (> 30 ns) is found with propidium, suggesting that the faster moving probes are those linked to acridine rings which thus appear as more loosely coupled to base pairs than propidium or ethidium ring.

3. Nature of the counter-ion and circular dichroism

This approach differ markedly from the preceding ones since the probe which is used displays no specific physical property (fluorescence or magnetism). One observes instead a perturbation of the CD spectrum of the DNA molecule when the counter-ion concentration is modified or at a given ionic-strength with changes in temperature in a range well below the melting temperature (Gennis and Cantor, 1972). This so-called "premelting" could also be followed by other techniques (see Palecek, 1976). This increase of the CD signal with temperature can be correlated either with the decrease of the number of conformational states which are explored by the B-helix or with a gradual decrease of the average twist angle between two successive base pairs. Evidences were given for the latter model (Ivanov et al, 1974) but any precise relationship between CD values and local structure is lacking. The model is also supported by the similarity of CD changes induced by dye binding (intercalated or non intercalated) to poly d(AT) poly d(AT) and those induced by increasing temperature in the premelting range (Das et al., 1980). It was thus interesting to look for the role of the counterion to control the thermal dynamics of the B-DNA. The problem was adressed by using tetramethylammonium ions (TMA^+) (de Murcia et al., 1978) and the results can be briefly summarized.

a) Molecular ellipticity at 280 nm (θ_{280}) is known to strongly decrease from about 8,000 in 10^{-2} M NaCl to 2,500 in 4 M NaCl. In contrast θ_{280} is almost insensitive to ionic strength in the range 10^{-2} M (θ_{280} = 8,600) to 10 Molal (θ_{280} = 8,200) TMA^+ ions (Table 2).

b) Premelting can be viewed as reflecting the capacity of B-DNA to modify its secondary structure (i.e. the helix parameters) according to external conditions (ionic strength, temperature, nature of the ions, etc.).

Deoxypolynucleotide	Solvent a)	λ (nm)	$\theta_{\lambda}^{25°C}$	$\Delta\theta_{\lambda}/\Delta T$
CT DNA	TMACl 0.01 M	280	8600 ± 200 c)	26
	TMACl 0.1 M	280	8400	12
	TMACl 1 M	280	8600	0
	TMACl 10 molal	280	8200	< 0
	b) ACl 1 M	280	4300	32
	c) MACl 1 M	280	7600	13
	d) TEACl 1 M	280	9200	44
Poly (dA) poly (dT)	TMACl 0.1 M	282	4200	83
Poly (dA.dT) poly (dT.dA)		262	7700	33
EC DNA		280	8700	4
ML DNA		280	9600	4
Poly (dG.dC) poly (dC.dG)		280	7600	65

TABLE 2. Effect of tetraalkylammonium salts on CD parameters of DNA's and deoxypolynucleotides

c) Only in presence of TMA ions and with DNA there is no more variation of θ_{280} with temperature (Fig. 7). Premelting, as viewed through CD measurements, is thus abolished and this specificity of TMA^+ to control the process was related to its unique property to fit snugly into the small groove and to be coordinated to the O_2 of thymine (de Murcia et al., 1978). NMR experiments have confirmed this model.

It is thus tempting to give it a role of strongly decreasing (if not blocking) the torsional fluctuations of the double helix (as described by ω').

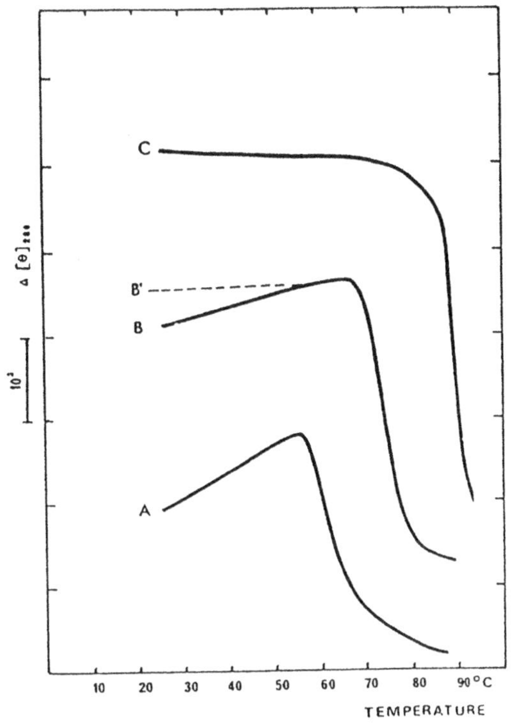

FIGURE 7. Temperature dependence of the molar ellipticity at 280 nm of CT DNA in different concentrations of TMACl and in the presence of 1 mM phosphate buffer pH 7, 0.2 mM EDTA : A (0.01 M); B (0.1 M); B' (0.1 M unbuffered); C (1 M). For the sake of clarity, CD melting profils have been arbitrarily translated along the θ axis. The scale on the ordinate represents a Δθ value of 10^3 degree cm^2 $dmole^{-1}$.

III. REACTION KINETICS BETWEEN MOLECULES AND DNA

The general idea is to take advantage of a chemical or physical binding to detect the dynamics of the DNA through a careful analysis of the reaction kinetics. We will examine successively two covalent bindings (formaldehyde and a carcinogen) and the physical binding of intercalated or non intercalated molecules.

1. Formaldehyde

Kinetic studies of the covalent binding to nucleic acids groups (sugar or bases) of reactive molecule as a probe of the dynamic state of the polymer was first developed in the case of formaldehyde (HCHO) (Utiyama and Doty, 1971, Von Hippel and Wong, 1971, Frank-Kamenetskii and Lazurkin, 1974, Mc Ghee and Von Hippel, 1975). According to a model initially proposed in the case of proton exchange (Printz and Von Hippel, 1965) HCHO can only react with an "open" state of the base pairs.

The term of "breathing" coined for this fast exchange between closed and open structure (Printz and Von Hippel, 1965) is represented by the simple conformational change

$$C \underset{k_c}{\overset{k_o}{\rightleftharpoons}} O$$

in which k_o and k_c are the rate constants for opening and closing respectively.

For the chemical reaction with the open state a series of assumptions have to be made :

1) Only one "open" state has to be considered.

2) The kind of chemical reactions observed when the reactive group is into the macromolecule is similar to that observed when the same group is fully accessible in solution.

3) It is possible to study the influence of external parameters (temperature, pH, ionic strength, ...) on either conformational or chemical step of the reaction.

Generally the concentration of the reactive molecule is in large excess relative to that of nucleotides and the kinetics of the binding step described by

$$O + F \underset{k_{-1}}{\overset{k_1}{\rightleftharpoons}} OF$$

can be assumed as a pseudo-first order reaction.

If, as a first approximation, k_{-1} is negligible before k_1 (the life of the derivative is much longer than the time of the reaction) the kinetic approach of the problem is similar to that derived in the case of proton exchange (Mc Connell and Von Hippel, 1970).

The rate constant k_{obs} of the pseudo first-order decay of base-pairing with time is given by

$$k_{obs} = \frac{k_o k_1}{k_o + k_c + k_1}$$

Generally and for temperatures much lower than the T_m, the concentration of the stationary open state is very small and therefore $k_c \gg k_o$ leading to

$$k_{obs} \sim \frac{(k_o/k_c\ k_1)}{1 + k_1/k_c}$$

If $k_c \gg k_1$, the chemical binding is the limiting step and then $k_{obs} = Kk_1$ where K is the equilibrium constant ($K \ll 1$) of the transconformation). If $k_c \ll k_1$, the opening reaction is the limiting step and then $k_{obs} = k_o$. Only in this case the value of k_o is directly measured by that of k_{obs}.

Actually as pointed out later (Mc Ghee and Von Hippel, 1977, Lukashin et al., 1976) it is very difficult to relate unambiguously the fluctuation opening of the base pair to the experimental data of the HCHO kinetics in view of

the complexity of the reaction (Mc Ghee and Von Hippel, 1977). One can also calculate (Lukashin et al., 1976) the fluctuational opening by using a classical theory of helix-coil transition which is supposed to remain valid below T_m (the so-called "premelting" range).
The two approaches yield data which differ of about 3 orders of magnitude for the average number of "opened" base pairs at a given temperature. Actually (Wartell and Benight, 1982) if interactions extending beyond neighboring base pairs are considered, a reconciliation between the two points of view can be made leading to about 10^{-2} and 10^{-3} as representing the average fractions of AT and GC base pairs respectively which are "opened" in the premelting range (∿ 50°C under the melting temperature). The rate constant opening is in the range 10 - 100 s^{-1}.

2. Aminofluorene derivatives

Fifteen years ago, a dynamic aspect of the DNA double helix was suggested in order to explain the covalent binding of the hepatocarcinogen N-acetoxy-N-2-acetylaminofluorene (N-Aco-AAF) to the poorly accessible C8 of the guanine residue (Kapuler and Michelson, 1971). Following the hypothesis of Printz and Von Hippel (1968) that exchange of interchain hydrogens occurs in a local, transiently-open conformational state, brought about by thermally-induced local fluctuations ("breathing" modes), a local conformational change of the double helix, allowing the C8 atom to be accessible during a short period of time, could be conceived.
Such an interpretation was already proposed as a tentative model (Daune, 1981 ; Westhof, 1983).

In order to obtain information about those dynamic states, the binding kinetics to native and denatured DNA of two closely related ultimate carcinogen, N-Aco-AAF and N-hydroxy-N-2-aminofluorene (N-OH-AAF) was studied. These two carcinogens were chosen because, although they both react with the C8 of guanine residues, they behave very differently in their reactions with nucleic acids. Thus, while N-Aco-AAF reacts better with denatured than with native DNA (Kapuler and Michelson, 1971), the opposite is true for N-OH-AF (Spodheim-Maurizot et al., 1979). Also, N-Aco-AAF binds equally well to both the B- and the Z-forms of poly d(G-C) but N-OH-AF binds only to the B-form (Rio and Leng, 1983). And finally, the B to Z transition is facilitated in AAF-modified poly d(G-C) (Sage and Leng, 1980 ; Santella et al., 1981) but not in the case of AF-modified poly d(G-C) (Sage and Leng, 1980). From those studies, it can be inferred that these two carcinogens react with guanine residues in different conformational states in the DNA double helix and, conceivably, could act as probes of the DNA dynamics governing the conformational transitions.

Covalent binding of the carcinogen to the C8 of guanine was proved to be a nucleophilic substitution (Miller, 1970) with the strong electrophilic nitrenium ion leading to a C-N bond between guanine and fluorene aromatic ring (arylamidation). We will assume that in most cases the C8 of G is only accessible to the reagent when the nucleotide conformation of the guanine residue differs from the classical B-structure. This new transient structure will be denoted G*. The electrophilic reagent can also react with water and this hydrolysis, leading to an inactive compound C_1, has to be taken into account as a competition reaction. In fact, several degradation products are issued from the hydrolysis reaction and the following equation (1) can be viewed as a coarse approximation but sufficient for our purpose. In view of the high concentration of water (55 M) compared to that of reagent (10^{-4} M), this hydrolysis can be treated as a pseudo first order reaction.

The whole process can thus be described by the three following reactions (where C designates the carcinogen and A the adduct) :

Hydrolysis	$C \xrightarrow{k_1} Cl$	(1)
Conformational change G	$G \underset{k_{-2}}{\overset{k_2}{\rightleftharpoons}} G^*$	(2)
Irreversible binding to give an adduct A with either G or G*	$G^* + C \xrightarrow{k_3} A$	(3)
	$G + C \xrightarrow{k_3} A$	(4)

Three types of experiments were made to determine the different kinetic parameters. Detailed calculations can be found elsewhere (Daune et al., 1985) and only final analytical expressions will be given here.

a) Measurement of the final amount of adduct

At zero time, $a = C_o/G_o$ is the ratio between the initial concentration of carcinogen and that of guanine. At the end of the process $R = A/G_o$ is the experimentally determined ratio between the adduct A present in the DNA molecule and the amount G_o of potential binding sites. When binding occurs only with G*, one has

$$m = \frac{k_3}{k_1} G_o \frac{K}{1 + K} = \frac{\ln(1 - R)}{R - a}$$

K is the equilibrium constant of the transconformation G <—> G* (K << 1) which is assumed to be much faster than the chemical reaction.
When binding occurs only with the "normal" G conformation

$$m = K \cdot \frac{\ln(1 - R)}{R - a}$$

The slopes of the linear plots of ln m versus 1/T are a linear combination of the activation energies $E_3^{\neq}$ (binding to G) $E_1^{\neq}$ (hydrolysis) and enthalpy change ΔH accompanying G <—> G* transconformation.

b) Determination of the initial rate

For experimental reasons, the concentration of carcinogen cannot be in large excess relative to that of nucleotides and the treatment of an open <—> closed process coupled to a chemical binding (Mc Connell and Von Hippel, 1970) cannot be used. Chemical binding can be followed spectrophotometrically in case of DNA and of free deoxyguanosine. Values of k_3, G* and $E_3^{\neq}$ can be readily obtained.

It is therefore possible to combine all the data to determine separately each of the parameters. Moreover, by comparing data obtained after binding of N-Aco-AAF and N-OH-AF with both native and denatured DNA it was possible to check the validity of our model according to which the rate of exchange between G and G* is fast relative to the rate of carcinogen binding.

c) Results

Tables 3 and 4 give the results obtained from determination of the final amount of adduct.

Table 3 Final Amount of AAF Binding to DNA at Different Temperatures As Expressed in Percentage of Modified Bases

buffer		t (°C) 4	10	20	37	50
2×10^{-3} M	nDNA	0.97	nd	1.58	2.01	3.66
citrate, pH 7	dDNA	2.95	nd	4.50	4.58	3.50
10^{-2} M	nDNA	0.13	0.19	0.25	0.32	0.48
citrate, pH 7	dDNA	0.60	0.77	1.20	1.00	nd

Table 4 Final Amount of AF Binding to DNA at Different Temperatures (in 10^{-2} M Citrate Buffer, pH 5.5) As Expressed in Percentage of Modified Bases

	t (°C) 4	10	15	20	37	50
nDNA	0.99	1.41	2.30	3.43	4.81	3.29
dDNA	0.98	1.51	2.06	2.18	2.18	nd

The unexpected result is the amount of G* found in native and denatured DNA at different temperatures (table 5).

Table 5 Values of Equilibrium Constant K of the G ⇄ G* Transconformation for Native and Denatured DNA at Different Temperatures with the Resulting Percentage Concentration of G*

	t (°C) 4	8	20	37	50
G^*/G_0	(0.04)	(0.056)	0.113	0.2	(0.3)
$K_N \times 10^2$	4.17	6.0	12.7	25.0	43
G^*/G_0	(0.236)	(0.265)	0.44	0.88	
$K_D \times 10^2$	30	36	72	730	

The thermodynamic parameters of the equilibrium G <——> G* are ΔH = 9 KCal mole^{-1} and ΔS = 27 Cal/K/mole. These values are close to those determined for base stacking (i.e. 9.4 and 24.8 respectively) (see Wartell and Benight, 1982). They compare well also with those obtained from infrared or circular dichroism studies of conformational changes of poly rA poly rU, namely ΔH_0 in the range 6.7 and 8.2 KCal/mole (Preisler et al., 1981).

3. Dynamics of "dye" binding to DNA

Interaction of DNA with molecules acting on the genetic expression was initiated more than thirty years ago with acridine dyes (Peacocke and Skerrett, 1956) which were used as antibiotics and mutagenic agents. Few years later the intercalation model was proposed (Lerman, 1961) and for now 25 years was extensively verified and improved. It is now usual to distinguish between intercalated and non-intercalated drugs and accordingly we have to successively examine their corresponding dynamics of binding.

A. Intercalated molecules

To the members of the acridine family, which were the first to be studied, were added many derivatives with aromatic rings (fluorene, anthracene, phenanthrene, phenoxazone, etc.).

Evidence for intercalation in double stranded DNA was tentatively related to specific modification of some physical properties of the DNA molecule or to a comparison between optical properties of bound and free ligand. Actually it appears that the best criteria rely on the two characteristic features of any intercalation complex : unwinding and stretching of the double helix.

Unwinding can be easily measured on a covalently closed circular DNA, the hydrodynamical behaviour of which sharply depends on its writhing number which in turn is correlated with the number of base pairs per turn.

Stretching is measured under the electron microscope provided convenient reference length could be determined on the same grid.

These two overall changes of DNA tertiary structure actually reflect the local structural changes occuring in the intercalation site. More details can be found in a recent discussion on probes of intercalation (Dougherty and Pilbrow, 1984). The structure of complexes between dinucleotides and an intercalated dye were determined by X-ray cristallography : ethidium bromide with ApU (Tsai et al., 1977) and CpG (Neidle et al., 1977, Tsai et al., 1977, Berman et al., 1979), acridine orange with CpG (Wang et al., 1979), proflavine with dCpG (Shieh and al., 1980).

The main features of the intercalation process are the conformational changes of the nucleotide backbone as displayed in figure 4 and 8 :

a) A mixed sugar pucker characterizes the two base pairs involved in the intercalation site with a C3' endo at the 5' end and often a C2' endo at the 3' end. Simultaneously the χ value at the 3' end is much higher (high anti) than that at the 5' end (low anti).

The ϕ angle is practically the only dihedral angle of the backbone which is modified.

Finally an unwinding angle dependent on the nature of the dye and the doubling of the distance between two adjacent base pairs achieve to describe the intercalation process.

b) These structural changes can also be viewed on a dynamical point of view : what could be the time scale of an intercalation process ? For more than 15 years fast kinetic experiments were designed to follow the process including chemical relaxation (T-jump and p-jump) and stopped-flow. In all cases, absorption or fluorescence changes were used to monitor the process.

Two general schemes were proposed in order to describe the two steps of intercalation.

In the first scheme, a local structural change of the DNA induced by a large fluctuation is supposed to be a prerequisite to create a convenient "cavity" for the drug. On can write accordingly :

$$P \underset{k_{-1}}{\overset{k_1}{\rightleftharpoons}} P* + L \underset{k_{-2}}{\overset{k_2}{\rightleftharpoons}} PL$$

The first step, a local transconformation, is followed by the binding step during which the ligand L can position itself into a preformed conformation of the polymer.

In the second scheme a diffusion-controled outside binding occurs first. The dye is sitting in one of the groove waiting for a fluctuation able to

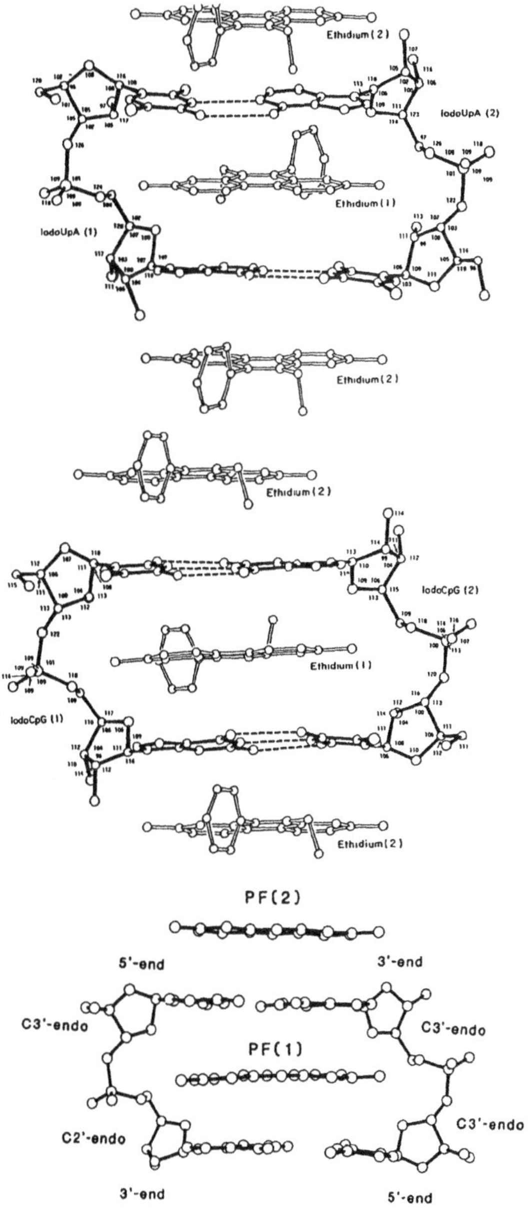

FIGURE 8. Position of ethidium bromide between two UpA and two CpG base pairs (Tsai et al., 1977) and of proflavine between two dCpG base pairs (Shieh et al., 1980)

create an intercalation site. One can write in this second case :

$$P + L \underset{k_{-1}}{\overset{k_1}{\rightleftarrows}} PL_1 \underset{k_{-2}}{\overset{k_2}{\rightleftarrows}} PL_2$$

in which PL_1 designates the outside binding and PL_2 the intercalation complex.

The first model was emphasized by Sobell (assuming the existence of "kinked" DNA ready to accomodate a planar dye molecule).
However, recent experimental evidence (Wilson et al., 1985) is in favor of the second model which also was largely used in the majority of experimental studies with sometimes the variant :

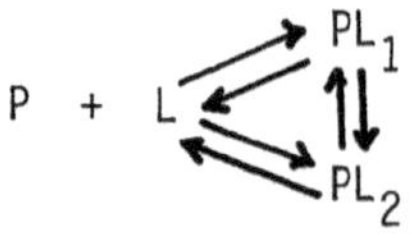

In any case we are only interested in the conformational change (either $P \rightleftarrows P^*$ or $PL_1 \rightleftarrows PL_2$) which will be assumed to be described by rate constants called k_2 and k_{-2} (both expressed in s^{-1}) or by the corresponding τ value ($1/\tau = k_2 + k_{-2}$). Values of log τ or log k_2^{-1} are given in table 6, with their corresponding activation energy when they have been determined.

From inspection of table 6, it is clear that k (or τ values) are at least 5 orders of magnitude higher than those corresponding to the internal fast torsional movement described in the preceding chapter.

Dye	Nucleic Acid	- log k_2	- log k_{-2}	$E^{\neq}_2$	$E^{\neq}_{-2}$	log τ		$E^{\neq}$		References
Proflavine	poly rA poly rU	- 2.35	- 1.8	15	23					Schmechel & Crothers, 1971
	DNA (CT)	- 3.84	- 2.62	16	14					Li & Crothers, 1969
	DNA (T_2)	- 2.87	- 2.2	13	17					
Ethidium bromide	DNA (CT)		- 1.74		18.5					Bresloff & Crothers, 1975
Tilorone	poly dA poly dT					- 1		28		Sturm, 1982
OHSA	poly dA poly dT					- 2.2	- 1.84	27.5	30	Sturm & al. 1975
	DNA					- 3.0	- 2.35	18	20	
	poly d(AT)					- 3.4	- 2.77	19.4	17	
NMR (hydrogen exchange)	AT					- 1.0		19.5		Lefèvre et al., 1985
	GC					- 0.3		18.1		
Base pair formation at the end of a helix		- 5.1		4						Pörschke, 1974

TABLE 6. Dye-nucleic acids interaction : kinetic parameters at 25°C
In the bottom are given data relative to direct measurements of DNA dynamics (see details in the text).

Before to go further into any interpretation it is worth to analyse the results obtained with non intercalating molecules.

B. Non intercalating molecules

A recent extensive review (Zimmer and Wahnert, 1986) was devoted to this kind of binding mode which was discovered in the case of steroidal diamines, aromatic diamidines, pyrrole amidine antibiotics like netropsin or more complex molecules like mithramycin and anthramycin. The non-planar geometry of the molecule preclude any intercalation, but instead many of these compounds are lying in the small groove of the B-DNA. Unfortunately few of them were studied from a dynamic point of view and we can only discuss of hydroxystilbamidine (OHSA) binding to DNA and synthetic polynucleotides, the only complete kinetic study (Festy and Daune, 1973, Festy et al., 1975, Sturm et al., 1975).OHSA, a trypanocidal drug, is made of a silbene ring with one phenolic group and two amidine group at each end. Binding to DNA and polydeoxyribonucleotides can be followed either by absorption or fluorescence changes. In the latter case and with DNA the quantum yield of the blue fluorescence (450 nm) is increased more than 50 times and the polarization reaches values between 0.4 and 0.45, i.e. similar to those determined in a rigid medium.

However striking differences appear in terms of base composition since the blue fluorescence is only increased 3 to 4 times in the case of poly d(GC) but 130 times with poly d(AT).

Upon binding to rod-shaped sonicated DNA molecule, no increase of intrinsic viscosity could be observed. In the case of closed circular DNA (form I) no change of hydrodynamic properties can be detected. These two results are strong evidences of a non-intercalating drug. The fitting of the molecule into the small groove which was assumed as a model (Festy et al., 1975) can be actually realized on a graphic display (Westhof and Daune, unpublished results) by using the coordinates of OHSA as deduced from a cristallographic study (Courseille et al., 1972) and those of a dodecanucleotide (Fratini et al., 1982) (Fig. 9).

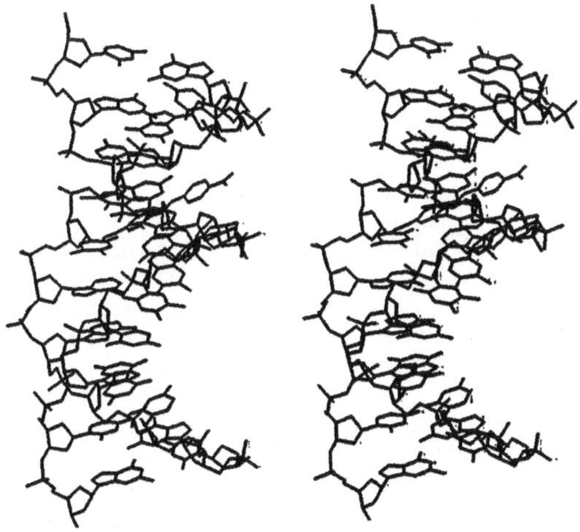

FIGURE 9. Stereoview of OHSA molecule lying in the small groove of a double stranded dodecadeoxynucleotide

Fluorescence monitored kinetic studies bring the most interesting data. Following the T-jump, a small (0.05% per °C) relaxational decrease of red and blue fluorescence takes place. Then in the longer time scale, while the "red" fluorescence levels off, the "blue" one increases slowly with an amplitude of 0.4% per °C. Normally any process related to binding would lead to a decrease of the fluorescence upon partial release of the bound dye. We experienced therefore a typical kinetic behaviour with the following characteristics :
1) The process is insensitive to DNA concentration (at constant value of dye/nucleotide ratio) and corresponds likely to some local transconformation. 2) The decay has to be fitted by a sum of two exponentials leading to two relaxation times τ_3 and τ_4. 3) The phenomenon is not observed with stilbamidine, in which the phenolic group is not present. The OH group must play a decisive role. 4) The process is not observed with OHSA alone in presence of glycerol, ATP or TMP to concentrations up to 10^{-1} M. The double helical structure appears therefore as a prerequisite. 5) Nothing comparable is observed with poly d(GC) pointing to the role of AT pairs.

Values of τ_3 and τ_4 with corresponding activation energies are displayed in table 6.

The types of kinetic experiments described above refer to reversible structural changes inside the B-DNA. Another type of chain dynamics occurs when dye intercalation triggers the passage from one helical structure to the other. This was observed with ethidium bromide and poly (dGdC) (Pohl et al., 1972). Intercalation is only possible into the B-helix and ethidium binding induces therefore a highly cooperative and large conformational change from a left-handed to a right-handed double helix. Recent experiments (Walker et al., 1985) show that for each bound drug, 3 to 4 base pairs of the Z-structure shifts to a B-structure and this local B-structure becomes a favored binding site for subsequent dye molecules. This nucleation process explains the high cooperativity of the transition.

It is also possible to convert B to Z and conversely only by changing the ionic strength (Pohl and Jovin, 1972). In that case the enthalpy change is close to zero and the activation energy is of the order of 22 KCal/mole for both forward and reverse conformational change. The τ value of the first order kinetics is in the range 10^2 - 10^3 s in the case of salt-induced as well as dye-induced transition. The conversion of B to Z structure is also triggered by divalent cations (Mg^{2+}, Zn^{2+}, Cu^{2+}) or trivalent complex ions ($Co(NH)_3)_6{}^{3+}$) (Peck et al., 1982, Chen et al., 1984, Woisard et al., 1985, Van de Sande et al., 1982).

IV. COMPARISON BETWEEN THE DATA AND MODELS

Afther this review of the dynamics of DNA molecule probed by bound molecules and ions or by the study of their binding process, the values of the mechanical parameters and of time constants have to be compared with the values of the corresponding parameters obtained from other experimental approaches. Indeed many experimental data are relevant to the internal dynamics of DNA :
a) Tritium or deuterium exchange were first studied by chromatographic or spectroscopic methods (Mc Connell and Von Hippel, 1970, Teitelbaum and Englander, 1975, Mandal et al., 1979, Woodward and Hilton, 1979, Markovits et al., 1983). Recently NMR offered a precise and detailed approach of the process (Kearns, 1984, Leroy et al., 1985, Lefèvre et al., 1985). The rate of opening the base pairs (a prerequisite for exchange of iminoprotons) was found initially to be of the order of 1 s. However recent NMR studies pointed to lower values, i.e. of the order of 10 to 100 ms, with an acti-

vation energy close to 20 KCal/mole.
b) Raman or microwave spectroscopy (Urabe and Tominaga, 1982, Van Zandt et al., 1982, Swicord and Davis, 1982) is only able to get information about fast movements. One has just to recall the correspondence between 1 cm^{-1}, a wavelength number which is at the end of the accessible spectra, and a time constant $\tau \sim 5$ ps.
c) In dynamic light scattering (Wilcoxon and Schurr, 1983), the diffusional behaviour of the DNA helix is analysed in terms of normal modes (Rouse's model) and again the corresponding relaxation times are very small. The presence inside the DNA helix of an average small number of opened base pairs (the transient state necessary for hydrogen exchange) would be however reflected in an increased flexibility of the molecule, which apparently does not correspond to experimental data.
d) Finally some direct experiments were made to determine the opening rate of base pairs at the end of a helix (Porschke, 1974) or the stacking and unstacking relaxation time in a single stranded polynucleotide (Dewey and Turner, 1979).

It was therefore interesting to compare all the data obtained from different experimental approaches.
Assuming an harmonic potential for the torsional oscillation as probed for example by FPA measurements, we can calculate the potential energy $E_p = 1/2\ C\theta^2$, where C is taken as the average (0.027 ± 0.014) of experimental determinations of the torsion constant and θ is the angular deviation between two adjacent base pairs from their equilibrium position. With thermal agitation θ could experience larger fluctuations, the probability of which could be assumed represented by a Boltzmann distribution. In other words the Boltzmann factor $\exp(-E_p/RT)$ will give the probability of occurrence of a local torsional movement leading to a mechanical energy E_p of the DNA molecule. According to Kao and Bobst (1985), we could assume a "zero" value of the order of 4 ns and then determine the corresponding time constant τ for each E_p value. Obviously the plot of log τ versus E_p is a straight line which can be considered as a simple reference in a log τ / Energy diagram. From inspection of Fig. 10 it is clear that the majority of movements of B-DNA are definitely outside, i.e. corresponding to higher energies than the pure torsional one.
Movements of double-stranded DNA thus belong to two different time and energy domains which are separated by at least 5 orders of magnitude in the time scale and by about 10 KCal in the energy scale.

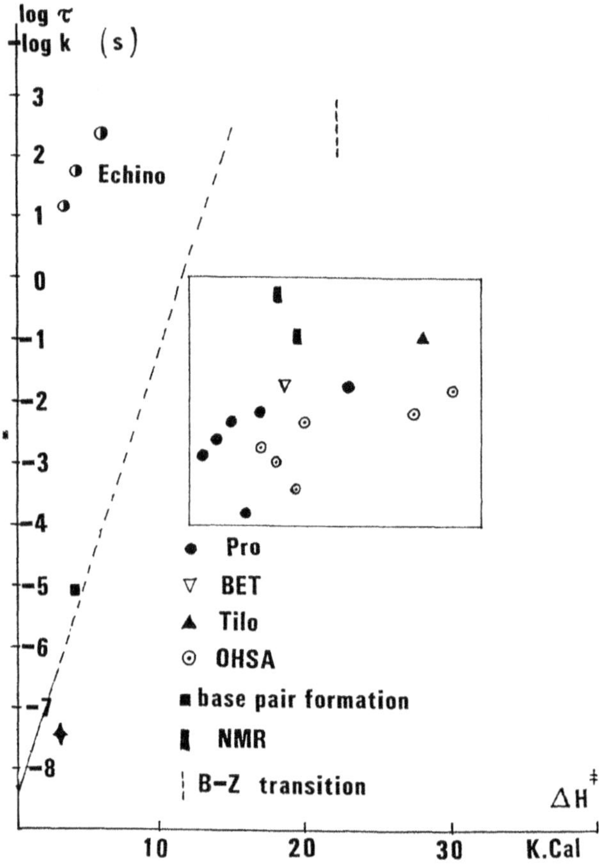

FIGURE 10. Plot of log τ (-log k) versus the activation enthalpy ΔH⁼. Correspondence between symbols and processes are given in the figure.

Three kinds of theoretical explanation can be proposed to describe the experimental situation : mathematical, structural and physicochemical. They will be given more as a matter of discussion than as definite modes of explanation.

A. Mathematical model

It is striking to find that the time limits of the domain (put into box in the figure) correspond to winding angles around 20 degrees i.e. of the order of the unwinding which accompanies for example dye intercalation. The simpler manner to describe such a behaviour is to assume a potential barrier between two wells A and B of the potential energy. Inside A would be localized the torsional oscillations of a DNA molecule. A larger thermal fluctuation could only allow the crossing of the barrier to go into a second well B. This could be an intercalation process and then torsional oscillations inside B were probed by FPA measurements in the time range of 10 to 100 ns. However the time to cross the barrier is much longer and depend not only of dynamics of DNA but also on the nature of the probe. Along the reaction path x the energy derivative dE/dx is represented by a cubic curve and in the domain where three roots (dE/dx = 0) are possible, two stable states could exist. This is a classical "bifurcation" in the catastrophe theory ans such a mathematical model could help to understand the two apparently distinct domains of movements of the DNA molecule.

B. Structural model

Such a model was already proposed (Daune et al., 1985) in the case of the covalent binding of a carcinogen molecule (AAF) to the guanine residue. When in DNA B-structure the O5'-C8 distance is plotted in function of P (the phase angle of pseudo-rotation) (Fig. 11) one observes in the "breathing" part of the curve (which covers the P fluctuation in a broad C2' endo region) a variation between 2.9 and 4.0 Å. Even for the highest value the acetyl group of AAF would collide with the phosphate group at O5'. Therefore the approach is only possible after a switch of ψ from g^+ to t conformation correlated with a switch of ω from g^- to t in order to maintain the double-stranded helical structure. This so-called "crankshaft" mechanism allows the O5'-C8 distance to reach a value close to 6 Å, and the nucleophilic substitution to take place without steric hindrance (Fig. 12).

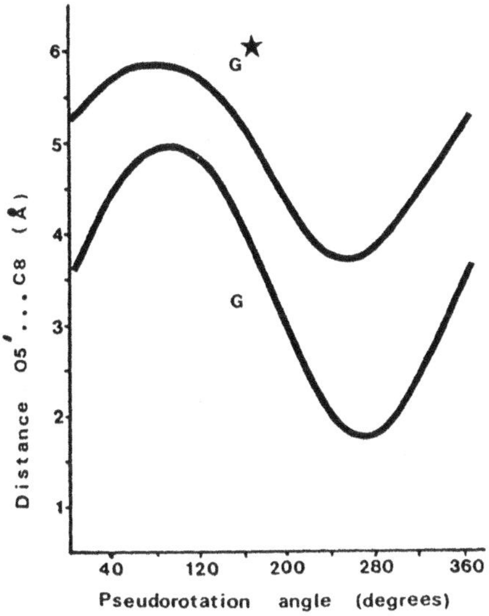

FIGURE 11. Plots of the distance between the C8 atom of guanine and the O(5') atom of the sugar-phosphate as a function of sugar pucker for two conformations of the torsion angle C(3')-C(4')-C(5') : first, the standard conformation (40°) (gauche +) and then the transconformation (180°), which is associated with G*. The glycosyl torsion angle, χ, has been left invariant (80°).

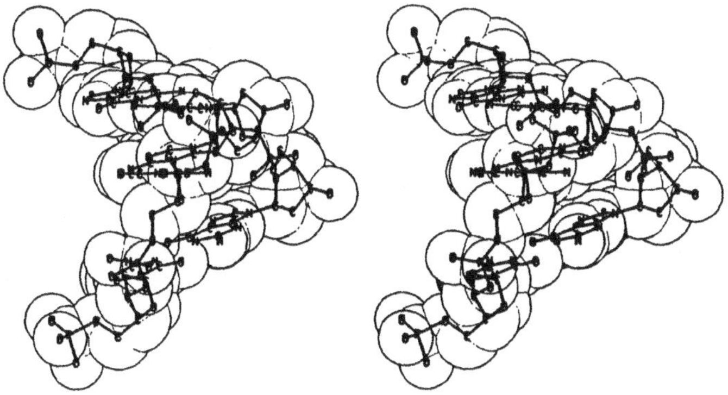

FIGURE 12. Stereo drawing of the trimer $d(G\text{-}C)_3$ incorporating a G* guanine residue (on the strand toward the viewer). Accessibility of C8 and minimal distortion of the DNA molecule are the main features of the model.

This double transconformation for ψ and ω was called G* and according to our data appears to represent a significant percentage of G. Some characteristics of this G $\leftrightarrow$ G* equilibrium have to be underlined :

a) The phosphate to phosphate distance remains practically unchanged, which explains why the conformational part of the energy change does not depend on ionic strength.

b) This local dynamics of guanine residue depends on neighboring base pairs since in a fragment of DNA the extent of the chemical modification of each of the 160 guanines may vary of more than one order of magnitude (Fuchs, 1984).

c) During the G $\leftrightarrow$ G* transition a transient opening of base pairs could occur and indeed the ΔH and ΔS values are very closed for the two processes. Nevertheless the opening time would represent only a small part of the life of the G* conformation and therefore the average values of open basepairs at any time would be much lower than the average number of G* states.

C. Physicochemical model

It relies on another striking experimental comparison between intercalated and non intercalated dyes and more precisely between ethidium bromide or proflavine in one hand and hydroxystilbamidine (OHSA) in the other. The similarity of kinetic parameters and activation energies raises the question of a common physicochemical process for the two different modes of binding to DNA.

In the model given previously (Fig. 9) of OHSA binding to DNA a hydrogen bond between OH group and the O2 of thymine was taken into account (Festy et al., 1975) and acceptable values of Van der Waals distances between atoms were only retained. In these conditions the positioning of OHSA in the small groove looks like that of netropsin in the same dodecamer deduced from X-ray analysis of the crystalline complex (Kopka et al., 1985). In the latter case obviously a greater number of hydrogen bonds can be formed but for each drug the peculiar feature (as pointed out by Dickerson's group) is the displacement of water molecules. Netropsin and OHSA are situated in the minor groove in place of the spine of hydration.

It is thus tempting to relate both the value of the thermodynamic parameters and the relaxation times to the modification of the hydration pattern of B-DNA. It could also characterize the intercalation process, explaining the relative similarity between two kinds of small molecules - DNA interactions.

A generalization to specific interactions, as those found in protein - DNA recognition processes, could provide a unifying manner to understand the role of molecular dynamics in many biological events.

REFERENCES

1. Allison SA and Schurr JM (1979) Torsion dynamics and depolarization of fluorescence of linear macromolecules. I. Theory and application to DNA, Chem. Phys. 41, 35-59.
2. Ashikawa I, Kinosita K, Ikegama A, Nishimura Y, Mitsuboi K, Watanabe K, Iso K and Nakano T (1983) Internal motion of DNA in chromatin. Nanosecond fluorescence studies of intercalated ethidium, Biochemistry 22, 6018-6026.
3. Ashikawa I, Kinosita K and Ikegami A. (1984) Dynamics of Z-form DNA, Biochim. Biophys. Acta 782, 87-93.
4. Barkley M and Zimm BH (1979) Theory of twisting and bending of chain macromolecules analysis of the fluorescence depolarization of DNA, Chem. Phys. 70, 2991-3007.
5. Berman HM, Stallings W, Carrell HL, Glusker JP, Neidle S, Taylor G and Achari A (1979) Molecular and crystal structure of an intercalation complex : proflavine-cytidylyl-(3'-5')-guanosine, Biopolymers 18, 2405-2429.
6. Bittman R and Blau L (1975) Stopped-flow kinetic studies of actinomycin binding to DNA, Biochemistry 14, 2138-2145.
7. Bresloff JL and Crothers DM (1975) DNA ethidium reaction kinetics : demonstration of direct ligand transfer between DNA binding sites, J. Mol. Biol. 95, 103-123.
8. Broersma S (1960) Rotational diffusion constant of a cylindrical particle, J. Chem. Phys. 32, 1626-1631.
9. Chen HH, Behe MJ and Rau DC (1984) Critical amount of oligovalent ion binding required for the B-Z transition of poly(dGm^{5}dC), Nucl. Acids Res. 12, 2381-2389.
10. Courseille C, Busetta B and Hospital M (1972) Etude cristallographique des composés se fixant sur les acides nucléiques. Structure cristalline du diiséthionate d'hydroxystilbamidine, C.R. Acad. Sci. Paris 274, 1921-1924.
11. Das GC, Mellet M and Daune MP (1980) Effect of dyes on the premelting conformational changes of synthetic polynucleotides, Studia Biophysica 81, 125-126.
12. Daune M, Sturm J and Zana R (1976) Temperature-jump study of tilorone binding on poly dA poly dT, Studia Biophysica 57, 139-144.
13. Daune M, Fuchs RPP and Leng M (1981) Structural modification and protein recognition of DNA modified by N-2-fluorenylacetamide, its 7-iodo derivative and by N-2-fluorenamine, Natl. Cancer Inst. Monogr. 58, 201-210.
14. Daune MP, Westhof E, Koffel-Schwartz N and Fuchs RPP (1985) Covalent binding of a carcinogen as a probe for the dynamics of DNA, Biochemistry 24, 2275-2284.
15. Davanloo P and Crothers DM (1976) Kinetic studies of drug-dinucleotide complexes, Biochemistry 15, 5299-5305.
16. Dewey TG and Turner DH (1979) Laser T-jump study of stacking in adenylic acid polymers, Biochemistry 18, 5757-5762.
17. Dougherty G and Pilbrow JR (1984) Physicochemical probes of intercalation, Int. J. Biochem. 16, 1179-1192.
18. Festy B and Daune M (1973) Hydroxystilbamidine : a non intercalating drug as a probe of nucleic acid conformation, Biochemistry 12, 4827-4834.
19. Festy B, Sturm J and Daune M (1975) Interaction between hydroxystilbamidine and DNA. I. Binding isotherms and thermodynamic of the association, Biochim. Biophys. Acta 407, 24-42.

20. Forster W and Stutter E (1984) Interaction of anthracycline antibiotics with biopolymers : 9. Comparative study of the interaction kinetics of daunomycin, adriamycin and iremycin with DNA, Int. J. Biol. Macromol. 6, 114-124.
21. Fox KR, Wakelin LPG and Waring MJ (1981) Kinetics of the interaction between echinomycin and DNA, Biochemistry 20, 5768-5779.
22. Fox KR and Waring MJ (1984) Kinetic evidence for redistribution of actinomycin molecules between potential DNA-binding sites, Eur. J. Biochem. 145, 579-586.
23. Frank Kamenetskii MD and Lazurkin AV (1974) Conformational changes in DNA molecules, Ann. Rev. Biophys. Bioeng. 3, 127-150.
24. Fratini AV, Kopka ML, Drew HR and Dickerson RE (1982) Reversible bending and helix geometry in a B-DNA dodecamer : CGCGAATT^{Br}CGCG, J. Mol. Chem. 257, 14686-14707.
25. Fuchs RPP (1984) DNA binding spectrum of the carcinogen N-acetoxy-N-2-acetylaminofluorene significantly differs from the mutation spectrum, J. Mol. Biol. 177, 173-180.
26. Fujimoto BS, Shibata JH, Schurr RL and Schurr JM (1985) Torsional dynamics and rigidity of fractionated poly(dGdC), Biopolymers 24, 1009-1022.
27. Genest D and Wahl P (1978) Fluorescence anisotropy decay due to rotational brownian motion of ethidium intercalated in double strand DNA, Biochim. Biophys. Acta 521, 502-509.
28. Genest D, Wahl P, Erard M, Champagne M and Daune M (1982) Fluorescence anisotropy decay of ethidium bromide bound to nucleosomal core particles, Biochimie 64, 419-427.
29. Genest D, Mirau P and Kearns DR (1985) Investigation of DNA dynamics and drug-DNA interaction by steady state fluorescence anisotropy, Nucl. Acids Res. 13, 2603-2615.
30. Genest D and Malfoy B (1986) Fluorescence decay of ethidium bromide in the presence of the Z conformation of poly dGdC and of poly (dG.dC) modified by chlorodiethylenediamine platinum (II) chloride, Biopolymers 25, 507-518.
31. Gennis RG and Cantor CR (1972) Optical studies of a conformational change in DNA before melting, J. Mol. Biol. 65, 381-399.
32. von Hippel PH and Wong KY (1971) Dynamic aspects of native DNA structure : kinetics of the formaldehyde reaction with calf thymus DNA, J. Mol. Biol. 61, 587-613.
33. Hogan ME and Jardetzky O (1980) Internal motions in deoxyribonucleic acid II, Biochemistry 19, 3460-3468.
34. Ivanov VI, Minchenkova LE, Minyat EE, Frank-Kamenetskii MD and Schyolkona AK (1974) The B to A transition of DNA in solution, J. Mol. Biol. 87, 817-833.
35. Kao SC and Bobst AM (1985) Local base dynamics and local structural features in RNA and DNA duplexes, Biochemistry 24, 5465-5469.
36. Kapuler AM and Michelson AM (1971) The reaction of the carcinogen N-acetoxy-N-2-acetylaminofluorene with DNA and other polynucleotides and its stereochemical implication, Biochim. Biophys. Acta 232, 436-450.
37. Kearns DR (1984) NMR studies of conformational states and dynamics of DNA, CRC Crit. Rev. Biochem. 15, 237-290.
38. Kopka ML, Yoon C, Goodsell D, Pjura P and Dickerson RE (1985) The molecular origin of DNA-drug specificity in netropsin and distamycin, Proc. Natl. Acad. Sci. USA 82, 1376-1380.
39. Lefèvre JF, Lane AN and Jardetzky O (1985) Nuclear magnetic resonance study of the proton exchange rate in the operator-promoter DNA sequence of the trp operon of Escherichia coli, J. Mol. Biol. 185, 689-699.

40. Lerman LS (1961) Structural consideration in the interaction of DNA and acridines, J. Mol. Biol. 3, 18-30.
41. Leroy JL, Broseta D and Guéron M (1985) Proton exchange and base-pair kinetics of poly(rA).poly(rU) and poly(rI).poly(rC), J. Mol. Biol. 184, 165-178.
42. Li HJ and Crothers DM (1969) Relaxation studies of the proflavine-DNA complex : the kinetics of an intercalation reaction, J. Mol. Biol. 39, 461-477.
43. Lukashin AV, Vologodskii AV, Frank-Kamenetskii MD and Lyubschenko YL (1976) Fluctuational opening of the double helix as revealed by theoretical and experimental study of DNA interaction with formaldehyde, J. Mol. Biol. 108, 665-682.
44. Mc Call PJ and Bloomfield VA (1976) Kinetics of proflavin binding to bacteriophages T2L and T4D, Biopolymers 15, 2323-2326.
45. Mc Connell B and von Hippel PH (1970) Hydrogen exchange as a probe of the dynamic structure of the DNA, J. Mol. Biol. 50, 297-316 and 317-332.
46. Mc Ghee JD and von Hippel PH (1975) Formaldehyde as a probe of DNA structure. I. Reaction with exocyclic amino groups of DNA bases. II. Reaction with endocyclic imino groups of DNA bases, Biochemistry 14, 1281-1296 and 1297-1303.
47. Mc Ghee JD and von Hippel PH (1977) Formaldehyde as a probe of DNA structure.III. Equilibrium denaturation of DNA and synthetic polynucleotides. IV. Mechanism of the initial reaction of formaldehyde with DNA, Biochemistry 16, 3267-3293.
48. Macgregor RB, Clegg RM and Jovin TM (1985) Pressure-jump study of the kinetics of ethidium bromide binding to DNA, Biochemistry 24, 5503-5510.
49. Mandal C, Kallenbach NR and Englander SW (1979) Base pair opening and closing reaction in the double helix. A stopped-flow hydrogen exchange study in poly rA poly rU, J. Mol. Biol. 135, 391-411.
50. Mandal C, Englander SW and Kallenbach NR (1980) Hydrogen-deuterium exchange analysis of ligand-macromolecule interactions : ethidium-DNA system, Biochemistry 19, 5819-5825.
51. Marcandalli B, Winzek C and Holzwarth JF (1984) A laser temperature jump investigation of the interaction between proflavine and calf-thymus DNA at low and high ionic strength avoiding electric field effects, Ber. Bunsenges Phys. Chem. 88, 368-374.
52. Markovits J, Ramstein J, Roques BP and Le Pecq JB (1983) Dynamic structure of DNA complexes - Fluorometric measurements of hydrogen-deuterium exchange kinetics of DNA-bound ethidium dimer and acridine-ethidium dimer, Biochemistry 22, 3231-3237.
53. Melchior WB and von Hippel PH (1973) Alteration of the relative stability of dAdT and dGdC base pairs in DNA, Proc. Natl. Acad. Sci. USA 70, 298-302.
54. Millar DP, Robbins RJ and Zewail AH (1982) Torsion and bending of nucleic acids studied by subnanosecond time-resolved fluorescence depolarization of intercalated dyes, J. Chem. Phys. 76, 2080-2094.
55. Miller JA (1970) Carcinogenesis by chemicals : an overview, Cancer Res. 30, 559-576.
56. de Murcia G, Wilhelm B, Wilhelm FX and Daune MP (1978) Effect of TMA ions on conformational changes of DNA in the premelting temperature range, Biophys. Chem. 8, 377-383.
57. Neidle S, Achari A, Taylor GL, Berman HM, Carrell HL, Glusker JP and Stallings WS (1977) Structure of a dinucleoside phosphate-drug complex as model for nucleic acid-drug interaction, Nature 269, 304-306.
58. Palecek E (1976) Premelting changes in DNA conformation, Progr. Nucl. Acids Res. 18, 151-213.

59. Peacocke AR and Skerrett JNH (1956) The interaction of aminoacridines with nucleic acids, Trans. Far. Soc. 52, 261-279.
60. Peck LJ, Nordheim A, Rich A and Wang JC (1982) Flipping of cloned $d(pCpG)_n$ $d(pCpG)_n$ DNA sequences from right to left-handed helical structure by salt, co(III) or negative supercoiling, Proc. Natl. Acad. Sci. USA 79, 4560-4564.
61. Pohl FM, Jovin TM, Baehr W and Holbrook JJ (1972) Ethidium bromide as a cooperative effector of a DNA structure, Proc. Natl. Acad. Sci. USA 69, 3805-3809.
62. Pohl FM and Jovin TM (1972) Salt-induced cooperative conformational change of a synthetic DNA : equilibrium and kinetic studies with poly (dGdC), J. Mol. Biol. 67, 375-396.
63. Pörschke D (1974) A direct measurement of the unzippening rate of a nucleic acid double helix, Biophys. Chem. 2, 97-101.
64. Preisler RS, Mandal C, Englander SW, Kallenbach NR, Howard FB, Frazier J and Toddmiles H (1981) in Biomolecular stereodynamics, ed. Sarma, Adenine Press, N.Y., vol. I, p. 405-415.
65. Printz MP and von Hippel PH (1965) Hydrogen exchange studies of DNA structure, Proc. Natl. Acad. Sci. USA 53, 363-370.
66. Printz MP and von Hippel PH (1968) On the kinetics of hydrogen exchange in DNA : pH and salt effects, Biochemistry 7, 3194-3206.
67. Ramstein J and Leng M (1975) Effect of DNA base composition on the intercalation of proflavine : a kinetic study, Biophys. Chem. 3, 234-240.
68. Ramstein J, Ehrenberg M and Rigler R (1980) Fluorescence relaxation of proflavine-DNA interaction - Kinetic properties of a base specific reaction, Biochemistry 19, 3938-3948.
69. Rio P and Leng M (1983) Preferential binding of the chemical carcinogen N-hydroxy-2-aminofluorene to B-DNA as compared to Z-DNA, Nucl. Acids Res. 11, 4947-4956.
70. Robinson BH, Lerman LS, Beth AH, Frisch HL, Dalton LR and Auer C (1980) Analysis of double helix motion with spin-labeled probes - Binding geometry and the limit of torsional elasticity, J. Mol. Biol. 139, 19-44.
71. Sage E and Leng M (1980) Conformation of poly (dG-dC) poly (dG-dC) modified by the carcinogens N-acetoxy-N-acetyl-2-aminofluorene and N-hydroxy-N-2-aminofluorene, Proc. Natl. Acad. Sci. USA 77, 4597-4601.
72. Sakoda M, Hiromi K and Akasaka K (1971) Kinetic studies of interaction between acridine orange and DNA, Biopolymers 10, 1003-1012.
73. Santella RM, Grunberger D, Weinstein IB and Rich A (1981) Induction of the Z conformation in poly (dGdC) poly (dGdC) by binding of N-2-acetylaminofluorene to guanine residues, Proc. Natl. Acad. Sci. USA 78, 1451-1455.
74. Schmechel DEV and Crothers DM (1971) Kinetic and hydrodynamic studies of the complex of proflavine with poly A - poly U, Biopolymers 10, 465-480.
75. Schurr JM (1984) Rotational diffusion of deformable macromolecules with mean local cylindrical symmetry, Chem. Phys. 84, 71-96.
76. Shibata JH, Fujimoto BS and Schurr JM (1985) Rotational dynamics of DNA from 10^{-10} to 10^{-5} seconds : comparison of theory with optical experiments, Biopolymers 24, 1909-1930.
77. Shieh HS, Berman HM, Dabrow M and Neidle S (1980) The structure of drug-deoxydinucleotide phosphate complex : generalized conformational behavior of intercalation complexes with RNA and DNA fragments, Nucl. Acids Res. 8, 85-97.
78. Spodheim-Maurizot M, Saint-Ruf G and Leng M (1979) Conformational changes induced in DNA by in vitro reaction with N-hydroxy-N-2-aminofluorene, Nucl. Acids Res. 6, 1683-1694.

79. Steenbergen J and Mohr SC (1973) A T-jump kinetic study of the binding of proflavine to poly rI poly rC, Biopolymers 12, 791-798.
80. Sturm J, Zana R and Daune M (1975) Interaction between hydroxystilbamidine and DNA. II. Temperature jump relaxation study - Dynamics of nucleic acids and polynucleotides, Biochem. Biophys. Acta 407, 43-60.
81. Sturm J (1982) Binding of ligands to a one-dimensional heterogeneous lattice. III. Kinetic model and relaxation study of the interaction of tilorone with DNA and polynucleotides, Biopolymers 21, 1189-1206.
82. Swicord ML and Davis CC (1982) Microwave absorption of DNA between 8 and 12 GHz, Biopolymers 21, 2453-2460.
83. Thomas JC, Allison SA, Appellof CJ and Schurr JM (1980) Torsion dynamics and depolarization of fluorescence of linear macromolecules. II. Fluorescence polarization anisotropy measurements on a clean viral ϕ29 DNA, Biophys. Chem. 12, 177-188.
84. Thomas JC and Schurr JM (1983) Fluorescence depolarization and temperature dependence of the torsion elastic constant of linear ϕ29 DNA, Biochemistry 22, 6194-6198.
85. Tritton TR and Mohr SC (1973) Kinetics of ethidium bromide binding as a probe of tRNA structure, Biochemistry 12, 905-914.
86. Tsai CC, Jain SC and Sobell HM (1977) Visualization of drug nucleic acids interactions at atomic resolution. I. Structure of an ethidium dinucleoside monophosphate crystalline complex, ethidium : 5-iodouridylyl(3'-5')adenosine. II. Structure of an ethidium dinucleoside monophosphate crystalline complex, ethidium : 5-iodocytidylyl(3',5')guanosine, J. Mol. Biol. 114, 301-315 and 317-331.
87. Urabe H and Tominaga Y (1982) Low-lying collective modes of DNA double helix by Raman spectroscopy, Biopolymers 21, 2477-2481.
88. Utiyama H and Doty P (1971) Kinetic studies of denaturation and reaction with formaldehyde on polydeoxyribonucleotides, Biochemistry 10, 1254-1264.
89. Van de Sande JH, Mc Intosh LP and Jovin TM (1982) Mn^{2+} and other transition metals at low concentration induced the right-to-left helical transformation of poly d(G-C), EMBO J. 1, 777-782.
90. Van Zandt LL, Kohli M and Prohofsky EW (1982) Absorption of microwave radiation by DNA double helix in aquo, Biopolymers 21, 1465-1468.
91. Wahl P, Paoletti J and Le Pecq JB (1970) Decay of fluorescence emission anisotropy of the ethidium bromide - DNA complex. Evidence for an internal motion of DNA, Proc. Natl. Acad. Sci. USA 65, 417-421.
92. Walker GT, Stone MP and Krugh TR (1985) Ethidium binding to left-handed (Z) DNAs results in regions of right-handed DNA at the intercalation site, Biochemistry 24, 7462-7471.
93. Wang AHJ, Quigley GJ and Rich A (1979) Atomic resolution analysis of a 2:1 complex of CpG and acridine orange, Nucl. Acids Res. 6, 3879-3890.
94. Wartell RM and Benight AS (1982) Fluctuational base-pair opening in DNA at temperatures below the helix-coil transition region, Biopolymers 21, 2069-2081.
95. Westhof E (1983) NATO-CECAM Workshop, Orsay, France.
96. Wilcoxon J and Schurr JM (1983) Temperature dependence of the dynamic light scattering of linear ϕ29 DNA : implication for spontaneous opening of the double helix, Biopolymers 22, 2273-2321.
97. Wilson WD, Krishna Moorthy CR, Wang YH and Smith JC (1985) Mechanism of intercalation : ion effects on the equilibrium and kinetic constants for the interaction of propidium and ethidium bromide with DNA, Biopolymers 24, 1941-1961.
98. Woisard A, Fazakerley GV and Guschlbauer W (1985) Z-DNA is formed by

poly(dCdG) and poly(dm^{5}C-dG) at micro or nanomolar concentrations of some Zn(II) and Cu(II) complexes, J. Biomol. Struct. Dyn. 2, 1205-1220.
99. Woodward CK and Hilton BD (1979) Hydrogen exchange kinetics and internal motions in proteins and nucleic acids, Ann. Rev. Biophys. Bioeng. 8, 99-127.
100. Zimmer C and Wähnert U (1986) Non intercalating DNA-binding ligands : specificity of the interaction and their use as tools in biophysical, biochemical and biological investigations of the genetic material, Prog. Biophys. Molec. Biol. 4, 31-112.

CIRCULAR DICHROISM AS A PROBE OF TERTIARY STRUCTURE

IRWIN TOBIAS

Department of Chemistry
Rutgers University, The State University of New Jersey
New Brunswick, New Jersey 08903, USA

1. INTRODUCTION

Chiral molecules, those which are not superimposable on their mirror images, have a distinctive property which depends critically on the details of their structure--they are optically active. That is, when linearly polarized light passes through a phase containing a uniform concentration of such molecules, the orientation of the plane of polarization varies smoothly with path length--the angle through which the plane rotates is simply proportional to the path length. Such behavior implies that the phase is circularly birefringent--right- and left-circularly polarized light are transmitted with different velocities. For light having a wavelength which falls within an absorption band of the chiral molecule, the medium will be circularly dichroic as well--circularly polarized light of one sense will be absorbed more than circularly polarized of the other sense. In this case not only does the plane of polarization of the linearly polarized light rotate as it passes through the phase, but the light also becomes elliptically polarized.

The nucleic acid molecules have an obvious characteristic causing them to be optically active, a helical-type chirality associated with the secondary structure. Furthermore, there are electronic transitions localized in the bases which happen to fall in a readily accessible wavelength region of the spectrum. Not surprisingly, therefore, circular dichroism has been one of the most popular tools for studying this class of molecules.

The space curve along which the helical axis lies may itself also be chiral. This leads one to speculate on the nature of the contribution that tertiary structure makes to circular dichroism. A quantum mechanical model of optical activity having a formalism particularly well suited to treat that part of the effect which may be traced to localized transitions, the so-called coupled oscillator model, was developed many years ago.[1] In this chapter we shall be concerned with questions such as--Does the coupled oscillator model adequately account for experimental results which seem to show the separate contributions to optical activity of secondary and tertiary structure? If not, what other approaches are possible?

We shall begin with a brief review of the parameters which are commonly used to describe the observed optical activity and circular dichroism of matter. A detailed description of the quantum mechanical origin of the phenomena will then be given, followed by a development of the coupled oscillator model. In order to be able to treat tertiary structure explicitly, we shall assume that the oscillators (chromophores) lie along an arbitrary space curve. A considerable quantity of experimental data is available for the case of nucleosomal DNA. Here this space curve is itself a helix, and so we shall apply the modified coupled oscillator model to the special case of the superhelix. Finally, we shall describe another, quite different, model, developed by the author, which has been successfully applied to the mononucleosome. No attempt will be made to review the rather extensive relevant literature. The discussion and the derivations will be almost completely self-contained, but references will be cited in which more details can be found.

2. PARAMETERS OF OPTICAL ACTIVITY AND CIRCULAR DICHROISM

In this section we review the most common parameters used to express optical activity and circular dichroism. By showing how they are related to the birefringence of the material under investigation, we will be able to see how they are related to each other.

Let the vector $\underline{E}$ be the electric field strength associated with a plane wave of light having a frequency Ω which is propagating in the positive z direction through some matter with an index of refraction n. $\underline{E}$ must always lie in a plane parallel to the x,y plane, the plane spanned by the orthogonal unit vectors $\underline{i}$ and $\underline{j}$. In elliptically polarized light of either sense (±) (with the major and minor axes oriented along the x- and y-axes), $\underline{E}$ is of the form

$$\underline{E}(\pm) = E^0_x\ \underline{i}\cos\Omega(t - nz/c) \pm E^0_y\ \underline{j}\sin\Omega(t - nz/c) \qquad (1)$$

where E^0_x and E^0_y are constants having the same sign, c is the speed of light in a vacuum, and t and z denote time and position, respectively. The upper(+) sign corresponds to left-elliptically polarized light and the lower(-) to right-. In the case that E^0_x and E^0_y have the same magnitude, the light is circularly polarized. Should one of these constants be zero, we have linearly polarized light. It will be convenient in what follows to use the formalism of complex numbers and to obtain $\underline{E}(\pm)$ as given by eq. (1) by taking the real part of the complex quantity (which we also call $\underline{E}(\pm)$)

$$\underline{E}(\pm) = \underline{E}^0(\pm)\exp\{-i\Omega(t - nz/c)\} \qquad (2)$$

where $\underline{E}^0(\pm)$ is the complex vector

$$\underline{E}^0(\pm) = E^0_x\ \underline{i} \pm iE^0_y\ \underline{j} \qquad (3)$$

It is useful to note that linearly polarized light can be thought of as a superposition of left- and right-circularly polarized light. Thus, for example,

$$[(\underline{i} + i\underline{j}) + (\underline{i} - i\underline{j})]\exp\{-i\Omega(t - nz/c)\} = 2\underline{i}\ \exp\{-i\Omega(t - nz/c)\} \tag{4}$$

If we assume that n is real (nonabsorbing medium), the real part of the right-hand-side of eq. (4) is $2\underline{i}\cos\Omega(t - nz/c)$, corresponding to light polarized along the x-axis.

We are now in a position to see what happens when linearly polarized light as given by eq. (4) impinges on a medium which is circularly birefringent. Again we think of the linearly polarized light as a superposition of left- and right-circularly polarized light but now the left-hand-side of eq. (4) must be modified appropriately

$$[(\underline{i} + i\underline{j})\exp\{-i\Phi(+)\} + (\underline{i} - i\underline{j})\exp\{-i\Phi(-)\} \tag{5}$$

where $\Phi(\pm) = \Omega(t - n(\pm)z/c)$. After the light has travelled a distance d through this medium, the real part of expression (5) is proportional to

$$(\underline{i}\cos\phi_r - \underline{j}\sin\phi_r)\cos\Phi \tag{6}$$

where $\Phi = \Omega(t - \tilde{n}d/c)$ and $\tilde{n}$ is the average of n(+) and n(-). Thus the light is still linearly polarized but its plane of polarization has been rotated through an angle

$$\phi_r = \Omega\ \delta n\ d/2c \tag{7}$$

with respect to the plane of polarization of the light entering the birefringent medium where $\delta n = n(+) - n(-)$.

The difference in refractive indices, δn, turns out to be proportional to the concentration of the active species and therefore, according to eq. (7), so is ϕ_r. When the concentration is constant, as we are assuming it to be, we see from eq. (7) that ϕ_r is also proportional to the path length of the light through the optically active medium. The various parameters used to express optical activity are defined so as to be independent of concentration and path length. For example, the specific rotation $[\phi_r]$ is the number of degrees the plane of polarization rotates for a path length of one decimeter in a phase in which the concentration of active material is one gram per cubic centimeter. That is, in terms of ϕ_r in units of radians,

$$[\phi_r] = 1800\ \phi_r/\pi dC \tag{8}$$

where the path length d is measured in centimeters and the concentration C in grams per cubic centimeter. A more convenient parameter for the biochemist is the molecular rotation $[\phi_r]$

$$[\phi_r] = 1.8 \times 10^4\ \phi_r/\pi dM \tag{9}$$

where M is the molarity of the active species in the phase. In the case of DNA the active species are not taken to be the whole molecules but the residues--each residue being associated with one base. The units of $[\phi_r]$ are degrees-cm²/decimol.

The above discussion can be easily broadened to include circular dichroism by considering the refractive index to be a complex quantity,

$$n = n_r + in_i \tag{10}$$

What we have been calling n up to this point is n_r, the real part of the complex refractive index. That is, we have been assuming n_i to be zero. When it is not, we see from eq. (2) that the electric field strength acquires $\exp\{-\Omega n_i z/c\}$ as a multiplicative factor, and the intensity of the light, therefore, which is proportional to $|\underline{E}|^2$, acquires the factor $\exp\{-2\Omega n_i z/c\}$. In this case, as the light propagates through the phase, it decreases in intensity, or, as one usually says, it is absorbed.

We have already considered the case of a circularly birefringent medium. In order to learn the effect on linearly polarized light of a medium which is also circularly dichroic, we return to expression (5), but now we use the complex n as given by eq. (9) and assume that $\delta n_i = n_i(+) - n_i(-) \neq 0$. Now after traversing d centimeters its real part is proportional to

$$\begin{aligned}\exp\{-\Omega\tilde{n}_i d/c\}[(\underline{i}\cos\phi_r - \underline{j}\sin\phi_r)\cosh\phi_i\cos\Phi \\ - (\underline{i}\sin\phi_r + \underline{j}\cos\phi_r)\sinh\phi_i\sin\Phi]\end{aligned} \tag{11}$$

where $\tilde{n}_i$, the average of $n_i(+)$ and $n_i(-)$, appearing in the first factor, determines how fast the intensity decreases as the light propagates. The rest represents light which is elliptically polarized in any plane perpendicular to the direction of propagation. As expected the major and minor axes have turned through an angle of ϕ_r radians after d centimeters. The angle ϕ_i is a measure of the degree of ellipticity

$$\phi_i = \Omega\ \delta n_i\ d/2c \tag{12}$$

Equation (12), which shows how the ellipticity is related to the circular dichroism is entirely analogous to eq. (7) which relates the optical activity to the circular birefringence. Also analogous to the definition of the molecular rotation (eq. (9)) is the definition of molar ellipticity

$$[\phi_i] = 1.8 \times 10^4\ \phi_i/\pi Md \tag{13}$$

Both $[\phi_r]$ and $[\phi_i]$ have the same units, degrees-cm²/decimol.

There are two quantities in common use for describing the degree of circular dichroism of a material. One of them is $[\phi_i]$, the molar ellipticity. The other is the difference between the molar extinction coefficients of the material for left- and right-circularly polarized light, $\delta\varepsilon = \varepsilon(+) - \varepsilon(-)$. To end this section we shall show how these two parameters are related.

We have already discussed the definition of the molar ellipticity: equations (12) and (13) summarize the results. The molar extinction coefficient is defined by the equation expressing Beer's law

$$\ln I(\pm)/I^0(\pm) = -2.303\ \varepsilon(\pm)\ Md \tag{14}$$

where $I^0(\pm)$ is the intensity of the incident light and $I(\pm)$ is the intensity of the transmitted light. The same equation in terms of n_i is

$$\ln I(\pm)/I^0(\pm) = -2\Omega n_i(\pm)d/c \tag{15}$$

from which it follows that

$$\delta\varepsilon = 2\Omega\ \delta n_i/2.303\ Mc \tag{16}$$

Comparing eqs. (12), (13), and (16) we can write

$$[\phi_i] = 1.8 \times 10^4 \times 2.303\ \delta\varepsilon/4\pi \tag{17}$$

Equation (17) expresses the sought-after connection between $\delta\varepsilon$ and the molar ellipticity. The constant which multiplies $\delta\varepsilon$ in eq. (17) has the value 3298.

Before we leave this section it will be useful to observe the following. Let us assume that light, linearly polarized along the $\underline{i}$ direction, is incident upon a circularly birefringent and dichroic medium. After travelling d centimeters, its polarization, and as a matter of fact, any general state of polarization $\underline{e}$, can be represented by a single complex expression

$$\underline{e} = \underline{i} \cos \phi - \underline{j} \sin \phi \tag{18}$$

where ϕ is a complex angle having real and imaginary parts given by ϕ_r and ϕ_i, eqs. (7) and (12). Expression (11) is just the real part of the product of (18) and $\exp\{-\Omega\bar{n}_i d/c\}\exp\{-i\Phi\}$.

3. QUANTUM MECHANICAL ORIGIN OF OPTICAL ACTIVITY AND CIRCULAR DICHROISM

It will become necessary to relate the experimental parameters we have been discussing in the previous section to the parameters which characterize the molecular system under consideration, and for that we must turn to quantum mechanics. The discussion of the previous section suggests that one can proceed by developing a quantum mechanical expression for the refractive index. Indeed this approach is possible, and, in fact, the early application of quantum mechanics to optical activity was carried out along these lines. Here we shall take a quite different tack. The method we shall develop affords further insight into the phenomena, and, in addition, is particularly well suited for arriving at the final equations needed for treating the oligonucleotides.

If are to avoid developing a molecular expression for ϕ by pursuing the index of refraction route, we need other equations, equivalent to (7) and (12), but which relate ϕ to some quantity other than δn. We shall do this using a method which is quantum mechanical in nature from the start.(2) We envision a two-state quantum mechanical system. Both states consist of a molecule in its ground state and a photon, but in one of the states the photon is linearly polarized along the $\underline{i}$ direction, and in the other it is linearly polarized along the $\underline{j}$ direction. In the absence of any interaction between the molecule and the photon, both of these states have precisely the same energy. To this approximation then, both of these states are stationary states--they do not change with time. Had we, for example, started out in the first state, the photon would forever remain linearly polarized along the $\underline{i}$ direction.

We now analyze, using elementary quantum mechanics, what changes may be expected to occur in this system when the interaction between the light and the molecule is taken into account. We let the symbol V denote the operator representing the interaction. It appears as an additive term in the total energy operator, the hamiltonian. Furthermore, we denote the two states described in the preceding paragraph as u_1 and u_2. Then, using u_1 and u_2, we construct the 2x2 matrix

V_{ij} (i,j = 1,2). It is known that its two eigenvalues give the correction to the energy. We shall see that should the matrix element V_{12} be nonzero, these eigenvalues will be unequal, that is, the inclusion of the interaction between photon and the molecule may result in a splitting of the degeneracy, and, furthermore, u_1 and u_2 would then no longer be states of definite energy, the stationary states, but, instead, some linear combination of them. (The stationary states can be constructed from the eigenvectors of the matrix.) In that case, if we started with the photon linearly polarized along the $\underline{i}$ direction, we would find that the polarization changed with time. This, of course, is precisely what we are looking for.

The eigenvalues of V_{ij} are

$$E(\pm) = \tfrac{1}{2}[(V_{11} + V_{22}) \pm \sqrt{\{(V_{11} - V_{22})^2 + 4V_{12}V_{21}\}}] \qquad (19)$$

It turns out that for the case of the particular system we are considering, certain relationships exist between the matrix elements which greatly simplify eq. (19), namely, $V_{11} = V_{22}$, and $V_{21} = -V_{12}$. These relationships arise after one takes into account the random orientation of the molecule with respect to the light by averaging over all orientations. We shall introduce them now into eq. (19). Later we shall have more to say about the orientational averaging. With these simplifications, eq. (19) becomes

$$E(\pm) = V_{11} \pm iV_{12} \qquad (20)$$

It is clear from eq. (20) that, as we stated above, when V_{12} does not vanish, the two-fold degeneracy is lifted. The splitting produced by the interaction is

$$\delta E = 2iV_{12}$$

In order to determine what happens to the polarization of a photon which starts out being linearly polarized along the $\underline{i}$ direction, the next step is to determine the explicit time dependence of the state which initially, at time t = 0, is u_1. This is done as follows. We first form those linear combinations of u_1 and u_2 which represent the two states having the energies E(+) and E(-) (relative to the unsplit energy). Then u_1 is expressed in terms of these two states. In this form it is a trivial matter to introduce the time dependence since stationary states having the energy E contain $\exp\{-2\pi iEt/h\}$ as a factor where h is Planck's constant, or, in the case of $u(\pm)$, $\exp\{-2\pi iE(\pm)t/h\}$. We now have a function which we call $U_1(t)$ which shows how the state which was initially u_1 evolves with time. Finally, we rewrite $U_1(t)$ as a linear combination of u_1 and u_2.

The two stationary states, the states of definite energy are

$$u(\pm) = (1/\sqrt{2})(u_1 \pm iu_2) \tag{21}$$

and, in terms of these,

$$u_1 = (1/\sqrt{2})(u(+) + u(-)) \tag{22}$$

When the time dependence described above is introduced into eq. (22) and eq. (21) is used to eliminate u(+) and u(-), we get

$$U_1(t) = \exp\{-2\pi i V_{11} t/h\}(u_1 \cos\beta - u_2 \sin\beta) \tag{23}$$

where $\beta = \pi\,\delta E\,t/h$.

Can we deduce from our knowledge of the time dependence of U_1 as given by eq. (23) the way the polarization of the photon changes? The general principles of quantum mechanics allow us to calculate the probability, w, that the photon associated with the state $U_1(t)$ has a polarization $\underline{e}$ which we can quite generally write in the same form as eq. (18). In this case

$$w = |\cos\beta|^2|\cos\phi|^2 + |\sin\beta|^2|\sin\phi|^2 + [(\cos\beta)(\sin\beta^*)(\cos\phi)(\sin\phi^*) + \text{c.c.}] \tag{24}$$

where the asterisk denotes the complex conjugate quantity and c.c. denotes a term which is the complex conjugate of the preceding term.

It is easy to see that w is identically unity for all time if ϕ is chosen to be equal to β, that is, if

$$\phi = \pi\,\delta E\,t/h \tag{25}$$

Given the physical significance of w, we are thus able to answer the question we have posed at the beginning of the preceding paragraph. Equation (25) tells us that the polarization, which, as we know, is determined by ϕ, changes uniformly with time. If δE is real, the photon remains linearly polarized but its plane of polarization smoothly rotates. If δE has an imaginary part as well, then as time goes on, the photon also becomes elliptically polarized, with the major and minor axes of the ellipse doing the rotating as the ellipticity keeps increasing.

This, then, is the picture of optical activity and circular dichroism which has quite naturally fallen out of our consideration of a simple quantum mechanical system. To get an expression for ϕ which is consistent with that of the previous section we have to replace t in eq. (25) by d/c, the time it takes for light to travel d centimeters, and to multiply by n, which now denotes the number of molecular units per cc.

$$\phi = n\pi \, \delta E \, d/hc \tag{26}$$

Now eq. (26) can be used in eqs. (9) and (13) for calculating the molecular rotation $[\phi_r]$ and the molar ellipticity $[\phi_i]$. In the next section we shall use it as a starting point for discussing the circular dichroism predicted by a specific model of a specific type of molecule, that is, we shall begin by deriving an expression for the imaginary part of δE applicable to a model representing an oligonucleitide.

4. A COUPLED OSCILLATOR MODEL OF AN OLIGONUCLEITIDE

4.1. Basic theory

We begin by reviewing the significance of δE which appears in eq.(26). We know that it represents the difference in energy between the two stationary states u(+) and u(-) given in eq. (21). It should come as no surprise that in both of these states the photon can be shown to be circularly polarized (as it turns out, right-circularly polarized in u(+), and left- in u(-)). Our theory should allow for δE to be complex, for, as we have seen, its real part is proportional to the molecular rotation, and its imaginary part, which we shall focus on, to the molar ellipticity.

We shall now present a detailed derivation of the coupled oscillator model of circular dichroism. In particular, we shall be determining the contribution to the circular dichroism which is associated with the geometrical spatial arrangement of a molecule containing N identical residues. Each residue is assumed to have dimensions much smaller than the wavelength of the light, but the separation between the residues can be any size. The appropriate form for V, the potential energy associated with the interaction between the molecule and the light, is then

$$V(\pm) = -\sum_{i=1}^{N} \underline{d}_i \cdot \underline{E}_i(\pm) \tag{27}$$

where $\underline{d}_i$ is the electric dipole moment operator of the ith residue and $\underline{E}_i(\pm)$ is the value of the electric field strength in the light evaluated at the position of the ith residue. As before, the + sign attached to $\underline{E}$ refers to left-circularly polarized light, and the - sign to right-circularly polarized light.

If H^0 is the energy operator (hamiltonian) of the molecule in the absence of any light, our task now is to determine the average value of the total energy, $\langle H^0 + V\rangle$, first when the light is right-circularly polarized, and then left-. The difference is δE. In order to derive these expectation values, the state of the system must be specified. There is a factor at work here which serves to simplify the form of the state function. When we assume that the frequency of the light is always close in value to the frequency of a certain absorption band, then the state function is, to a first approximation, a linear combination of the two states involved in the transition--the ground state r_g and the appropriate excited state r_e. Thus we write for R, the state of the system under the influence of the light

$$R(\pm) = C_g(\pm)\ r_g + C_e(\pm)\ r_e \tag{28}$$

As we have indicated with the notation, the coefficients will depend on the state of polarization of the light. The functions r_g and r_e are eigenfunctions of H^0. We shall take them to be time independent so that the necessary time dependence of $R(\pm)$ is contained solely in the coefficients. Furthermore, the excited states we will be treating are the ones in which one of the N identical residues has been electronically excited. There are, of course, N such excited states all of the same energy, that is, this particular energy level is N-fold degenerate: in this class of excited states any one of the N residues may be excited but the energy associated with each residue excitation is the same. This property of the excited level will become important at a later point in our discussion.

We can introduce the natural lifetime of the states by making the eigenvalues complex, $E'_{g(e)} = E_{g(e)} - ih\Gamma_{g(e)}/4\pi$, where $E_{g(e)}$ and $h\Gamma_{g(e)}/4\pi$ are the energies and widths of the states, respectively. The lifetimes themselves are proportional to Γ^{-1}. In the case of the ground state, the lifetime is infinite, or, in other words, the state is not broadened. Accordingly, we set Γ_g equal to zero, and from here on, call Γ_e simply Γ.

We now use the state function, eq. (28), to compute the expectation value of the operator $H^0 + V$. One can immediately deduce the form of the imaginary part of δE

$$\mathrm{Im}\{\delta E\} = -(h\Gamma/4\pi)[\,|C_e(-)|^2 - |C_e(+)|^2] \tag{29}$$

Equation (29) shows that the imaginary part of δE, and therefore the molar ellipticity itself, is proportional to the product of the linewidth of the excited state and the difference in the populations of the excited state induced by the light in its two forms of circular polarization. We thus proceed by deriving the expression for this difference predicted by quantum mechanics for a molecular system under the influence of a monochromatic plane wave of either of the appropriate polarizations.

The function R(±), eq. (28), must satisfy the Schrödinger wave equation

$$(ih/2\pi)\,\partial R(\pm)/\partial t = (H^0 + V(\pm))\,R(\pm) \tag{30}$$

When R(±), as given by eq. (28), is substituted into eq. (30), and use is made of the fact that r_g and r_e are eigenfunctions of H^0, we obtain the first-order differential equation that each coefficient obeys

$$(ih/2\pi)dC_{g(e)}/dt = E'_{g(e)}C_{g(e)} + V_{ge}C_{e(g)} \tag{31}$$

where $V_{ge} = \int r_g V r_e dx$.

(For the moment we shall omit the light polarization label, ±, on V, E, and on the coefficients.) From eq. (31) one can readily prove that

$$d|C_e|^2/dt + \Gamma|C_e|^2 = -(4\pi/h)V_{ge}\mathrm{Im}\{C_e C_g^*\} \tag{32}$$

Averaging this equation over time gives as the steady-state value of $|C_e|^2$

$$|C_e|^2 = -\langle(4\pi/h\Gamma)V_{ge}\mathrm{Im}\{C_e C_g^*\}\rangle_t \tag{33}$$

It is clear from eq. (33) that in order to proceed we need an expression for $C_e C_g^*$. Just as it was possible to derive eq. (32) from eq. (31), so we can use eq. (31) to derive a differential equation for $C_e C_g^*$

$$d(C_e C_g^*)/dt + (i\Omega_0 + \tfrac{1}{2}\Gamma)C_e C_g^* = -(2\pi i V_{ge}/h)(|C_g|^2 - |C_e|^2) \tag{34}$$

where Ω_0 is $2\pi(E_e - E_g)/h$. Consistent with the low-intensity limit in which we are working, on the right-hand-side of eq. (34), we must set $|C_g|^2$ equal to one and $|C_e|^2$ equal to zero. It is evident from eq. (27) that the interaction integral V_{ge} appearing in eqs. (33) and (34) is

$$V_{ge} = -\Sigma_{i=1}^{N} (\underline{d}_i)_{ge} \cdot \underline{E}_i \qquad (35)$$

The integral vector quantities $(\underline{d}_i)_{ge}$ are the so-called transition moments associated with each residue.

We now have all that we need to determine Im{δE} according to eq. (29). The molar ellipticity, $[\emptyset_i]$, is then constructed with the help of eqs. (13) and (26). We have already noted that in order to determine the steady-state value of $|C_e|^2$, we need the steady-state $C_eC_g^*$. This is readily obtained from eq. (34) after the explicit form of the electric field strength in the light, $\underline{E}_i(\pm)$, is substituted into V_{ge} (eq. (35)). The nature of the time dependence of the right-hand-side of eq. (34) now becomes apparent--there are terms which vary as $\exp\{i\Omega t\}$ and terms which vary as $\exp\{-i\Omega t\}$. When $\Omega \approx \Omega_0$, as is the case here, the properties of eq. (34) are such that only the latter are important and need be retained. By inspection we find that

$$C_eC_g^*(\pm) = \pi E^0 \Sigma_{i=1}^{N} (\underline{d}_i)_{ge} \cdot \underline{e}(\pm) \exp\{i\Omega(z_i/c - t)\}/hL \qquad (36)$$

where $\underline{e}(\pm) = (1/\sqrt{2})(\underline{i} \pm i\underline{j})$, $L = (\Omega_0 - \Omega - \frac{1}{2}i\Gamma)$, and z_i is the component of the position vector of the ith residue along the direction of light propagation.

The imaginary part of eq. (36) is needed in eq. (33). At this point, some simplification is achieved if we first average the right-hand-side of eq. (33) over time and drop all terms which do not carry a ± sign since they will disappear anyway when the difference indicated in eq. (29) is taken. We are then left with the task of carrying out the orientational average on terms of the form $[(\underline{d}_i)_{ge} \cdot \underline{i}][(\underline{d}_j)_{ge} \cdot \underline{j}]\exp\{ik(z_i - z_j)\}$ where k is Ω/c. Such an expression is thereby converted into $\frac{1}{2}ij_1(kR_{ij})\underline{R}_{ij} \cdot (\underline{d}_i)_{ge} X (\underline{d}_j)_{ge}/R_{ij}$ where $\underline{R}$ is a vector which points from residue j to residue i and has a magnitude R_{ij} equal to the distance between these residues.[3] The function $j_1(kR_{ij})$ is $(kR_{ij})^{-2}(\sin kR_{ij} - kR_{ij}\cos kR_{ij})$. The details of this averaging process are explained in ref. 3, and will not be repeated here. Now it is a simple matter to pick out the required imaginary part of the resulting complex expression. One finally arrives at the following expression for $\emptyset_i$, the imaginary part of eq. (26)

$$\emptyset_i = (\pi^2 nk\Gamma/h)\Sigma_{i,j} j_1(kR_{ij})\underline{n}_{ij} \cdot (\underline{d}_i)_{ge} X (\underline{d}_j)_{ge}/T \qquad (37)$$

where $T = [(\Omega_0 - \Omega)^2 + (\Gamma/2)^2]$ and $\underline{n}_{ij}$ is a unit vector pointing in the same direction as R_{ij}.

As we mentioned earlier, the excited level of the type we are considering--excitation of a single residue--is N-fold degenerate, where N is the number of residues per molecule. The right-hand-side of eq. (37) must also be summed over these N states: we shall denote them with the label α. Each transition moment as well as Ω_0 should carry this label. In the absence of any interaction between the residues, and, as we shall see, ϕ_i is zero. In order to proceed, therefore, the interaction between the residues must be considered. Accordingly, we now introduce as a perturbation the operator v which describes this interaction. If u_i and u'_i represent the ground and excited states of the isolated ith residue, then the zeroth order molecular ground state wave function G_0 is

$$G_0 = \prod_i u_i \tag{38}$$

The N zeroth-order excited molecular states are linear combinations of the states $G_i = (u'_i/u_i)G_0$,

$$M_\alpha = \Sigma_{i=1}^{N} b_{i\alpha} G_i \tag{39}$$

When the functions given by eqs. (38) and (39) are used to compute the transition moments in eq. (37), each one is of the form

$$(\underline{d}_i)_{ge} = \underline{d}_i b_{i\alpha} \tag{40}$$

where $\underline{d}_i = \int u'_i \underline{d} u_i dx$.

The sums on the right-hand-side of eq. (37) then become

$$\Sigma_{i,j} j_1(kR_{ij}) \underline{n}_{ij} \cdot \underline{d}_i X \underline{d}_j \Sigma_\alpha b_{i\alpha} b_{j\alpha} / [(\Omega_{\alpha 0} - \Omega)^2 + (\Gamma/2)^2]$$

The coefficients $b_{i\alpha}$ and the energies of the now split level $\Omega_\alpha = \Omega_{\alpha 0} - \Omega_0$ are, of course, the eigenvectors and eigenvalues of the matrix $\underline{v}$ whose elements are the quantities v_{ij} where

$$v_{ij} = \int G_i v G_j dx \tag{41}$$

We have noted above that the particular type of circular dichroism we are considering here vanishes in the absence of any interaction between the residues. In the absence of such interaction, the transition frequencies $\Omega_{\alpha 0}$ are independent of α, and, therefore, the sum over α on the right-hand-side of eq. (37) reduces to $\Sigma_\alpha b_{i\alpha} b_{j\alpha}$. But because of the orthonormality of the eigenvectors, this sum just gives the elements of the unit matrix $\underline{1}$--all elements with i is not equal to j are zero, and each of the diagonal elements, those with i is equal to j, is one. But eq. (37) is still zero because the factor $\underline{n}_{ij} \cdot \underline{d}_i \times \underline{d}_j$ is zero whenever $i = j$.

It is evident from eq. (37) that according to this model, the circular dichroism can be expressed as a sum of terms, each of which corresponds to the contribution of a pair of residues. All pairs are represented. We also see from eq. (37) that each of the pairwise contributions, say the one coming from the i,jth pair, $(\phi_i)_{ij}$, can be written in the form

$$(\phi_i)_{ij} = (2\pi^2 nk\Gamma/h)A_{ij}B_{ij} \tag{42}$$

where

$$A_{ij} = j_1(kR_{ij})\underline{n}_{ij} \cdot \underline{d}_i \times \underline{d}_j \tag{43}$$

and

$$B_{ij} = \Sigma_\alpha b_{i\alpha} b_{j\alpha}/[(\Omega_{\alpha 0} - \Omega)^2 + (\Gamma/2)^2] \tag{44}$$

The factor B_{ij} contained in each term is a sum over the states associated with the excited level. To evaluate this sum the elements of the residue interaction matrix must be specified, the so-called $\underline{v}$ matrix. In fact the sum can be evaluated directly in terms of $\underline{v}$. There is no need to determine its eigenvectors and eigenvalues; the sum is just the i,jth element of the matrix $[(\underline{v}')^2 + (\Gamma/2)^2\underline{1}]^{-1}$, where $\underline{v}' = \underline{v} + (\Omega_0 - \Omega)\underline{1}$.

4.2. The effect of tertiary structure

Equation (37) is the starting point for a discussion of the effect of changes in tertiary structure on circular dichroism as predicted by the coupled oscillator model. We shall choose a specific distribution of point transition moments which serves as a model for a superhelix. To establish a reference structure, however, we first imagine the transition moments placed along a line, equally spaced, and each perpendicular to the line. Furthermore, as we progress from moment to moment, the moments rotate uniformly about the axis--each moment rotates through some constant angle with respect to the previous moment. (The total number of turns made by the moments about the axis is called the twist.)

We had previously calculated the circular dichroism predicted by eq. (37) for this distribution of moments, and, in addition, when only nearest-neighbor interactions are included.[3] Figure 1 shows the the type of frequency dependence predicted for ϕ_i/N in the limit of very large N for a right-handed turning of the moments. The units chosen for the ordinate and the abscissa, as well as the values chosen for the other parameters which must be specified are discussed in the reference.

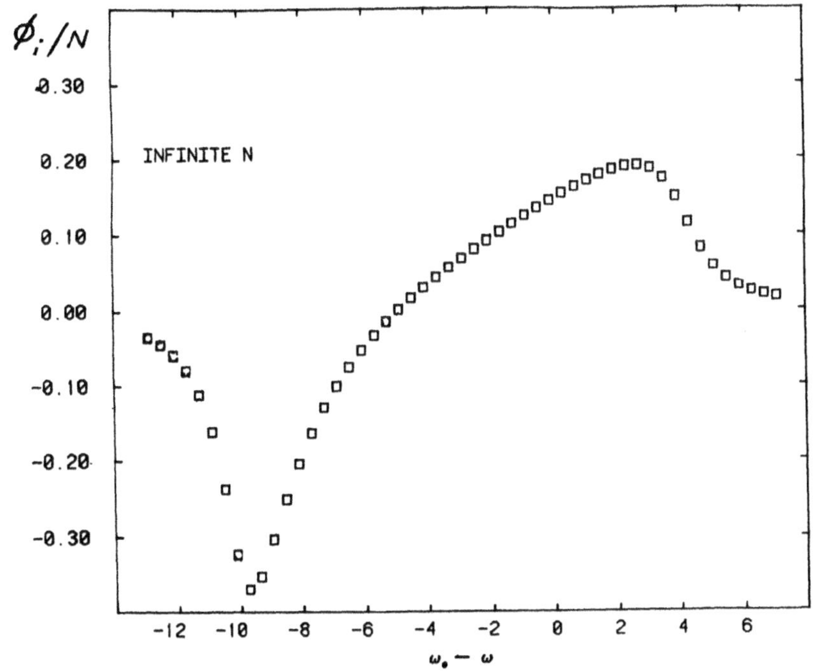

FIGURE 1. ϕ_i/N plotted as a function of frequency.

We now relax the condition that the moments lie on a line. Instead, we have them lie equally spaced on a helix. Each moment is now directed so as to be contained completely in the plane normal to the tangent to the helix at its position. In addition, the moments wind around uniformly as before. The equal spacing, perpendicularity, and uniform winding of the moments characterize the secondary structure of this model structure, while its tertiary structure is characterized by the radius and pitch of the superhelix along which the moments are situated. We thus wish to estimate the sensitivity of the circular dichroism, as predicted by eq. (37), to changes in the radius and pitch.

We first consider the evaluation of the purely geometrical factors in these terms, A_{ij}, eq. (43), for our new distribution of transition moments. The radius vector from the origin to any point P on a helix with its axis passing through the origin and lying along the z-direction is given by

$$\underline{X}(s) = \underline{i}a \cos \tau s + \underline{j}a \sin \tau s + \underline{k}b\tau s \tag{45}$$

where a and b are constants, ($2\pi b$ is the pitch of the helix, and a is the radius of the cylindrical surface containing the helix.) $\tau = |a^2 + b^2|^{-\frac{1}{2}}$, and $|s|$ is the distance traversed along the helix from the point where the helix intercepts the x,y plane to the point P. For right-handed-helices b is positive, and for left-handed ones, it is negative. Then $|\underline{X}(s_i) - \underline{X}(s_j)|$, which is R_{ij}, the distance between moments i and j, turns out to be

$$R_{ij} = 2[a^2 \sin^2 \tfrac{1}{2}(k - 1)\tau \, \delta s_0 + \tfrac{1}{4}(k - 1)^2 b^2 \, \delta s_0{}^2]^{-\frac{1}{2}} \quad (46)$$

and the unit vector which points from the jth to the ith moment is

$$\underline{n}_{ij} = R_{ij}{}^{-1}[\underline{X}(s_i) - \underline{X}(s_j)] \quad (47)$$

where δs_0 is the constant separation, measures along the helix, between adjacent moments.

For computing the vector cross product, $\underline{d}_i$ X $\underline{d}_j$, it is convenient to introduce still another coordinate system, a local right-handed cartesian system associated with each point on the helix. The three mutually orthogonal unit vectors characterizing this system are $\underline{t}$, the tangent to the helix at the point in question, $\underline{n}$, the so-called principal normal, and $\underline{b}$, the so-called binormal. These three vectors are illustrated at a point of an arbitrary space curve in fig. 2.

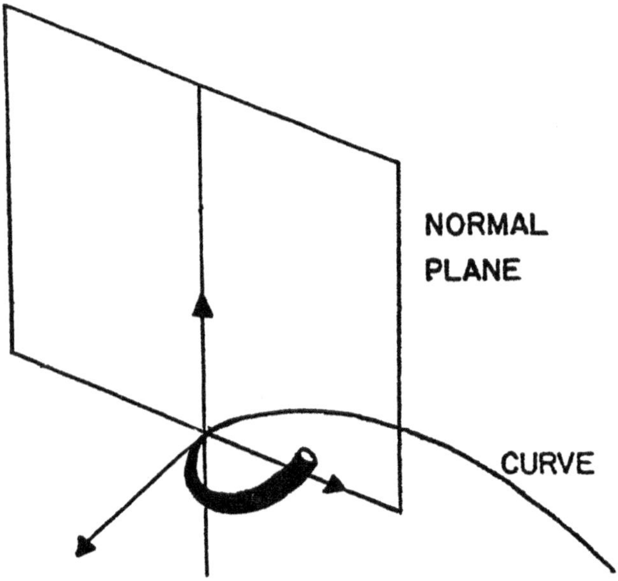

FIGURE 2. The tangent, principal normal, and binormal at a point of an arbitrary space curve.

The tangent is readily obtained from $\underline{X}(s)$ by differentiating with respect to s

$$\underline{t} = d\underline{X}(s)/ds = -\underline{i}a\tau\sin \tau s + \underline{j}a\tau\cos \tau s + \underline{k}b\tau \qquad (48)$$

Similarly, the principal normal is obtained from $\underline{t}$

$$\underline{n} = ad\underline{t}/ds = -\underline{i}\cos \tau s - \underline{j}\sin \tau s \qquad (49)$$

The binormal, which is orthogonal to $\underline{t}$ and $\underline{n}$, is

$$\underline{b} = \underline{t} \times \underline{n} = \underline{i}b\tau\sin \tau s - \underline{j}b\tau\cos \tau s + \underline{k}a\tau \qquad (50)$$

In the case we are treating, each transition moment, for example the ith, can be simply expressed in terms of $\underline{n}$ and $\underline{b}$

$$\underline{d}_i = d(\underline{n}_i\cos \alpha_i + \underline{b}_i\sin \alpha_i) \qquad (51)$$

where d is the common magnitude of the transition moments and α_i is the angle that this moment makes with the binormal to the helix at the point where the moment is located. This angle itself depends on the tertiary structure. When the moments lie on the line, we can express α_i as follows

$$\alpha_i = \alpha^0 + 2\pi i H_S/N' \qquad (52)$$

where α^0 is an arbitrary phase, and N' is the number of moments per turn of the secondary structure. The factor H_S allows us to specify the handedness of the structure. We assign it to be +1 when the structure is right-handed and -1 when it is left-handed. When we distort the line of moments into a helix, we do so in such a way so as not to change the secondary structure. In particular, we keep the twist constant. Even so the angle α_i changes in form. (We shall not discuss this point in any detail here except to say that it turns out that as the local coordinate frame we have introduced moves along the helix, the principal normal and the binormal themselves rotate about the tangent at a rate determined by the quantity $b\tau^2$. Thus the angle of rotation of the moments, measured with respect to this coordinate frame, appears to change as the helical parameters are changed.) The angle is now

$$\alpha_i = \alpha^0 + i\phi \qquad (53)$$

where

$$\phi = 2\pi H_S/N' - b\tau^2\ \delta s^0 \tag{54}$$

Using eqs. (45), (47), (48), (49), (50), (51), and (53) we obtain for A_{ij}, eq. (43)

$$\begin{aligned} A_{ij} = R_{ij}^{-1}j_1(kR_{ij})\{C_{ij}b\tau(i-j)\ \delta s_0 \sin(i-j)\tau\ \delta s_0 \\ - D_{ij}[b\tau^2(i-j)\ \delta s_0 \cos(i-j)\tau\ \delta s_0 + a\tau^2 \sin(i-j)\tau\ \delta s_0] \\ + E_{ij}[b^3\tau^3(i-j)\ \delta s_0 \sin(i-j)\tau\ \delta s_0 \\ + 4a^2b\tau^2 \sin^2 \tfrac{1}{2}(i-j)\tau\ \delta s_0]\} \end{aligned} \tag{55}$$

where

$$C_{ij} = \cos\alpha_i \cos\alpha_j$$

$$D_{ij} = \sin(\alpha_i - \alpha_j) = \sin(i-j)\phi$$

and $E_{ij} = \sin\alpha_i \sin\alpha_j$

When we average eq. (55) over the arbitrary phase α_0, D_{ij} is unchanged since it is independent of α_0, but C_{ij} and E_{ij} both become equal to $\frac{1}{2}\cos(i-j)\phi$.

In the cases of physical interest, it can be shown that the second factor, B_{ij}, making up the contribution to the circular dichroism of the pair of moments labeled i and j falls off sharply in magnitude as the magnitude of the difference $i - j$ increases, that is, the dominant contribution comes from closely situated pairs. We thus need to examine eq. (55) only for the case of $|i - j| \approx 1$. It is also true for the cases of physical interest that the radius of curvature and pitch of the superhelix are both large compared with the corresponding parameters of the helix characterizing the secondary structure. We can thus treat $(i - j)\tau\ \delta s_0$ as a small parameter in eq. (55) and expand the right-hand-side in terms of it. (Note that when $|i - j| \approx 1$ we can set $R_{ij}^{-1}j_1(kR_{ij})$ equal to $k/3$.) If we truncate this expansion after the terms linear in the small parameter we obtain

$$\begin{aligned} A_{ij} = -[k(i-j)\ \delta s_0/3][\sin 2\pi H_S(i-j)/N' \\ - b\tau^2(i-j)\ \delta s_0 \cos 2\pi H_S(i-j)/N'] \end{aligned} \tag{56}$$

The first term on the right-hand-side of eq. (56) is independent of a and b. It depends entirely on the secondary structure and corresponds to the value of A_{ij} when the transition moments lie on a straight line. The second term, which does contain the parameters of the tertiary structure, differs in magnitude from the first by a factor $b\tau^2(i - j)\ \delta s_0$. For the molecular systems which we are concerned with here, this is very much less than unity. We expect that the situation is similar in the case of the factor B_{ij} because in the $\underline{v}$ matrix the elements v_{ij} for which $|i - j|$ is large, those we expect to be most sensitive to tertiary structure, are negligible compared with the ones with $|i - j| \approx 1$, but the latter are little changed from the straight-line case.

We are thus able to conclude that for the usual DNA-type molecular systems in which the pitch and radius of curvature of the superhelix are large compared the pitch and radius of curvature of the helix associated with the secondary structure, the coupled-oscillator model predicts that the circular dichroism is not very sensitive to changes in the tertiary structure--indeed, the prediction is that one should be hard-pressed to tell from this kind of measurement whether the DNA were linear or supercoiled. We must now turn to experiment to test the validity of this prediction.

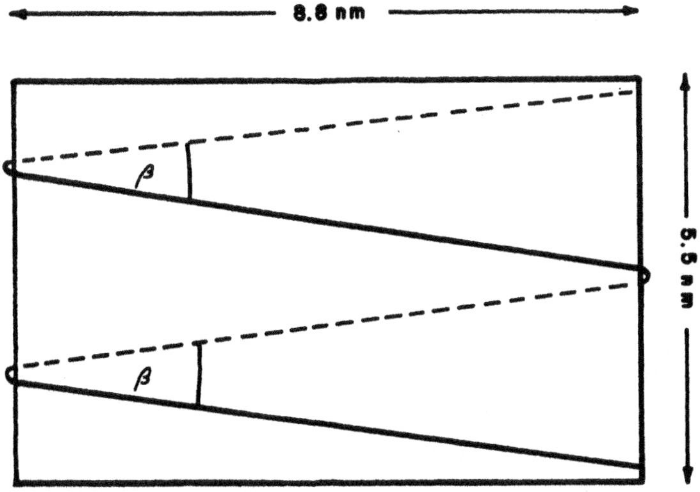

FIGURE 3. Schematic representation of core DNA in a mononucleosome.

One system on which detailed measurements of circular dichroism have been made is the aqueous solution of mononucleosomes. This is an attractive system to study because the tertiary structure of the B-DNA associated with the histone core in nucleosomes has been well-characterized. The core DNA, shown schematically in fig. 3, is believed to consist of somewhat under two turns of a left-handed superhelix with a radius, a, of 44 Å and a pitch of 28 Å, corresponding to a value of b of 4.5 Å. For these values of a and b, and with δs_0 set equal to 3.4 Å, the spacing between base pairs, the parameter $b\tau^2 \delta s_0$ discussed above has a value of 0.008. We would expect, therefore, that CD bands associated with molecular transitions of the type to which the coupled oscillator model is applicable would be virtually identical for linear DNA and for core DNA in mononucleosomes.

Such a band is the one in the ultraviolet centered at a wavelength of 258 nm. The absorption at this wavelength is believed to be associated with a localized π to π^* transition in the bases, with the corresponding transition moments oriented in the plane of the base pairs. Contrary to expectations, however, the experimental molar ellipticity in the region of the 258 nm band is markedly different for DNA bound to the core from that for free DNA.(4) The CD band for the free DNA shows the same qualitative shape predicted by the coupled oscillator model, (See fig. 1.), but in the case of the core DNA although the negative lobe on the short-wavelength side of the transition is pretty much unchanged, the positive lobe on the long-wavelength side is all but gone. The authors of ref. 4 show that to get the molar ellipticity of the core DNA from that of free DNA one must add to the latter a negative band centered at 275 nm. The authors suggest that this band is characteristic of the tertiary structure of the core DNA. From the discussion here, it is clear that, if so, we must look beyond the coupled oscillator model to find it.

5. ANOTHER MODEL OF CIRCULAR DICHROISM: THE MONONUCLEOSOME CORE AS A TRAIN OF TWO OPTICAL ELEMENTS

The schematic diagram of the tertiary structure of the core DNA, fig. 3, suggests another model which leads to circular dichroism.(5) Each planar projection of the front and rear surface of the cylindrical nucleosomal core is a projection of oriented DNA. Oriented DNA is <u>linearly</u> birefringent--we shall call $n_r(\sigma)$ the real part of the index of refraction for linearly polarized light having an E-field perpendicular to the secondary helical axis and $n_r(\pi)$ the real part of the index of refraction when it is directed parallel to this axis. In the linearly birefringent medium the difference $\delta n_r = n_r(\sigma) - n_r(\pi)$ is not zero. When the wavelength of the light is around the molecular transition centered at 258 nm., the oriented DNA is also linearly dichroic, for when the E-field is parallel to the transition moments, which lie in the plane of the base pairs, the light is absorbed, and when the E-field is perpendicular to the plane of the base pairs, and, therefore, also perpendicular to the transition moments, the light is not absorbed. As we have previously seen, the imaginary part of the index of refraction accounts for absorption. In this case we have $n_i(\sigma)$ and $n_i(\pi)$ with $\delta n_i = n_i(\sigma) - n_i(\pi)$ not zero.

How does the circular dichroism arise? Each cylindrical nucleosomal core, oriented as in fig. 3, appears to the light like two planar complex linear retarders, that is, planar optical elements which are both linearly birefringent and linearly dichroic. Furthermore, the two retarders are rotated with respect to each other through an angle β. Such an arrangement of optical elements, it can be shown, is circularly dichroic, and if we call, as before, I(+) the intensity of transmitted left-circularly polarized light and I(-) the intensity of right circularly polarized light, it turns out that

$$I_-/I_+ = (ks)^2\ \delta n_r\ \delta n_i \sin 2\beta \qquad (57)$$

where $I_- = I(-) - I(+)$, I_+ is the sum of these quantities, and s is the thickness of each retarder.[5] The wave number $k = \Omega/c$ has been used previously.

Equation (57), which is applicable to one pair of planar retarders completely intercepting a light beam, cannot be used as it stands. What we are dealing with in an aqueous solution of mononucleosomes is a collection of many randomly oriented cylinders, each intercepting only a tiny fraction of the light. We begin the modification of eq. (57) by multiplying its right-hand-side by the total area cross-sectional area of of core structures, oriented as shown in fig. 3, which intercepts a light beam of unit cross-sectional area. There are $MNl/1000n_B$ such structures, where M is the molarity of the bases, N is Avogadro's number, n_B is the number of bases per mononucleosome, and l is the path length of the light through the solution. If a is the cross-sectional area of a single core structure in cm², eq. (57) then becomes

$$I_-/I_+ = (aMNl/1000n_B)(ks)^2\ \delta n_r\ \delta n_i \sin 2\beta \qquad (58)$$

We must also correct for the effect of random orientation of the mononucleosomes with respect to the incident light. For light incident perpendicular to the plane of fig. 3, clearly nothing changes when we rotate the cylindrical core about axis of the cylinder. It not as clear, but nevertheless true, that nothing changes either when we rotate the core about a line which is perpendicular to the cylindrical axis and also perpendicular to the plane of the figure. The situation will change, however, if we rotate about the axis which is perpendicular to the cylindrical axis and lies in the plane of the figure. But for light incident on the core directed along the cylindrical axis there is no birefringence or dichroism. We thus correct eq. (58) for the effects of random orientation by multiplying its right-hand-side by 2/3.

There is one final correction to be made. We have so far assumed that the values of δn_r and δn_i are not functions of position on the retarders. In fact, this is not so. The orientation of the base pairs is such that, for each, the normal to the aromatic plane is perpendicular to the normal to the cylindrical surface. For the light, however, the angle between the direction of propagation and the surface normal, ϕ, the angle of incidence, varies between $-\frac{1}{2}\pi$ and $\frac{1}{2}\pi$. If we now specify that δn_r and δn_i are the values of these quantities for the case of an angle of incidence of zero, (light incident perpendicular to the cylinder), then, it can be shown, $\delta n_r \cos^2\phi$ and $\delta n_i \cos^2\phi$ are their values for any other angle of incidence. We take this effect into account by multiplying $\delta n_r\, \delta n_i$ in eq. (58) by the average value of $\cos^4\phi$, 3/8. Both corrections may be taken into account by multiplying the right-hand-side of eq. (58) by 1/4. We finally obtain for the contribution of the tertiary structure to the molar ellipticity of this solution

$$[\phi_i]^T = (0.72Na/n_B)(ks)^2\, \delta n_r\, \delta n_i \sin 2\beta \qquad (59)$$

We now discuss the predictions of eq. (59). The sign and wavelength dependence of $[\phi_i]^T$ are determined by the product $\delta n_r\, \delta n_i \sin 2\beta$. The angle β can be deduced from the known geometry of the mononucleosome and turns out to be -0.31 rad. With this value of β, $\sin 2\beta$ is -0.59. Furthermore, in the neighborhood of an isolated absorption band, it is possible to say enough about the general form of the wavelength dependence of $\delta n_r\, \delta n_i$ to decide whether this model can predict the observed negative band apparently associated with the tertiary structure. In dilute media the connection between the complex index of refraction and the complex polarizability here takes the form

$$\begin{aligned} \delta n_i &= -B\, \mathrm{Im}\{\alpha_\sigma(\lambda - \lambda_0)\} \\ \delta n_r &= \delta n_0 + B\, \mathrm{Re}\{\alpha_\sigma(\lambda - \lambda_0)\} \end{aligned} \qquad (60)$$

where δn_0 and B are real constants ($B > 0$), and α_σ is the complex polarizability associated with the transition ($\alpha_\pi = 0$ for this transition.) which is centered at λ_0 (= 258 nm). The wavelength dependence of the complex polarizability in the vicinity of a Gaussian-shaped band is given by the complex error function

$$\alpha \sim \exp\{-x^2\}[2\pi^{-\frac{1}{2}}\int_0^x \exp\{t^2\}\, dt - i] \qquad (61)$$

where $x = 2(\ln 2)^{\frac{1}{2}}(\lambda - \lambda_0)/\Delta\lambda$ and $\Delta\lambda$ is the full-width at half-height of the absorption band.

Curves I and II of fig. 4 show δn_i and $\delta n_r - \delta n_0$ with B chosen so as to normalize δn_i to unity at the line center $\lambda = \lambda_0$. To get δn_r, curve II must be shifted along the ordinate by an amount δn_0. This constant has the significance of being the value of δn_r at a given wavelength if we were to omit the contribution of the contribution of the band centered at λ_0. Since π electrons are much more polarizable in the plane of the base pairs, we expect δn_0 to be positive, and, therefore, to get δn_r, curve II must be raised. If it is raised just enough to be everywhere positive, curve III is the result. Then the product $-\delta n_r\ \delta n_i$, curve IV, which represents the effect of the tertiary structure of the mononucleosome on the molar ellipticity, is in the form of a single band,negative at all wavelengths, much like the experimental observations. If the wavelength scale of fig. 4 is normalized so that the maximum of curve I is set at 258 nm, and its full-width at half-height is set at 46nm, the experimental values, then curve IV has a minimum at 269nm and a linewidth of 32nm. These should be compared to the experimental values of 275 and 33 nm, respectively.

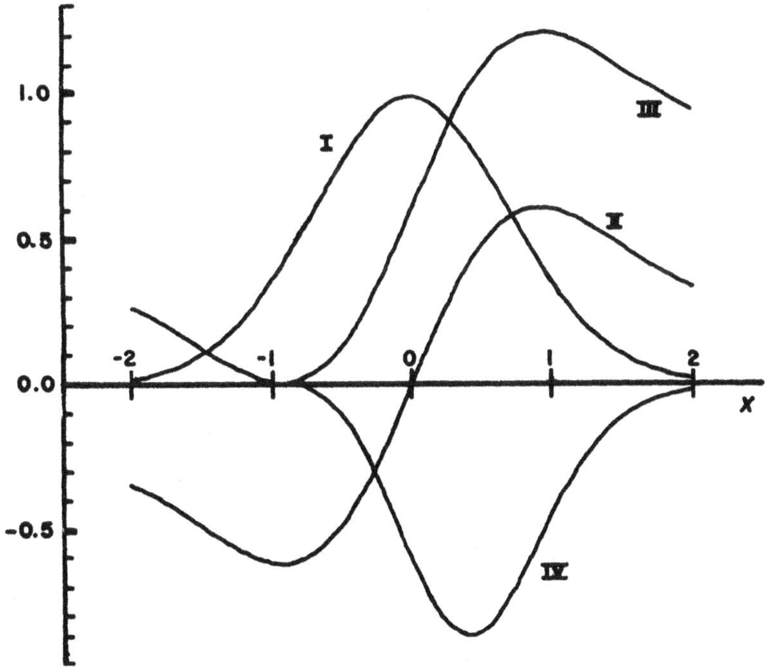

FIGURE 4. Curve I, δn_i vs $x = 2(\ln 2)^{\frac{1}{2}}(\lambda - \lambda_0)/\Delta\lambda$. Curve II, $\delta n_r - \delta n_0$ vs x. Curve III, δn_r vs x. Curve IV, $-\delta n_r\ \delta n_i$ vs x. Curves are normalized so that $\delta n_i = 1$ at x = 0.

Finally, using eq. (59), we shall make an estimate of the predicted magnitude of $[\emptyset_i]^T$ at the wavelength of its minimum. We first express $\delta n_r\ \delta n_i$ at the wavelength of the absorption maximum in terms of ε_{258}, the measured molar extinction coefficient of B-DNA. We have (see eq. (16))

$$\delta n_i = n_\sigma = (2.3/2k)\varepsilon_\sigma M_B \tag{62}$$

where ε_σ is the molar extinction coefficient of a film of oriented B-DNA for light linearly polarized perpendicular to the secondary helical axis, and M_B is the molarity of the bases in the film. From the dimensions of the mononucleosome we can estimate this to be

$$M_B = 1000 n_B/2Nab \tag{63}$$

Furthermore, $\varepsilon_\sigma = (3/2)\varepsilon_{258}$. We can also determine from fig. 4 that $\delta n_r = 0.61\ \delta n_i$ and that the value of $\delta n_r\ \delta n_i$ at its minimum is 1.42 times larger in magnitude than its value at the wavelength of the absorption maximum. We thus get for $[\emptyset_i]^T$ at its minimum

$$[\emptyset_i]^T = 0.46 \times 10^6 (n_B/aN)(\varepsilon_{258})^2 \sin 2\beta \tag{64}$$

With $\varepsilon_{258} = 6600 M^{-1}/cm$, $a = 8.8 \times 5.5 \times 10^{-14} cm^2$, $n_B = 280$, and $\beta = -0.31$ rad, eq. (64) gives $[\emptyset_i]^T = -11.4 \times 10^3$ deg cm²/dmol. The experimental band has minimum value of -8.5×10^3 in these units. This model, then, seems to account for the principle features of the observed change in the molar ellipticity of B-DNA when it is associated with the histone core in nucleosomes.

In the argument presented above, the quantity δn_0 was treated like an empirical parameter, that is, we found that it was possible to find a value for it such that the wavelength dependence of $[\emptyset_i]^T$ as given by eq. (59) matched the experimental dependence, and it was this value that we chose. We have pointed out that δn_0 has physical significance, so it is important to ascertain whether the chosen value is reasonable. From the fact that when λ = 258nm, $\delta n_r = \delta n_0$, and that for the chosen value of δn_0, $\delta n_r = 0.61\ \delta n_i$, we learn from eq. (62) that $\delta n_0 = 0.031$ with b taken as 4.4 nm, the radius of the mononucleosome.

A quantity $\delta\alpha = \alpha_\sigma - \alpha_\pi$, the anisotropy of polarizability associated with a base pair, occurs in the theory relating flow birefringence to molecular parameters. This can be related to δn_0 by the equation

$$\delta n_0 = 2\pi(n_B/4ab)\ \delta\alpha \tag{65}$$

where the quantity in parentheses represents the number density of base pairs in the film. Estimates of $\delta\alpha$ vary between 12.5 and 19 Å^3. Equation (65) shows that this implies a range in δn_0 from 0.026 to 0.039, consistent with the value 0.031 chosen here.

One lesson to be learned from the success of this model is that it is not generally possible to explain all observations in this type of measurement in terms of changes in secondary structure alone. As we have just seen, a CD spectrum may also be sensitive to tertiary structure, and we expect, to higher order structure as well.

REFERENCES

1. Moffitt W, Fitts DD, and Kirkwood JG: Proc. Nat. Acad. Sci. USA 43, (1957) 723
2. Tobias I, Brocki TR, and Balazs NL: J. Chem. Phys. 62, (1975) 4181
3. Tobias I and Balazs NL: Chem. Phys. 44, (1979) 9
4. Cowman MK and Fasman GD: Proc. Nat. Acad. Sci. USA 75, (1978) 4759
5. Tobias I: Biopolymers 23, (1984) 1315

ZEEMAN LASER SCATTERING (ZLS): A NEW LIGHT SCATTERING TECHNIQUE

ROGER G. JOHNSTON, SHERMILA B. SINGHAM, AND GARY C. SALZMAN

Experimental Pathology Group
Life Sciences Division
Los Alamos National Laboratory
Los Alamos, New Mexico USA

1. INTRODUCTION

Measurements of the phase and polarization properties of elastically scattered light have received considerable attention in recent years [1-11]. They provide more information about the size, morphology, and internal structure of the scattering particle than do simple measurements of the scattered irradiance [1].

Polarization properties of the scattered light can be determined by measuring the Mueller scattering matrix [1,6-11]. Phase information can be obtained from optical heterodyne techniques [2,3]. The instrumentation in either case is fairly complex. The measurements are highly susceptible to experimental artifacts [12,13].

The recent commercial availablility of the two-frequency, Zeeman Effect laser [14] permits the development of a new light scattering technique, which we call **Zeeman Laser Scattering (ZLS)**. ZLS allows access to much of the information contained in the Mueller scattering matrix or available from optical heterodyne measurements. The instrumentation, however, is much simpler and the measurements are relatively free of experimental artifacts.

2. EXPERIMENTAL IMPLEMENTATION

The existing two-frequency, Zeeman Effect laser [14] uses the Zeeman effect [15] to produce two coherent, colinear laser beams having orthogonal polarizations. The polarizations may be linear or circular, but only linear polarizations will be considered here. The two beams have a power of 0.5 mw each at approximately 632.8 nm. They differ in frequency by only 250 kHz. This frequency difference is stable to 0.1 parts per million.

When the two colinear beams produced by the Zeeman Effect laser interfere on a photodetector, a 250 kHz beat frequency is detected. The phase of this 250 kHz sine wave depends on the optical phase difference between the two laser beams. The phase difference between two orthogonally polarized electromagnetic waves oscillating at half a million GHz can thus be measured at an experimentally accessible 250 kHz frequency.

The instrument we have used to make ZLS measurements on aqueous suspensions of particles is shown schematically in Figure 1. The output of the Zeeman Effect laser (ZEL) passes unfocused into a scattering cell containing the sample of interest. The light scattered at each scattering angle θ is detected with a photomultiplier tube (PMT) mounted on an arm that rotates under computer control about the scattering cell. A lockin amplifier is used to record the phase and amplitude of the 250 kHz beat frequency.

Before reaching the photomultiplier tube, the scattered light must pass through an analyzing polarizer (POL) and two apertures (A1 and A2) that define the scattering volume. The angle of the polarizer's transmission axis (0, 45, or 90 degrees) determines the nature of the recorded ZLS information.

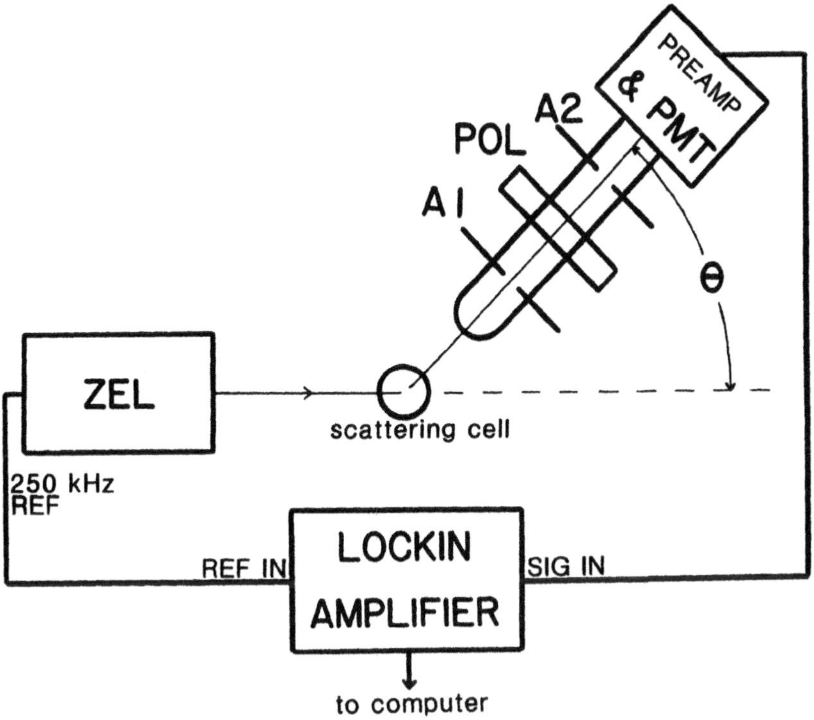

Figure 1. The ZLS instrument. Note its simplicity. There is only one optical component to orient: a polarizer (POL). Electronics within the Zeeman Effect laser (ZEL) produce a 250 kHz reference sine wave which is used by the lockin amplifier as a phase and frequency reference.

3. ZLS THEORY

ZLS is sensitive to three different scattering mechanisms. Each depends on the nature of the scatterer. A particular scattering particle can invoke all three mechanisms simultaneously.

At a given scattering angle, the scattering particle may scatter the incident horizontal and vertical polarizations with different efficiencies. This can be measured by setting the transmission axis of the analyzing polarizer at 45 degrees to the horizontal and recording the **amplitude** of the 250 kHz beat frequency.

The scattering particle may also retard one of the two scattered polarizations relative to the other. This will be apparent from the **phase** of the 250 kHz beat frequency when the polarizer is oriented at 45 degrees.

The scatterer can also partially convert one orthogonal polarization into the other during the scattering process. The extent and relative retardation of this "mixing" of polarizations depends on the asymmetry of the particle. It can be characterized by measuring the **amplitude** and **phase**, respectively, of the 250 kHz beat frequency when the polarizer is at 0 and 90 degrees.

A quantitative theory for ZLS follows from the scattering amplitude matrix formalism for light scattering [1]. The scattering amplitude matrix consists of four complex matrix elements S_j (j=1,2,3,4), where $S_j = A_j e^{i\psi_j}$ with A_j and ψ_j real. For scattering of the Zeeman Effect laser light by a single scattering particle with fixed orientation, the relationship between the scattered and incident electric fields is given by

$$\begin{pmatrix} E_{\parallel s} \\ E_{\perp s} \end{pmatrix} = \frac{e^{ik(r-z)}}{-ikr} \begin{pmatrix} S_2 & S_3 \\ S_4 & S_1 \end{pmatrix} \begin{pmatrix} E^o_{\parallel i}\ e^{-i\omega_1 t} \\ E^o_{\perp i}\ e^{-i\omega_2 t} \end{pmatrix} \tag{1}$$

where $E_{\parallel i} = E^o_{\parallel i}\ e^{-i\omega_1 t}$ is the incident electric field parallel to the scattering plane , $E_{\perp i} = E^o_{\perp i}\ e^{-i\omega_2 t}$ is the incident electric field perpendicular to the scattering plane, $E_{\parallel s}$ and $E_{\perp s}$ are the scattered electric fields parallel and perpendicular to the scattering plane, r is the distance from the scatterer to the detector, z is the position of the scatterer, k is the wavenumber, and ω_1 and ω_2 are the output frequencies of the Zeeman Effect laser with $\Delta\omega/2\pi = (\omega_2-\omega_1)/2\pi = 250$ kHz.

After the scattered light passes through the analyzing polarizer (POL in Figure 1) the scattered intensity, I, at the photodetector has the form:

$$I = DC + \Gamma \cos(\Delta\omega t + \gamma) \tag{2}$$

where DC is a relatively uninteresting time-independent constant and Γ and γ are the amplitude and phase, respectively, of the 250 kHz beat frequency.

The ZLS phase, γ, in Equation (2) for a single scatterer with the analyzing polarizer oriented at 0, 45, and 90 degrees, respectively, can be shown from Equation (1) to be:

$$\gamma(0^\circ) = \psi_2 - \psi_3 \tag{3}$$

$$\gamma(45^\circ) = \tan^{-1}\frac{A_2\sin\psi_2 + A_4\sin\psi_4}{A_2\cos\psi_2 + A_4\cos\psi_4} - \tan^{-1}\frac{A_1\sin\psi_1 + A_3\sin\psi_3}{A_1\cos\psi_1 + A_3\cos\psi_3} \tag{4}$$

$$\gamma(90^\circ) = \psi_4 - \psi_1 \tag{5}$$

where $S_j = A_j e^{i\psi_j}$.

These equations may also be expressed in terms of the more familiar Mueller matrix elements (S_{ij}):

$$\gamma(0^\circ) = \tan^{-1} \frac{S_{14} + S_{24}}{S_{13} + S_{23}} \tag{6}$$

$$\gamma(45^\circ) = \tan^{-1} \frac{S_{14} + S_{34}}{S_{13} + S_{33}} \tag{7}$$

$$\gamma(90^\circ) = \tan^{-1} \frac{S_{14} - S_{24}}{S_{13} - S_{23}} \tag{8}$$

where the Mueller scattering matrix elements S_{ij} are defined by the following relation between the incident and scattered Stokes parameters [1]:

$$\begin{pmatrix} I_s \\ Q_s \\ U_s \\ V_s \end{pmatrix} = \frac{1}{k^2 r^2} \begin{pmatrix} S_{11} & S_{12} & S_{13} & S_{14} \\ S_{21} & S_{22} & S_{23} & S_{24} \\ S_{31} & S_{32} & S_{33} & S_{34} \\ S_{41} & S_{42} & S_{43} & S_{44} \end{pmatrix} \begin{pmatrix} I_i \\ Q_i \\ U_i \\ V_i \end{pmatrix} \tag{9}$$

The amplitude, Γ, of the 250 kHz beat frequency also can be expressed in terms of the Mueller scattering matrix elements. For the analyzing polarizer at 0, 45, and 90 degrees, Γ is proportional to the following functions of the S_{ij}'s:

$$\Gamma(0^\circ) \propto \mathrm{SQRT}[(S_{13} + S_{23})^2 + (S_{14} + S_{24})^2] \tag{10}$$

$$\Gamma(45^\circ) \propto \mathrm{SQRT}[(S_{13} + S_{33})^2 + (S_{14} + S_{34})^2] \tag{11}$$

$$\Gamma(90^\circ) \propto \mathrm{SQRT}[(S_{13} - S_{23})^2 + (S_{14} - S_{24})^2] \tag{12}$$

Equations (3)-(12) are valid for scattering from a **single** particle having a **fixed** orientation. The equations are also valid for scattering from a collection of **spheres** if the spheres have an isotropic dielectric constant and can be treated as independent scatterers.

To model scattering from an aqueous suspension of **non-spherical** particles, it is necessary to rotationally average the scattering from individual particles. The rotationally averaged values for S_{13}, S_{23}, S_{14}, and S_{24} will be zero for non-chiral particles [1]. As a consequence, $\Gamma(0^\circ)$ and $\Gamma(90^\circ)$ in Equations (10) and (12) will be zero for rotationally averaged particles, unless the particles are chiral. ZLS is thus sensitive to the chemical or morphological chirality of aqueous suspensions when the analyzing polarizer is oriented at 0 or 90 degrees.

4. EXPERIMENTAL RESULTS

The ZLS data for Figures 2-9 were taken with the analyzing polarizer at 45 degrees. Figures 2-5 show the ZLS phase and amplitude as a function of scattering angle for polystyrene microspheres. Figures 6-9 demonstrate discrimination between different bacterial species in pure aqueous suspensions.

Figure 10 shows the ZLS phase for three biological specimens when the polarizer is at 0 degrees. No ZLS phase or amplitude can be measured above the noise for polystyrene microspheres when the analyzing polarizer is at 0 or 90 degrees. This is to be expected from theory since the microspheres are non-chiral.

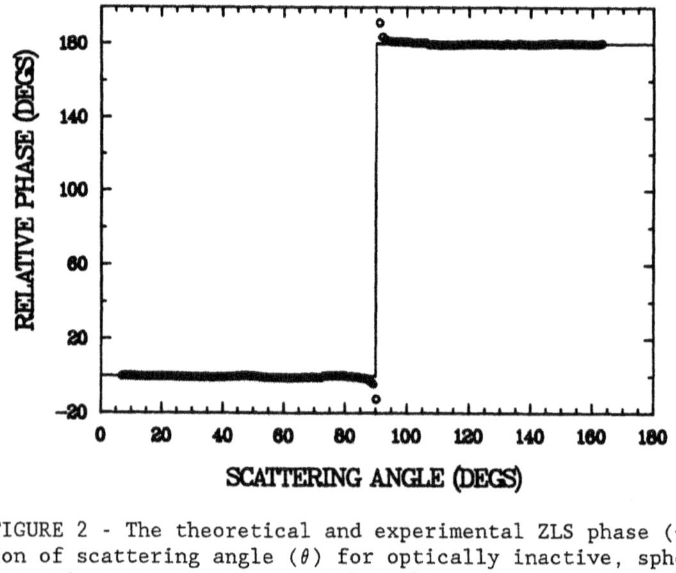

↑FIGURE 2 - The theoretical and experimental ZLS phase (γ) as a function of scattering angle (θ) for optically inactive, spherical dipole scatterers in aqueous suspension when the analyzing polarizer is at 45 degrees. The experimental data (circles) are for scattering from polystyrene microspheres of diameter 85 nm. The data are independent of microsphere concentration. The 180 degree phase shift between forward and backward scattering angles can be predicted analytically [1] and is analogous to the phase shift seen in specular reflection from a dielectric surface.

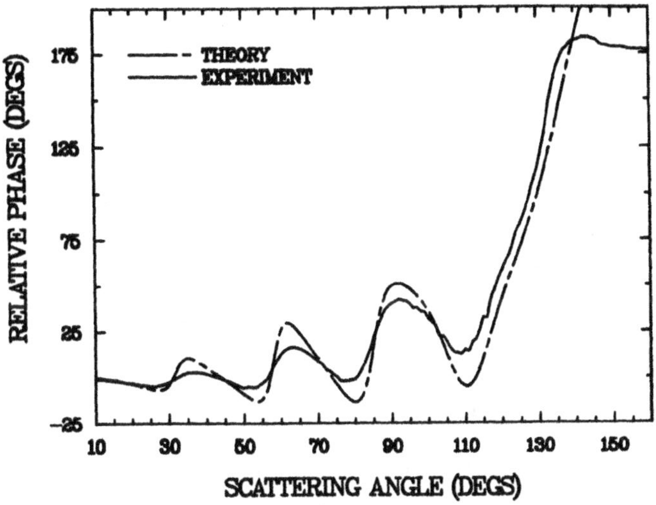

†FIGURE 3 - ZLS phase as a function of scattering angle for 1.09 μm diameter polystyrene microspheres with the analyzing polarizer at 45 degrees. The theoretical curve was determined from ZLS theory and Mie scattering calculations. The theoretical curve has no adjustable parameters and is independent of concentration. The general shape of the experimental curve is independent of concentration but, as is the case for all scatterers larger than dipoles, the amplitude of the ripples increases strongly with decreasing concentration. The experimental data were taken every 1 degree of scattering angle. The microsphere concentration was 6 X 10^7/ml.

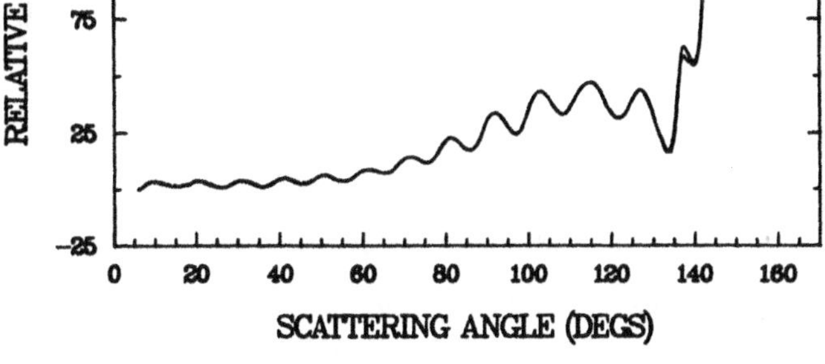

†FIGURE 4 - Typical reproducibility of the ZLS phase for polystyrene microspheres with the polarizer at 45 degrees. The two data runs nearly

overlap. The microspheres were 2.95 μm in diameter and 4 X 10^6/ml.

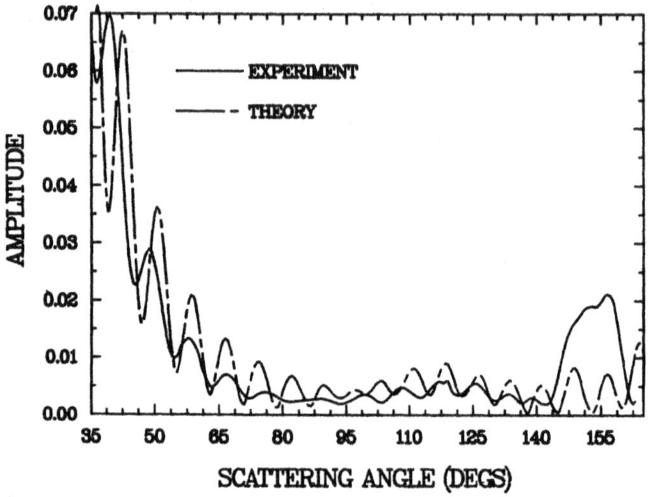

†FIGURE 5 - Relative amplitude (Γ) of the ZLS beat frequency as a function of scattering angle for 3.53 μm diameter polystyrene microspheres with the polarizer at 45 degrees. The microsphere concentration for the experimental curve was 8 X 10^5/ml.

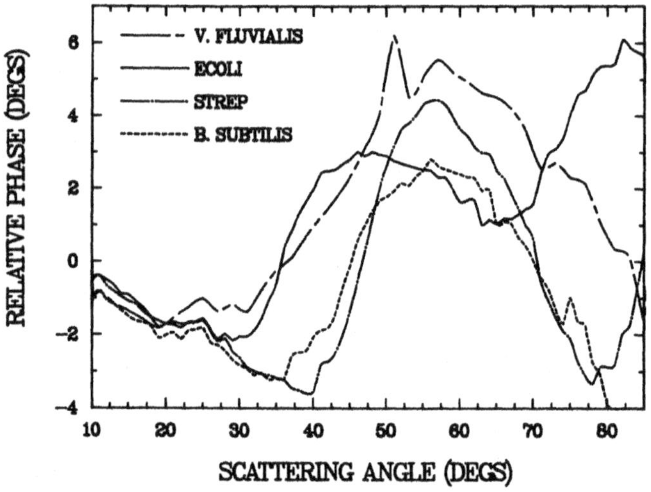

†FIGURE 6 - ZLS phase discrimination between pure suspensions of bacteria with the polarizer at 45 degrees for forward scattering angles. The concentration of each sample was approximately 1 X 10^7/ml.

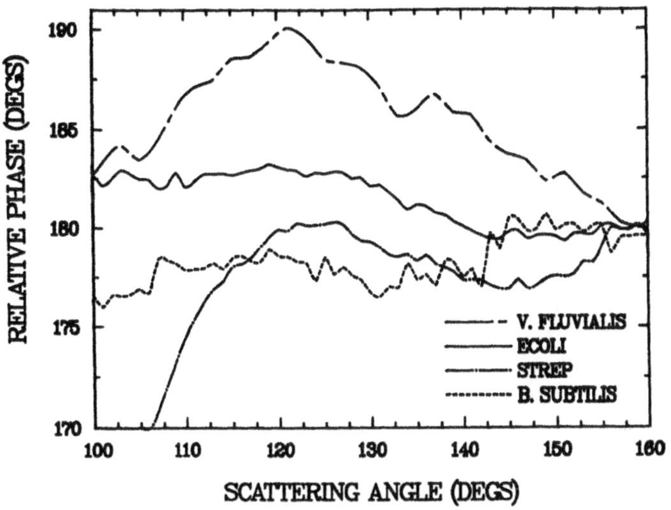

↑FIGURE 7 - Discrimination at backward scattering angles between the same bacterial samples used for the previous figure. The analyzing polarizer was at 45 degrees.

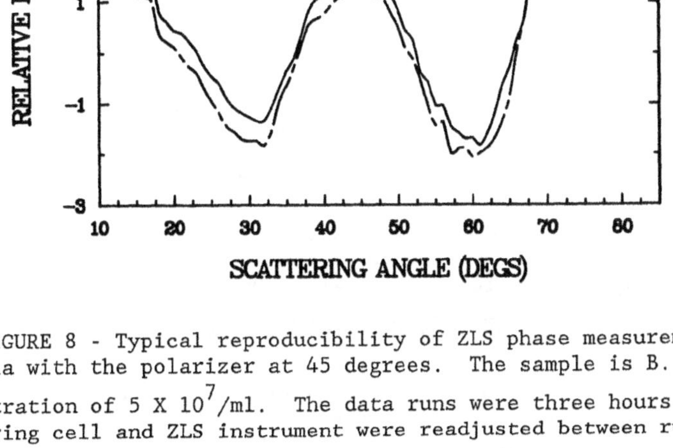

↑FIGURE 8 - Typical reproducibility of ZLS phase measurements on bacteria with the polarizer at 45 degrees. The sample is B. Globigii at a concentration of 5 X 10^7/ml. The data runs were three hours apart. The scattering cell and ZLS instrument were readjusted between runs.

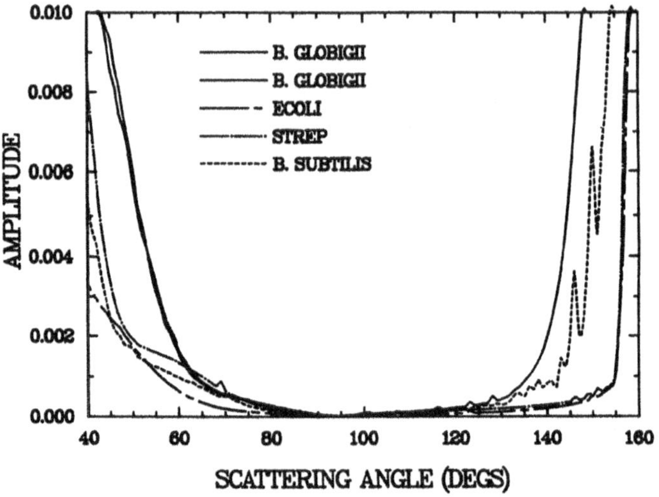

†FIGURE 9 - Discrimination between bacterial species on the basis of the relative ZLS beat frequency amplitude, Γ.

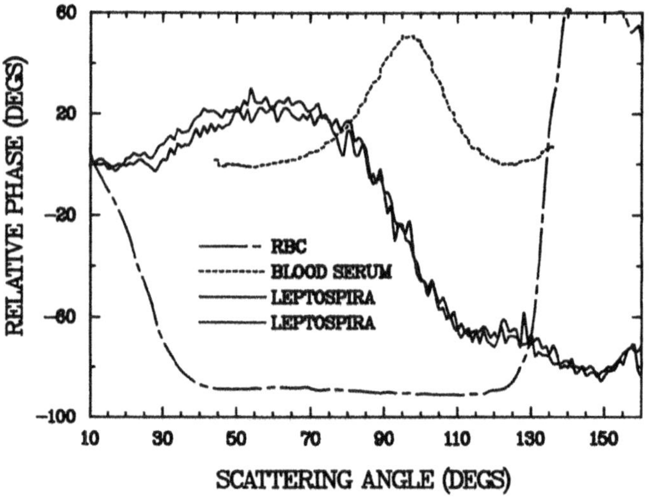

†FIGURE 10 - ZLS phase, γ, with the polarizer at 0 degrees for suspensions of human red blood cells, human blood serum, and the bacteria Leptospira biflexa at concentrations of 3 X 10^6/ml, 50% dilution, and 1 X

10^7/ml, respectively. All three samples had strong beat frequency amplitudes (Γ) with the polarizer at 0 or 90 degrees, indicating the samples are chiral. Indeed, the Leptospira bacteria have a helical shape with a pitch of approximately 0.5 μm.

5. CONCLUSIONS

Zeeman Laser Scattering (ZLS) appears to be a promising new light scattering technique. It does not provide as much information about the scatterer as measurements of the complete Mueller scattering matrix [1], nor does it currently offer the phase resolution theoretically possible with optical heterodyne techniques [2,3]. ZLS does, however, provide more information that conventional irradiance measurements and it does so using simple instrumentation that is relatively free of artifacts.

ZLS currently relies on a 1 mw helium-neon Zeeman Effect laser at 632.8 nm. This is adequate for a number of applications. An argon-ion version of the Zeeman Effect laser is expected within two years [14]. Its higher power and shorter wavelength should greatly increase the usefulness of ZLS.

REFERENCES

1. CF Bohren and DR Huffman. Absorption and Scattering of Light by Small Particles. New York: Wiley, 1983.
2. SF Jacobs. Optical Eng. 17, 544 (1978).
3. DR Pettitt and TW Peterson. Aerosol Sci. Tech. 2, 351 (1983); 3, 305 (1984).
4. JM Greenberg, NE Pedersen, and JC Pedersen. J. Appld. Phys. 32, 233 (1961).
5. RW Hart and EP Gray. J. Appl. Phys. 35, 1408 (1964).
6. AC Holland and G Gagne. Appl. Optics 9, 1113 (1970).
7. RJ Perry, AJ Hunt, and DR Huffman. Appl. Optics, 17, 2700 (1978).
8. RC Thompson, JR Bottiger, and ES Fry. Appl. Optics, 19, 1323 (1980).
9. JE Katz, S Wells, D Ussery, C Bustamante, and MF Maestre. Rev. Sci. Instru. 55, 1574 (1984).
10. GC Salzman, JK Griffith, and CT Gregg. Applied and Environmental Microbiology 44, 1081 (1982).
11. SB Singham and GC Salzman. J. Chem. Phys., 84, 2658 (1986).
12. Y Shindo and M Nakagawa. Applied Spectr. 39, 32 (1985).
13. Y Shindo and M Nakagawa. Rev. Sci. Instru. 56, 32 (1985).
14. Optra, Inc. Peabody, Massachusetts, USA.
15. BW Shore and DH Menzel. Principles of Atomic Spectra. New York: Wiley, 1968.

Structure and Dynamics of Double Helices in Solution: Analysis of DNA Bending by Electro-optical Experiments.

Dietmar Porschke

Max Planck Institut für biophysikalische Chemie

3400 Göttingen, Federal Republic of Germany

Abstract

The long range structure of DNA and its dynamics has been studied for restriction fragments with 43 to 4362 base pairs and for long viral DNA by electro-optical procedures. The data obtained both for short and long DNA molecules indicate bending of the double helix by ligands like spermine or spermidine. Part of this effect is attributed to inherent bending of natural DNA. Inherent bending explains the spontaneous self-organization of long DNA molecules into toroidal structures. The observed high rate of toroid formation suggests the analogy of a spring, which is kept under tension by electrostatic repulsion: the spring collapses as soon as the repulsion is turned off. The molecular basis of DNA bending may be some phasing of A clusters at a period of about 11 base pairs detected by autocorrelation analysis in e.g. λ-DNA. Probably inherent bending of relatively short DNA segments is sufficient to serve as a nucleus for toroid formation.

Introduction

Our present knowledge on the structure of biological macromolecules mainly comes from x-ray analysis of crystallyzed samples. Although this method has been extremely successful even for very large macromolecules, there has always been a strong demand for methods providing information on the structure of macromolecules in solution, because the structure found in a crystal is not necessarily identical to that in solution. Another, even more important argument for the investigation of structures in solution results from the fact that the rigid form of macromolecules seen in the crystal would not be suitable for biological function. Function requires mobile structures and the required mobility is usually much more than small fluctuations around an equilibrium position. Unfortunately there is no ideal method to approach this problem. However, some of the problems associated with structure, dynamics and function may be analysed by electro-optical procedures (1-4), which prove to be particular useful in the case of nucleic acids.

Electro-optical procedures in general imply techniques for the induction of molecular processes by electric field pulses and the characterization of these processes by spectroscopic measurements. The field pulses may induce two different types of processes: 1) orientation of molecules with their induced or permanent dipole in the direction of the field vector; 2) reactions of different nature including dissociation of ligands, conformation changes and interactions between macromolecules due to induced dipoles. These different processes may be distinguished for example by UV spectroscopy using polarized light or by careful analysis of time constants including their concentration dependence. The results obtained by a quantitative analysis of orientation processes are the anisotropy of the macromolecule providing information on the orientation of the chromophore, the nature and magnitude of the molecular dipole, the dimensions of the molecule and also its flexibility. The information obtained from a quantitative analysis of field induced reactions include thermodynamic and kinetic parameters of ligand binding and of conformation changes. A major advantage of electro-optical procedures is their particularly broad time range extending over many orders of magnitude from seconds to nanoseconds. In the present contribution a short summary will be given on results obtained by electro-optical methods on the structure and dynamics of DNA double helices.

The orientation of DNA bases in the double helix

The results obtained by electro-optical measurements of DNA can only be interpreted quantitatively, if the samples are homogeneous. Thus, the detection of restriction nucleases and their application to the preparation of strictly homogeneous DNA fragments has been very useful for electro-optical investigations. A large number of these fragments have been analysed by measurements of the linear dichroism as a function of the electric field strength (5, 6). The linear dichroism indicates the orientation of the bases with respect to the helix axis, when the DNA helices are aligned parallel to the electric field vector. Since the alignment is not complete even at the highest field strengths applied to the sample, a model has to be used for extrapolation of the observed degree of orientation at different field strengths up to the limit of infinite field strength. Orientation models are well established for the standard classical cases of induced and permanent dipoles. The orientation of DNA fragments proved to be more complex (6): short DNA fragments with chain lengths $\leq$ 50 bp are aligned according to the expected induced dipole mechanism up to field strengths around 50 kV/cm, whereas fragments of e.g. 100 bp are aligned already at field strengths in the range above 10 kV/cm according to a permanent dipole mechanism. The observation of permanent dipoles for DNA double helices has been confusing, since - a priori - a molecule like DNA should not have a permanent dipole. Apparently the polarizability of DNA approaches saturation,

Table I

Limit values of the linear dichroism ξ_∞ according to a saturating induced dipole model (from ref. 6; buffer: 1 mM NaCl, 1 mM Na-cacodylate pH7.1, 0.2 mM EDTA)

Number of base pairs	ξ_∞
43	-0.96±0.05
64_a	-1.25±0.02
64_n	-1.29±0.02
69	-1.27±0.03
76	-1.11±0.03
84	-1.21±0.02
95	-1.21±0.02
118	-1.04±0.02
134	-1.19±0.03
180±3	-1.10±0.02
194	-1.08±0.02
245±4	-1.08±0.02
258	-1.17±0.02
430 pool	-1.23±0.02
436	-1.15±0.02
603	-1.01±0.02
880±13	-1.25±0.02
1450±50	-1.45±0.02
4362	-1.38±0.04

when the electric field strength increases beyond some limit value and finally the DNA is oriented as if associated with a permanent dipole. Thus, the dichroism data have been analysed by a saturating induced dipole model (6), assuming saturation of the polarizability within a given range of the electric field strength, which is dependent upon the DNA chain length. Using this model, the dichroism data measured for a large number of DNA fragments over a broad range of field strengths can be represented with high accuracy. The limit values of the linear dichroism ξ_∞ corresponding to complete alignment of the fragments according to the saturating induced dipole model are given in Table I. All ξ_∞-values are different from -1.5 expected for perpendicular orientation of the base planes with respect to the helix axis. Furthermore, the experimental values are not at all constant for the different fragments and demonstrate a rather large sequence dependence variation of the helix structure. These sequence dependent variations should be averaged for the long DNA molecules and thus a mean ξ_∞-value may be expected for long DNA chains. However, the ξ_∞-values for long DNA chains are not at all an average of those observed for short chains, but approach the limit value -1.5 expected for perpendicular orientation of the bases with respect to the helix axis. This result suggests an interpretation of the dichroism data based on field induced changes of the long range structure: electric field pulses do not only align DNA fragments along the field vector, but also stretch DNA helices, which are partly bent at zero field strength. The stretching force is expected to increase with chain length and thus long DNA fragments are apparently stretched to a linear form at high field strength, whereas short fragments remain partly bent. Since bending reduces the dichroism, the relatively low ξ-values observed for short fragments may be explained, at least partly, by bending of the double helix.

Rotation time constants: sensitive indication of dimensions

In addition to the amplitude, an electric dichroism experiment also provides time constants, which reflect the transition between the partly aligned and the random orientation by rotation diffusion. In the limit of low chain lengths, DNA helices behave like rigid rods (5, 7-9) and their rotation diffusion coefficient decreases with the third power of the chain length (10). In this range the dichroism time constants are particularly sensitive to changes of the length. For example, intercalation of a single aromatic residue between the base pairs of a helix with 95 bp can be detected (11). Thus, dichroism time constants can be very useful to analyze the structure of e.g. protein-DNA complexes. Recently the method has been used to demonstrate bending of specific DNA fragments upon binding of the cyclic AMP receptor protein (12).

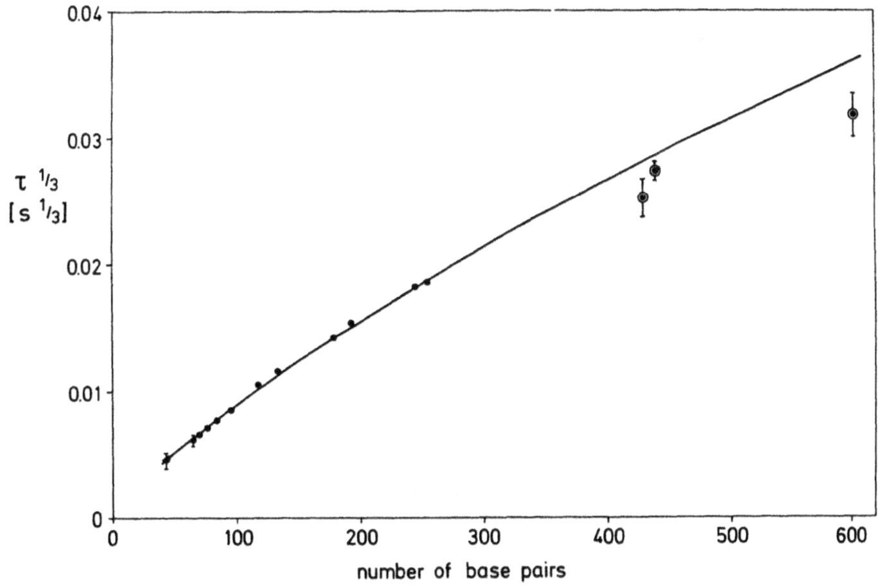

Fig. 1. Third root of dichroism decay time constants $\tau^{1/3}$ as a function of the chain length. The continuous line represents a fit according to the weakly bending rod model (13) with the parameters: rise per base pair 3.4 Å, axial diameter 26 Å, persistence length 500 Å, bead-diameter 50 Å (from ref. 8)

Since DNA molecules are not completely rigid, the increase of the rotation time constants with the third power of the chain length is not valid anymore for DNA molecules with more than ~100 bp. A quantitative interpretation of the chain length dependence by an appropriate hydrodynamic model provides the "persistence length", which is a measure for the DNA stiffness. As shown in Fig. 1, the rotation time constants observed for a large number of DNA fragments can be fitted with high accuracy by a "weakly bending rod model" (13) using a persistence length of 500 Å corresponding to approximately 150 bp.

DNA condensation: well ordered retreat of long molecules in a short time

Natural DNA molecules are exposed to an environment with many factors like enzymes, metal-ions or simply hydrodynamic shear gradients, which may lead to degradation. Thus, DNA has to be protected by some packing mechanism. A particularly simple packing reaction has been detected by Gosule and Schellman (14, 15), who added spermine or spermidine to long DNA at a very low DNA concentration and found that under these conditions DNA "condenses" in an intramolecular reaction to a well ordered toroidal structure (16). Subsequent investigations of this reaction demonstrated that condensation can also be induced by ligands like $[Co(NH_3)_6]^{3+}$ (17) and that the main function of the ligands is reduction of the polyelectrolyte charge density below some limit value (18-20). Obviously the repulsion between adjacent DNA segments has to be reduced before they can be packed together. However, ligand binding alone does not yet explain the self-organization of DNA into well-ordered toroids, since various factors are against the transition to the toroidal form, including the entropy of the long DNA chain, the stiffness of the double helix and also some electrostatic repulsion remaining even after ligand binding. Thus, the condensation of DNA represents a remarkable case of molecular self-organization against several factors which stabilize the disordered state and appear to be dominant. The spontaneous self-organization is particularly surprising in view of the extremely long DNA chains and the topological problem of winding these chains without entanglement.

The kinetics of this reaction has been analysed recently by stopped flow and electric field jump measurements (21, 22). These measurements demonstrate that two reaction phases can be distinguished: 1) ligand binding and 2) folding of DNA strands into toroids. The binding of ligands like spermine to DNA is a relatively complex reaction because of the fact that spermine covers more than a single nucleotide residue (23). The association of spermine molecules leads to a distribution of ligands along the DNA, which excludes occupation of further binding sites, unless the ligands are rearranged. A simulation of the binding process including the rearrangement demonstrates that spermine molecules in the associated state are mobile along the DNA: migration from one binding site to the next one (i.e. shift by one nucleotide residue) takes about 5 ms (21). This mobility is important to achieve a reasonable high rate of ligand association up to high degrees of ligand binding. As soon as the degree of ligand binding arrives at or exceeds the threshold value, the DNA chain is folded into the toroidal form in a relatively fast reaction, which cannot be resolved by the stopped flow technique because of its limited time resolution.

For a more detailed investigation of the DNA folding process, electro-optical techniques proved to be useful owing to their potential for the induction and observation of reactions at a particularly high time resolution.

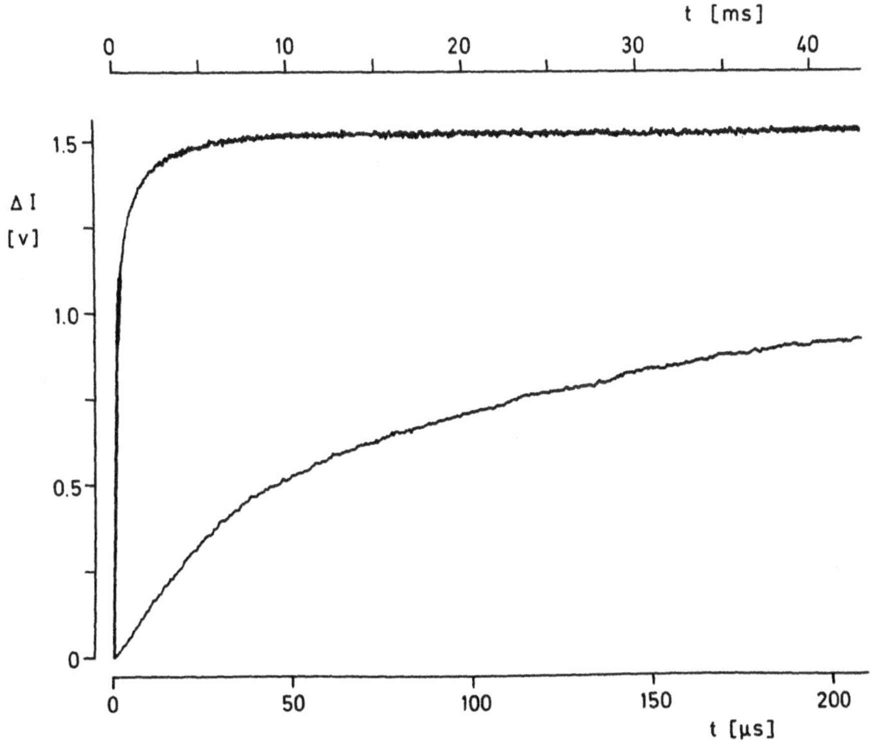

Fig. 2. Scattered light intensity (arbitrary units) of 5 μM λ-DNA in the presence of 600 μM spermine as a function of time after a field pulse of 21 kV/cm (average value) and 130 μs given at two different time scales. Fitting of exponentials provides time constants of 25 μs, 200 μs and 2 ms with relative amplitudes of 31%, 41% and 28% respectively (from ref. 17; 1 mM NaCl, 1 mM Na-cacodylate pH6.5, 50μM EDTA; 20°C).

Condensed DNA was exposed to high electric field pulses to induce dissociation of spermine molecules via a 2. Wien effect (22). When enough spermine molecules are pushed from the DNA, the toroids dissociate. After pulse termination the reaction is reverted: spermine molecules bind again to the DNA and the DNA is folded into toroids. DNA folding is associated with a large change of the light scattering intensity and thus can be followed relatively easily. Owing to the high time resolution of the field jump technique, the condensation reaction could be characterized at very high ligand concentrations (e.g. 600 μ spermine; cf. Fig. 2), where the ligand binding is fast compared to the DNA folding reaction. Under these conditions, the condensation of λ-DNA is reflected by a broad spectrum of time constants ranging from 25μs to 2ms. Apparently, the spectrum of time constants reflects the distribution of the internal mobility of DNA segments in the wormlike coil.

The electric field jump measurements demonstrate a very high rate of the DNA condensation reaction, which is remarkable in view of the great topological problem to arrange an extremely long molecular chain into a well ordered toroid. These results suggest the analogy of a spring, which is kept under tension by electrostatic repulsion and collapses as soon as the repulsion is turned off. A similar analogy has been used by Manning (20).

Gradual reduction of electric dichroism upon spermine binding

DNA toroids have been observed by electron microscopy in various laboratories. Nevertheless, it should be useful to demonstrate their existence in solution by some independent method. The existence of toroids may be tested by measurements of the electric dichroism. As discussed above the limit dichroism of long linear DNA molecules is approximately -1.4. Folding of this DNA in a circular arrangement should lead to a clear reduction of the limit dichroism. The theoretical value for DNA circles with the base planes oriented perpendicular to the helix axis is -3/8, if the circles are aligned with their diameter parallel to the field vector, as should be expected according to the direction of maximal polarizability. Measurements at different degrees of spermine binding demonstrated that the limit dichroism found above the condensation threshold is indeed very close to that expected for toroids (24). Furthermore, the orientation time constant observed for condensed DNA is also consistent with a toroidal organization. [cf. also (25)].

Another result of the dichroism measurements was completely unexpected: below the condensation threshold, the limit dichroism decreases linearly with the spermine concentration (cf. Fig. 3) and indicates a gradual transition to the condensed state without any sign for a cooperative transition (24). This is in contrast to the cooperative appearance of the condensation reaction reflected by measurements of the light scattering intensity under exactly corresponding experimental conditions (21). Thus, there are two complementary aspects of the condensation reaction. The

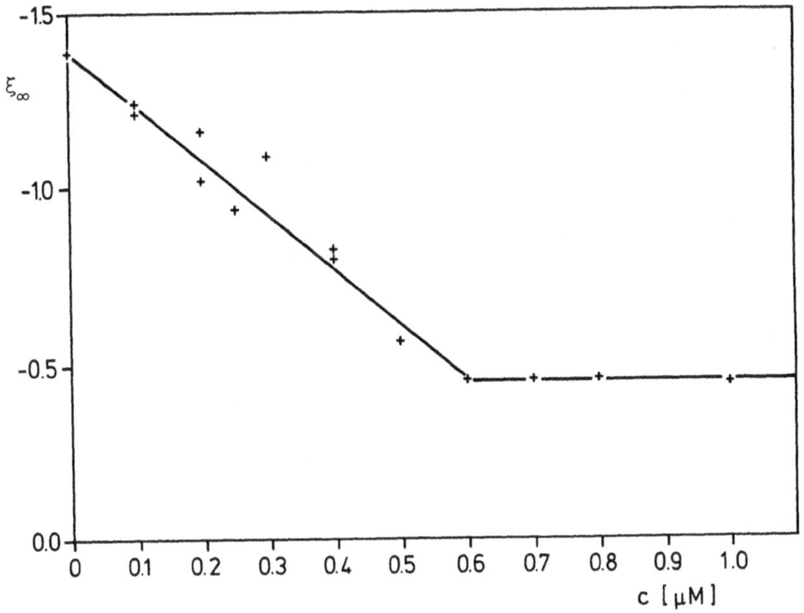

Fig. 3. Limit electric dichroism ξ_∞ of 2 µM λ-DNA as a function of the spermine concentration (from ref. 24; 1 mM NaCl, 1 mM Na-cacodylate pH6.5, 50 µM EDTA; 20°C).

gradual change of the dichroism indicates the existence of precursors on the way to the toroidal state. Apparently the folding of the DNA chain approaches almost smoothly the toroidal organization. Probably the intermediates have some superhelical structure, which is formed due to spermine induced bending of the double helix. Superhelices would be optimal precursors for fast formation of toroids and thus would explain the high rate of DNA condensation.

Bending amplitudes and time constants in short DNA fragments

Bending of DNA double helices may result from two different sources: 1) according to the classical view the double helix is inherently straight and is bend only due to thermal fluctuations; 2) more recently evidence has been accumulated that double helices may be inherently bent provided that small bending effects caused by certain sequences of base pairs are added by appropriate phasing (26). The usual description of DNA bending by a persistence length does not consider the possibility of inherent DNA bending, partly because the standard methods for the characterization of the DNA long range structure in solution cannot distinguish between different molecular origins of bending.

The problem may be analyzed again by electro-optical methods. As mentioned above electric field pulses do not only induce orientation of DNA fragments, but also lead to stretching. The reverse bending process may be observed after pulse termination. Measurements on DNA fragments of different chain lengths showed that the bending process can be observed directly for helices with more than 100 base pairs (8). As should be expected both bending amplitudes and time constants increase with the chain length: for a fragment with 179 bp the bending amplitude relative to the overall dichroism amplitude is 30% and the corresponding number for a fragment with 256 bp is 43%; the bending time constants for these fragments are 180 ns and 400 ns respectively (1 mM NaCl, 1 mM Na-cacodylate pH 7.0, 100 μM $MgCl_2$, 20°C).

Ligand induced bending may be demonstrated directly by an analysis of the overall rotation time constants for short DNA fragments. As shown in Table II for two different restriction fragments, the rotation time constants decrease with increasing spermine concentration. Since the local structure of the DNA fragments remains in the B-form as judged from CD spectra, the reduction of the rotation time constants must be attributed to bending of the double helix.

Table II

Rotation time constants (ns) from dichroism decay as a function of the spermine concentration (1 mM NaCl, 1 mM Na-cacodylate pH 7.0, 0.2 mM EDTA; 20°C)

Spermine (μM)	0.0	1.5	3.0	4.0
69 bp fragment	317	298	289	261
84 bp fragment	525	460	454	423

Measurements of both bending amplitudes and persistence lengths under different solvent conditions demonstrate that the experimental data can hardly be explained by thermal bending alone and indicate a contribution due to inherent bending (27). A semiquantitative analysis of bending amplitudes and their chain length dependence according to an "arc model" is consistent with an inherent curvature of the fragments similar to that observed in the toroidal state. A combination of various results suggests that inherent bending prevails at low temperatures, whereas thermal bending dominates at high temperatures.

<u>Inherent bending of natural DNA and toroid nucleation</u>

The results discussed above indicate ligand induced bending of DNA. The molecular basis of this effect appears to be a local discontinuity of the ion atmosphere around the double helix. Due to the relatively high number of positive charges associated with ligands like spermine, the electrostatic repulsion between adjacent phosphates at the binding site is shielded more effectively than elsewhere and

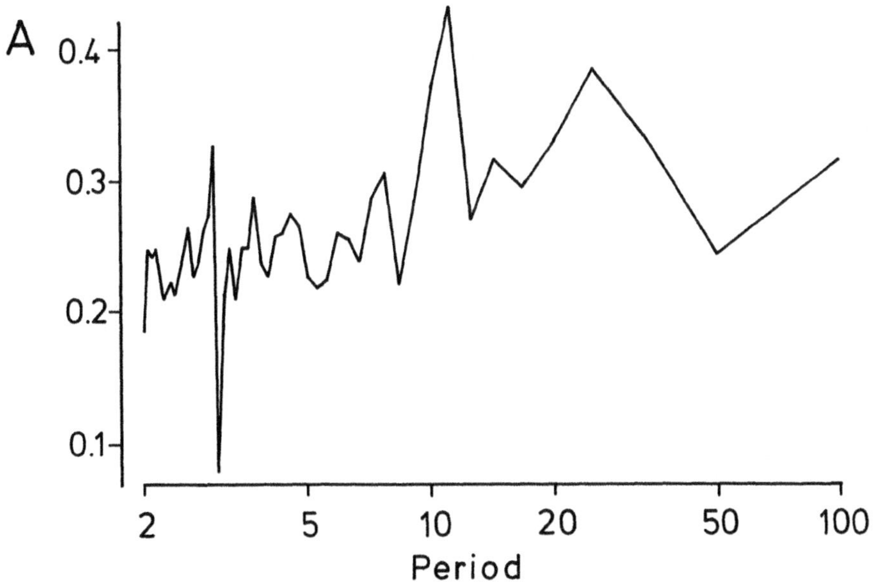

Fig. 4. Fourier analysis of an autocorrelation of A dimers in the complete sequence of λ-DNA taken from the EMBL library. An enhanced contribution is observed at a period of about 11 base pairs.

thus the helix axis is bent at this site. The question arises, however, whether these bends cancel each other over long stretches of DNA. Obviously local bends will not compensate each other, when there is an appropriate phasing of bending sites. As first suggested by Trifonov (26), bending may be enhanced and become detectable by a periodic arrangement of certain base sequences. An extreme case of strong helix bending has been detected in a natural "kinetoplast" DNA and has been attributed to the presence of adenine clusters in a period corresponding to the turn of the double helix (28-31). Thus, it should be useful to look for a periodic arrangement of A clusters in the DNA used for condensation experiments. An analysis of A dimers in the complete λ-phage genome revealed a detectable autocorrelation of these dimers at a distance of about 11 base pairs close to the period of B-DNA (cf. Fig. 4). This result suggests that λ-DNA is inherently bent, when the electrostatic repulsion between phosphate residues is reduced. Some correlation of A dimers has been found for other natural DNA molecules. Thus, the spontaneous formation of toroids may be explained by inherent bending of natural DNA.

Inherent bending will also be useful for packing of DNA in phage heads. In many cases including λ-phages, the DNA in the phage heads appears to be wound like a thread on a spool. Although the packing of DNA in phage heads is supported by proteins, the nature of DNA packing in phage heads and in toroids may be very similar.

Since the sequence correlations observed in natural DNA molecules are not at all perfect, it cannot be expected that inherent bending leads to the formation of any perfect superhelix in solution. Some DNA's may have very limited sequence correlations leading to bending only for relatively short segments. Nevertheless, these limited bends can be very useful for nucleation of toroids, since inherent bending supports the formation of a first circle of DNA. Subsequent growth of toroids by winding of more DNA along a first circle is expected to be a rather fast process, which is not inhibited by serious activation barriers, when the free energy associated with contact formation between DNA helices is favorable due to reduction of phosphate repulsion by ligand binding.

Acknowledgements:

The author is indebted to M. Jung and J. Lange for their adaptations of autocorrelation software and to J. Ronnenberg for his technical assistance.

References

1. Fredericq, E. & Houssier, C., (1973) Electric Dichroism and Electric Birefringence (Clarendon, Oxford)
2. O'Konski, C.T., Ed. (1976) Molecular Electro-optics (Dekker, New York)
3. Krause, S., Ed. (1981) Molecular Electro-optics (Plenum, New York)
4. Porschke, D. (1985) Ann. Rev. Phys. Chem. 36, 159-178
5. Hogan, M., Dattagupta, N. & Crothers, D.M. (1978) Proc. Natl. Acad. Sci. USA 75, 195-199
6. Diekmann, S., Hillen, W., Jung, M., Wells, R.D., & Porschke, D. (1982) Biophys. Chem. 15, 157-167
7. Elias, J.G. & Eden, D. (1981) Macromolecules 14, 410-419
8. Diekmann, S., Hillen, W., Morgeneyer, B., Wells, R.D. & Porschke, D. (1982) Biophys. Chem. 15, 263-270
9. Hagerman, P.J. (1981) Biopolymers 20, 1503-1535
10. Broersma, S. (1960) J. Chem. Phys. 1626-1631
11. Porschke, D., Geisler, N. & Hillen, W., (1982) Nucleic Acids Res. 10, 3791-3802
12. Porschke, D., Hillen, W. & Takahashi, M. (1984) EMBO J. 3, 2873-2878
13. Hearst, J.E. (1963) J. Chem. Phys. 38, 1062-1065 and personal communication
14. Gosule, L.C. & Schellman, J.A. (1976) Nature 259, 333-335
15. Gosule, L.C. & Schellman, J.A. (1978) J. Mol. Biol. 121, 311-326
16. Marx, K.A. & Ruben, G.C. (1983) Nucleic Acids Res. 1, 1839-1854
17. Widom, J. & Baldwin, R.L. (1980) J. Mol. Biol. 14, 431-453
18. Wilson, R.W. & Bloomfield, V.A. (1979) Biochemistry 18, 2192-2196
19. Benbasat, J.A. (1984) Biochemistry 23, 3609-3619
20. Manning, G.S. (1985) Cell Biophysics 7, 57-89
21. Porschke, D. (1984) Biochemistry 23, 4821-4828
22. Porschke, D. (1985) Biopolymers 24, 1981-1993
23. Epstein, I.R. (1979) Biopolymers 18, 2307-2050
24. Porschke, D. (1986) Biomolecular Stereodynamics IV, 11-20
25. Marquet, R., Houssier, C. & Fredericq, E. (1985) Biochim. Biophys. Acta 825, 365-374
26. Trifonov, E.N. (1985) CRC Crit. Rev. Biochem. 19 89-106
27. Porschke, D. (1986) submitted
28. Marini, J.C., Levene, S.D., Crothers, D.M. & Englund, P.T. (1982) Proc. Natl. Acad. Sci. USA 79, 7664-7668
29. Diekmann, S. & Wang, J.C. (1985) J. Mol. Biol. 186, 1-11
30. Koo, H.S., Wu, H.M. & Crothers, D.M. (1986) Nature 320, 501-506
31. Ulanovsky, L., Bodner, M., Trifonov, E.N. & Choder, M. (1986) Proc. Natl. Acad. Sci. USA 83, 862-866

NMR STUDIES OF LOOPFOLDING IN A DNA HAIRPIN MOLECULE

M.J.J. BLOMMERS, C.A.G. HAASNOOT and C.W. HILBERS

Laboratory of Biophysical Chemistry, Faculty of Science, University of Nijmegen, Toernooiveld, 6525 ED Nijmegen, The Netherlands.

J.H. VAN BOOM and G.A. VAN DER MAREL

Laboratory of Organic Chemistry, Gorlaeus Laboratories, University of Leiden, P.O. Box 9502, 2300 RA Leiden, The Netherlands.

1. INTRODUCTION

The last decade has witnessed an outburst of NMR studies of nucleic acid structure and dynamics. Initially such investigations were aimed at the study of exchange properties of imino protons involved in hydrogen bonding in basepairs, at the determination of the melting properties of double helical molecules and at resolving the conformational properties of the sugar moieties in small oligonucleotides. Detailed structural studies were hampered by the inability to interpret the complicated spectra which are obtained for medium sized and larger nucleic acid molecules. In the last five years this situation has changed dramatically. The introduction of NMR spectrometers operating at 500 MHz and the development of two dimensional Fourier Transform methods have increased the resolution of the NMR method tremendously. The concommitant development of assignment procedures and the possibility to measure short range distances, albeit in a qualitative way, between protons in a molecule, have made the NMR technique one of the most powerful methods to study the structure of molecules in solution.

Assignments of resonances in nucleic acid molecules have concentrated on the identification of imino proton resonances and on the interpretation of pyrimidine and/or purine H6 and H8 resonances via mutual NOE connectivities with H1' and H2'/H2'' resonances. Once one has succeeded in assigning these resonances the remaining proton signals may follow relatively easy. Assignments of the first category have solved a long standing problem in the NMR spectroscopy of nucleic acids namely the identification of the imino proton resonances in tRNA's (1-6). The second approach (7) has played a crucial role in detailed structural studies of well defined DNA fragments in solution (8,9).

In this contribution the application of the aforementioned methods will be illustrated for the partly selfcomplementary oligonucleotide d{ATCCTA$(T)_4$TAGGAT}. Elsewhere we have established that this molecule forms a hairpin structure with a melting temperature of 53°C at 0.2 M NaCl (10-12). In the next section we will first consider 2D NOE spectra obtained for this molecule and use these for the interpretation of the spectrum. Subsequently, the assignment of the resonances and NOE cross peaks will be used to derive a preliminary structure for the hairpin. The results will be compared with the folding principle we derived for loop formation in hairpin molecules and with the results of a molecular mechanics calculation carried out for d{ATCCTA$(T)_4$TAGGAT} (13).

2. ASSIGNMENT OF NON EXCHANGEABLE PROTONS IN d{ATCCTA(T)$_4$TAGGAT}

Assignment of sugar backbone and base ring protons is possible on the basis of 2D NOE spectra recorded in D_2O. Standard sequential assignment procedures are available, which make use of the fact that the distances between the base ring protons H6 and H8 of pyrimidines and purines and the sugar H1' and H2'/H2'' are short enough (i.e. < 4 Å) that NOE effects can be observed between the resonances of these protons (14-16). This is illustrated in Fig. 1. which displays the distances between the mentioned ring protons and the H1' and H2'/H2'' sugar protons of the sugar attached to the base and of the sugar of its 5'-neighbour found in a B-DNA helix.

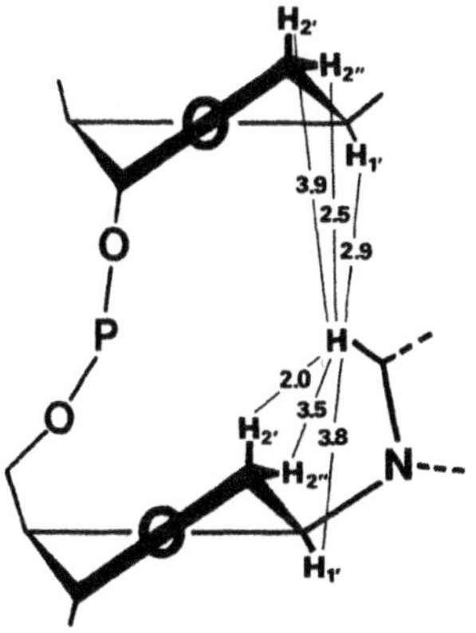

FIGURE 1. Average distances between the purine H8 and pyrimidine H6 proton and the H1'/H2'/H2'' deoxyribose protons in B-DNA helices (Arnott et al., 1976, ref. 17). These distances should not be interpreted in an absolute way as the Arnott model is necessarily only an approximation to the true helix conformation at the atomic level. The data serve to show that the purine H8 on pyrimidine H6 proton is not only close to its intranucleotide 1', 2' and 2'' protons but also to the 1', 2' and 2'' protons of its 5'-neighbouring nucleotide. Therefore in a 2D NOE experiment intra- as well as internucleotide cross peaks will be observed for the base protons of an oligonucleotide (see Fig. 2).

This observation is used in the sequential assignment procedures. Briefly, starting with the cross peak between the base ring proton (H6 or H8) and the sugar H1' proton resonance of the residue at the 5'-end of the sequence one proceeds by looking for the cross peak between this H1' sugar resonance and the ring proton resonance of the next residue in the 3'-direction. Once this cross peak is found the procedure is repeated and in this way one proceeds through the nucleotide chain. A similar approach is possible when starting from the 3'-end of the molecule. This procedure allows the assignment of the sugar H1' and base H6 and H8 resonances to particular residues in the sequence.

For d{ATCCTA(T)$_4$TAGGAT} this approach is illustrated in Fig. 2. Starting with the H8 - H1' cross peak for A1, we can assign the ring and sugar H1' proton resonances until we arive at the loop residue L3 for which we can assign the H1' resonance, but we are not able to make the connection to the H6 resonance of the loop residue L4.

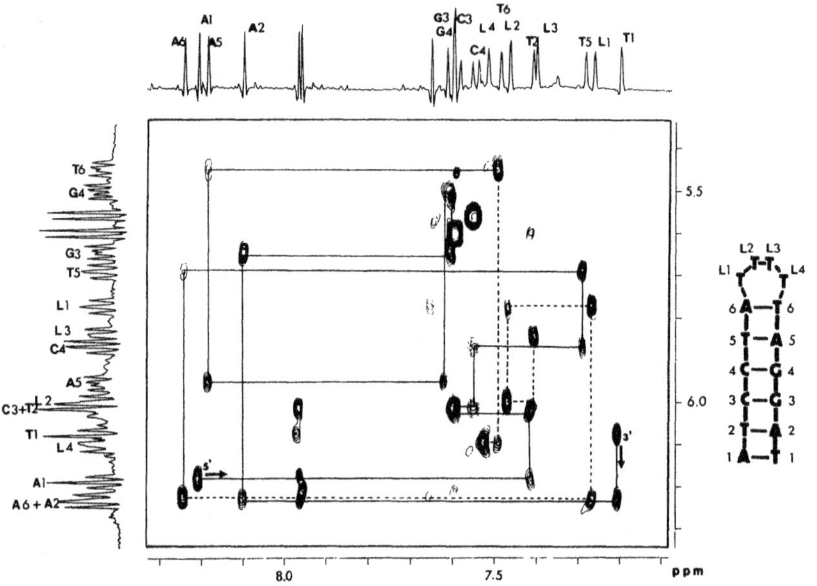

FIGURE 2. Part of 500 MHz ^{1}H NMR NOESY spectrum of d{ATCCTA(T)$_4$ TAGGAT} recorded with a mixing time of 0.3 s at 27°C. Cross peaks between aromatic base proton resonances (7.1 - 8.3 ppm) and H1' proton resonances (5.4 - 6.3 ppm) are shown. In the assignment of the resonances the numbering is used as indicated for the hairpin structure at the right hand side. The sequential assignment of the cross peak is indicated. Starting with the cross peak between the 5'-terminal A1H8 and A1H1' resonances, a solid line connects all cross peaks belonging to neighbouring residues in the stem. The sequential assignment can be extended up to the third base in the loop as indicated by the dashed line. Analogously, the cross peaks from the other strand in the stem can be assigned by starting from the 3'-end as indicated. The sequential assignment can be extended up to the base L4. The connectivity pathway in the loop is again indicated by a dashed line.

On the other hand when we start at the 3'-end with the resonances of T1 we are able to proceed to the H6 resonance of L4 mutatis mutandis we are not able to proceed to the H1' resonance of L3. The results suggest that the stacking of the bases present in the stem is transferred to the bases in the loop, i.e. proceeding in the 3'-direction L1 and L2 stack in a more or less helical fashion on A6 and L4 on T6. The situation with respect to L3 is somewhat less clear since the cross peak between the H1' resonance of L2 and the H6 resonance of L3 strongly overlaps with the cross peak between the H1' and H6 resonance of T2 (c.f. Fig. 2).

A similar approach can be followed with respect to the connectivities between the base ring proton and the H2'/H2'' sugar resonances. Because of overlap of cross peaks the analysis cannot be extended into the loop as far as for the H1' resonances, but by combination of the results with those obtained for the H1' resonances it is nevertheless possible to assign all H2' and H2'' resonances. Discrimination between H2' and H2'' resonances can be achieved by means of COSY (J-splitting patterns of the cross peaks) and NOESY spectra (intensity differences of cross peaks at short mixing times): all H2' resonances are found at higher field compared with the H2'' resonances of the same sugar residue.

Having identified the pyrimidine H6 resonances it is now straightforward to analyse the spectrum of the methylgroups of the thymidine residues. In turn these resonances can be used to support the structural features of the hairpin already apparent from the analysis of the H1' and H2'/H2'' resonances, i.e. in B-DNA the distance between the thymine methyl groups and the H6 or H8 proton of the residue at the 5'-side is sufficiently short to give rise to cross peaks in a 2D NOE spectrum. The observed cross peaks corroborate the findings already available from the sequential assignments described above, i.e. the stem of the hairpin forms a B-type double helix, the thymidine L1 is stacked upon A6, L2 is stacked upon L1 and T6 on L4. There is no connectivity between L4 and L3 indicating that the disruption of the regular stacking pattern in the loop is between L3 and L4 as was already indicated by the results of the sequential assignment procedure. The results of these assignments are summarized in Fig. 3, where interresidual connectivities obtained for the hexadecanucleotide are schematized.

FIGURE 3. Summary of the sequential NOE's (solid lines) observed in the NOESY spectra of d{ATCCT$(T)_4$TAGGAT} between base ring proton resonances and sugar proton or methyl resonances. The dashed lines indicate situations in which cross peaks between resonances of neighbouring nucleotides could be observed but some uncertainty as to the assignment remained due to overlap. Note the absence of cross peaks between the loop residues L3 and L4, which indicates a perturbation of the B-type stacking pattern in this part of the molecule.

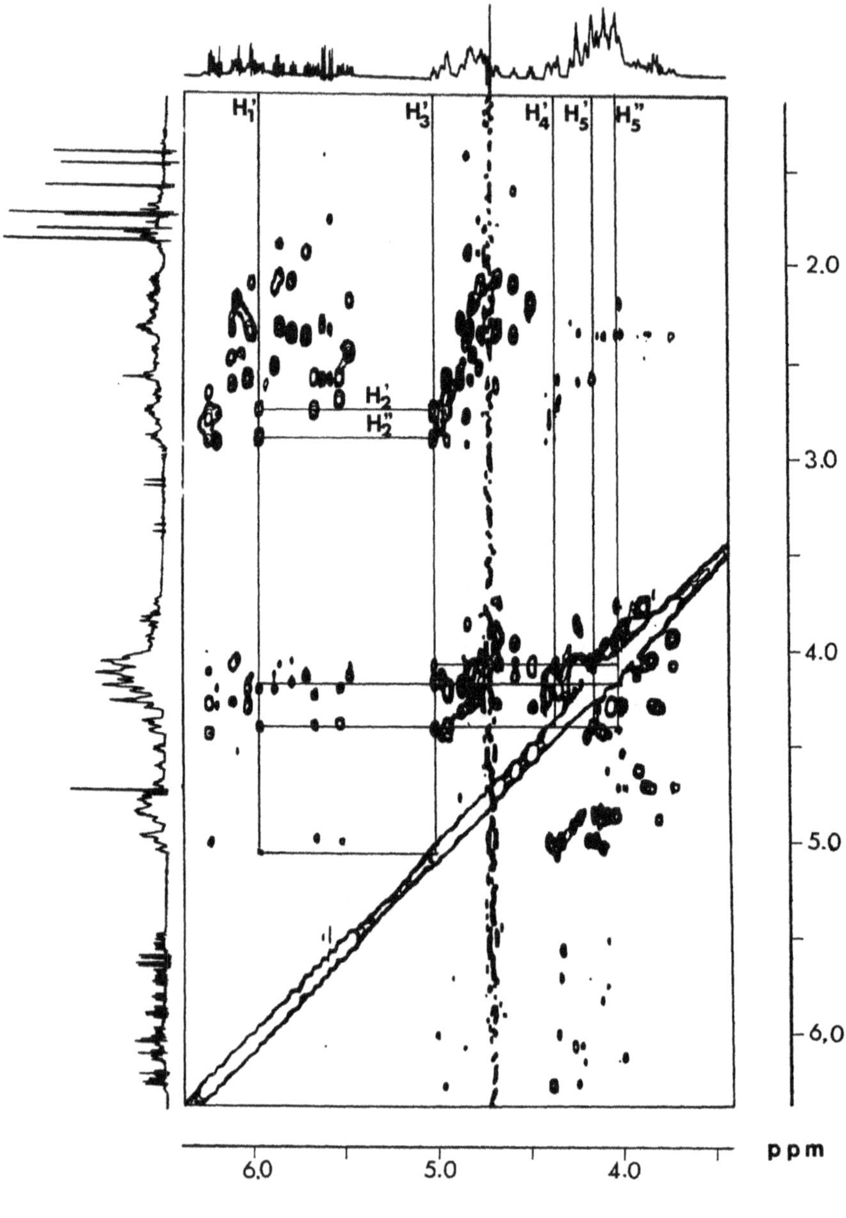

FIGURE 4. Part of the 500 MHz ^{1}H NMR NOESY spectrum (ω_1 = 1.15 to 6.28 ppm, ω_2 = 3.43 to 6.37 ppm) of d{ATCCTA(T)$_4$TAGGAT} recorded with a mixing time of 0.3 s at 27°C. The assignment of the sugar protons of residue A5 is indicated.

Having obtained an identification for the H1' and H2'/H2'' sugar resonances one may attempt to assign the rest of the proton resonances to the individual sugar residues. This is of some interest because eventually this may lead to a definition of the sugar conformation once after the interpretation of the sugar resonances the J-couplings between the proton spin can be determined. In most instances we succeeded in obtaining a complete identification of the sugar proton resonances. An example of the assignment of the sugar resonances of residue A5 is presented in Fig. 4. It is noted that spin diffusion effects between H1' and the other spins such as H3', H4' and H5' or H5'' in the sugar ring very often facilitate the assignment procedure; this is also clear for the A5 sugar (see Fig. 4).

The results obtained so far cannot but give a qualitative description of the loop structure. For a more detailed characterisation additional data are required. In this respect it is interesting to mention the occurrence of some additionally observed NOE cross peaks between protons of loop residues. It turned out that the cross peaks between H6 of L4 and its H5'/H5'' sugar proton resonances persist even at a mixing time of 125 ms when such effects are no longer observed for the other nucleotides. This indicates that the torsional angle (γ) of the (O5'-C5'-C4'-C3') bond of this sugar moiety corresponds to a g^- or t conformation instead of to the normally observed g^+ conformation. Two other connectivities are worth mentioning. At ω_1, ω_2 = 1.95, 5.82 ppm a cross peak is observed between the methyl resonance of residue L4 and the H1' resonance of L3. This is the only cross peak observed between a methyl resonance and an H1' resonance and forms an indication of the peculiar structure in the loop formed by the L3 and L4 residues (vide infra). Furthermore we note the occurrence of a cross peak between L3 CH3 and L2 H3' indicating that although L3 is stacking on L2 the stacking pattern may deviate from that found in normal B-DNA.

3. EXCHANGEABLE PROTONS

When d{ATCCTA(T)$_4$TAGGAT} folds into a hairpin six canonical Watson Crick base pairs are formed giving rise to six imino proton resonances when the molecule is dissolved in a H_2O solution. Six such resonances are indeed observed but in addition four extra imino proton resonances are present at ~ 11 ppm which must arise from the imino protons of the thymidines in the loop (see Fig. 5). Assignment of the imino protons in the stem were made by means of 1D NOE experiments (18). Presaturation of one particular imino proton resonance should give rise to Overhauser effects at the imino proton resonances of the neighbouring base pairs. An example is provided by the experiment presented in Fig. 5.

The resonance from GC4 at 12.7 ppm is irradiated and Overhauser effects are observed at 13.7 ppm for the resonance from AT5 and at 12.6 ppm for the resonance from GC3. Using the same approach we have tried to assign the imino proton resonances of the loop thymidines. Presaturation of the high field resonance (intensity corresponding to one proton) at 10.7 ppm gives rise to an Overhauser effect at the imino proton resonance position of base pair AT6 and a larger one at the position of the other loop imino proton resonances (at 10.8 ppm). This suggests that the presaturated resonance is from one of the thymidines (L1 or L4) neighbouring the AT6 base pair. The sequential assignments of the non-exchangeable protons discussed above indicated that both L1 and L4 are stacked on the AT6 base pair so that indeed an Overhauser effect between the imino proton resonances of AT6 and that of L1 and/or L4 may be expected. L1 and L4 being stacked on basepair AT6 might easily lead to the formation of a T-T basepair involving two hydrogen bonds as indicated in Fig. 5. For such a base pair we expect a large, i.e. 10 - 20% Overhauser effect when one of the imino proton resonances is irradiated. This is indeed what is observed in Fig. 5 indicating that such a base pair is present in the loop structure of the hexadecanucleotide. This interpretation is nicely confirmed when an NMR melting experiment is conducted (see Fig. 6).

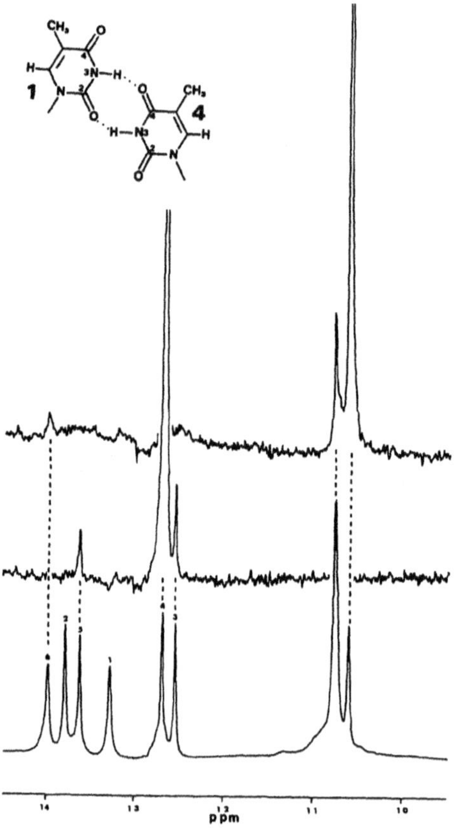

FIGURE 5. 500 MHz ^{1}H NMR 1D NOE experiments performed with d{ATCCTA(T)$_4$TAGGAT} at 0°C. The lower trace is the reference spectrum exhibiting the imino proton resonances arising from the base pairs (between 12 and 14.5 ppm) and from the thymidines in the loop (between 10 and 11 ppm). The numbers refer to the basepair numbering given in Fig. 1. The middle trace represents the NOE difference spectrum obtained after presaturation (during 0.3 s) of the imino proton resonance of base pair GC4. NOE's are seen for the imino proton resonances of GC3 and AT5. The upper trace represents the NOE difference spectrum obtained after presaturation of the high field resonance from the loop residues. A weak NOE is observed for basepair AT6 and a much larger NOE (~15%) for the imino proton resonance from another loop residue. This result is interpreted to indicate the existence of a T-T basepair between residues L1 and L4 stacked upon the double helical stem of the hairpin (see text).

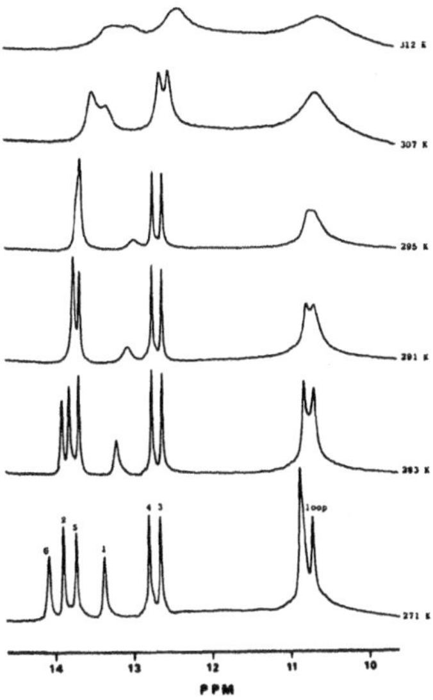

FIGURE 6. The 500 MHz imino proton spectrum of d{ATCCTA(T)$_4$TAGGAT} recorded as a function of temperature. We note that at 291 K the intensity of the loop imino proton signal has decreased by an amount corresponding to two proton resonances. The remaining signals assigned to the T-T pair disappear more or less concomitantly with the other imino proton signals (see text).

Elevation of the temperature leads to broadening and shifting of the imino proton resonances. The interpretation of these effects has been discussed elsewhere (10); here we want to point out that at 291 K we obtain two resonances of equal intensity which originate from signals with an intensity ratio of 3:1 at 271 K. In view of the results presented above the most plausible interpretation of this observation is that the imino protons of the thymidines L2 and L3 exchange rapidly with H_2O at 291 K while the T-T pair remains intact. As judged from the spectra measured at higher temperatures this basepair melts concomitantly with the base pairs in the stem.

At this point the only resonances that have not been assigned are those from the adenine C2H moiety and those arising from the amino groups. In one dimensional NOE difference spectroscopy these signals can be identified in principle after presaturation of assigned imino proton resonances (1,19).

NOESY experiments of the sample dissolved in H_2O allow a more systematic approach. For the hexadecanucleotide this is shown in Fig. 7. For the GC pairs, cross peaks can be observed between imino proton resonances and cytidine amino proton resonances. In turn we see cross peaks between the cytidine amino proton signals and between the amino proton resonances and the C5H resonance. Note that also a connectivity is observed between the imino protons and the C5H resonance which is caused by spindiffusion. The latter cross peaks provide an independent check on the correctness of the assignments since the position of the CH5 resonances was already established (vide infra). The assignment of three adenine amino proton resonances is also indicated in Fig. 7. It turns out that assignment of the guanine amino proton signals is not very well possible. In these base pairs the amino group still has an appreciable amount of rotational freedom which at 5°C leads to a strong broadening of the amino proton signals. Apart from the mentioned exceptions we have now come to the point that a complete assignment of the proton spectrum of d{ATCCTA$(T)_4$TAGGAT} has been achieved.

4. CONFORMATIONAL FEATURES OF THE HAIRPIN FORMED BY d{ATCCTA$(T)_4$TAGGAT}

In earlier studies (10-12) we have found that the stability of DNA hairpin loops as a function of loop size is at its maximum for hairpin molecules with four or five residues (i.e. thymidines) in the loop. This is in contrast to RNA hairpins for which an optimal stability is found for six or seven nucleotides in the loop (20,21). To provide a structural explanation for this behavior, we have recently advocated that the loop folding in hairpin structures is guided by the propagation of the base stacking pattern of the stem into the loop. Then the "natural" way to fold a loop in B-DNA hairpins is to extend the stacking of the nucleotides at the 3'-end of the duplex stem. The hexadecanucleotide d{ATCCTA$(T)_4$ TAGGAT} has been subjected to conformational studies in order to uncover whether this mechanism is operative in this molecule.

It is well established that the conformational characteristics of the sugar moiety may serve to expose the structural genus of the nucleic acid molecule under study, i.e. broadly speaking whether it has A or B type structure. Given enough J-coupling constants between the proton spins on the sugar ring a detailed conformational analysis of the sugar ring is possible (21). For the hexadecanucleotide sufficient data for such an analysis are not available. However, in COSY spectra cross peaks between H2'' and H3' are lacking for all sugars, which indicates that the J-coupling between these protons is small (~1 Hz). In addition the sum of the J-couplings, i.e. $J1'2' + J1'2'' > 13.6$ Hz for all nucleotides except the terminal ones. This is characteristic of a predominantly S-type sugar conformation which is representative of B-type folding of the DNA fragment (23). This conclusion is augmented by the observation that in 2D NOE spectra only an intranucleotide cross peak between H6/H8 and H2' and not between H6/H8 and H2'' is observed at short mixing times (75 ms). Therefore we may conclude that the double helical stem has a B-type structure which is propagated into the loop region.

A more detailed picture is obtained if we incorporate the results of the preceding section. There it was shown that the stacking of the bases in the stem is extended in the 3'-direction into the loop involving L1 and L2 and possibly L3. The fourth thymidine L4 stacks on top of T6 and favors the formation of a T-T base pair.

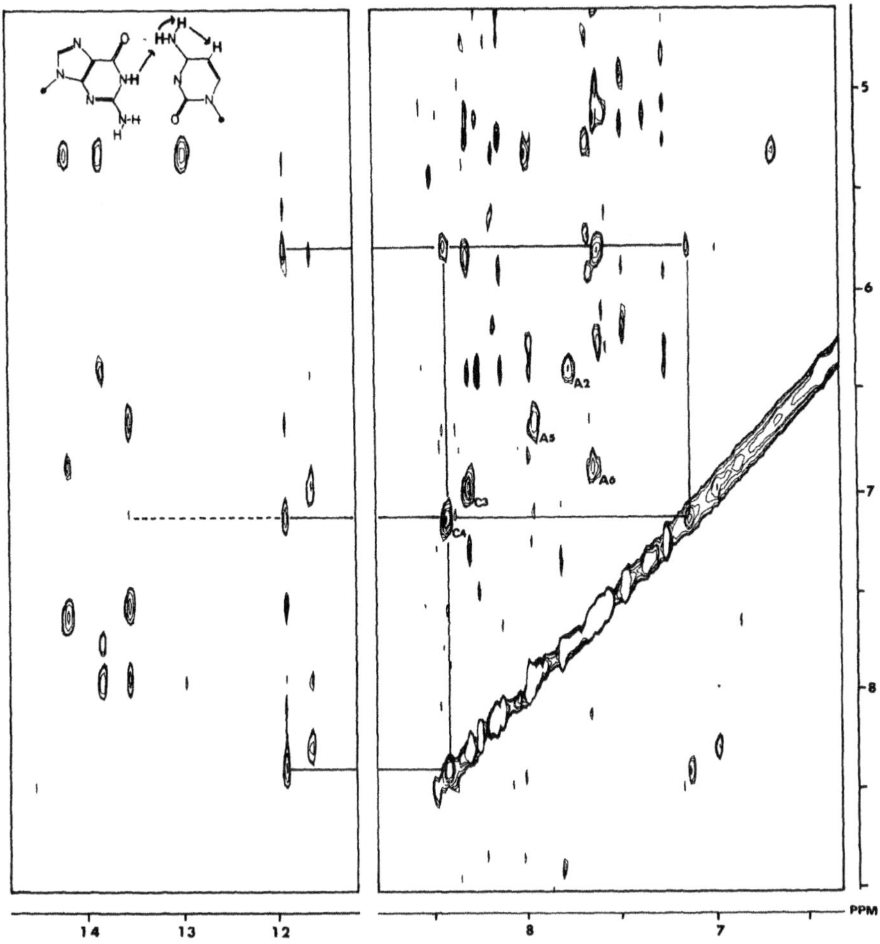

FIGURE 7. Part of the 500 MHz ^{1}H NMR NOESY spectrum of d{ATCCTA(T)$_4$TAGGAT} dissolved in H O recorded with a mixing time of 0.3 s. at 0°C. Water suppression methods to obtain this spectrum have been described elsewhere (18). Two spectral regions are connected, i.e. ω_1 = 4.5 to 9.0 ppm, ω_2 = 11.2 to 14.8 ppm, containing imino proton cross peaks, and ω_1 = 4.5 to 9.0, ω_2 = 6.3 to 8.8 ppm, containing amino and base ring proton cross peaks. The solid lines connect connectivities between the imino proton resonance of G4 (see Fig. 1) and the amino- and C5H resonances of C4. The latter connectivities can only be observed when sufficient spin diffusion is permitted; the spin diffusion pathway is indicated in the base-pair (top left). The cross peaks observed for the amino proton resonances of C3, A2, A5 and A6 are also indicated.

To check whether this base overlap is energetically feasible we performed a series of molecular mechanics calculations. In the hairpin structure used to start the calculations the bases L1 and L2 were placed on top of a regular B-DNA stem so that the stacking pattern of the double helical stem was extended in the 3'-direction as prescribed by our folding principle which is visualized in Fig. 8A (13). The dinucleotide formed by L3 and L4 was docked into the remaining gap between L2 and T6.

After an all atom energy minimization the structure sketched in Fig. 8B is obtained. It is clear that the base stacking scheme derived from the NMR data is also found after the minimization procedure. The bases L1 and L2 continue to stack upon the double helical stem in a B-type fashion. Base L3 is partly stacked upon L2 and partly on L4; the latter base is turned inward to form a T-T pair with L1. This base pair formation nicely reproduces the NMR results. Generally, the structure of the loop (Fig. 8) is rather tight; all bases are turned inward. This is in good agreement with the observation that the exchange of the imino protons is considerably slowed down with respect to the exchange of imino protons of free thymidines. That the loop thymidines are turned inward also follows from binding experiments between $d(A)_4$ and the hexadecanucleotide (10). Even at NMR concentrations and low temperature no evidence of binding between these molecules can be obtained. Only after the loop has been extended to 6 thymidines complex formation becomes readily observable.

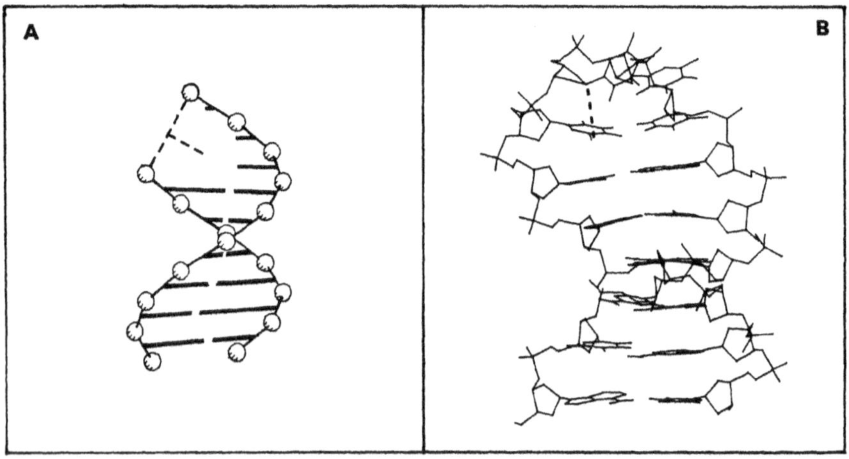

FIGURE 8. Left: Scheme of B-type helix used to clarify the folding of bases in a DNA loop. The stacking of the bases in the loop is continued in the 3'-direction. When three bases are added the distance between the 3'- and 5'-terminal phosphate is reduced to 11 Å (dashed line between circles). This distance can be spanned by one nucleotide. Right: Structure obtained after energy minimization; use was made of the AMBER program developed by Kollman and colloborators (24,25). Note that after energy minimization the stacking pattern of the double helical stem is extended into the loop structure in the 3'-direction. The first two bases in the loop, L1 and L2, stack in a more or less B-type helical fashion; the third base L3 is partly stacked on L2 and partly on L4. Base L4 is turned inward into the loop thereby forming a T-T base pair with L1, as exhibited in Fig. 5. The T-T base pair in the simulated structure is somewhat buckled. The dashed line indicates the short distance between L4 CH3 and L3 H1' (see text).

In order to allow a further comparison between the simulated structure in Fig. 8 and the NMR results, coordinates derived from the energy minimized structure were used to predict a 2D NOE spectrum for the hexadecanucleotide, using the program NOESIM (to be published). Part of the results of this calculation (i.e. the spectral region ω_1 = 1.5 to 3.2 ppm and ω_2 = 5.4 to 6.4 ppm) is presented in Fig. 9 side by side with the experimental results. Scrutiny of the two spectra shows that the cross peaks are reproduced by the theoretical calculation, albeit that a perfect match of the intensities has not yet been obtained. The prediction of a cross peak at ω_1 = 1.95, ω_2 = 5.82 ppm, which arises from the close proximity of the methyl group of L4

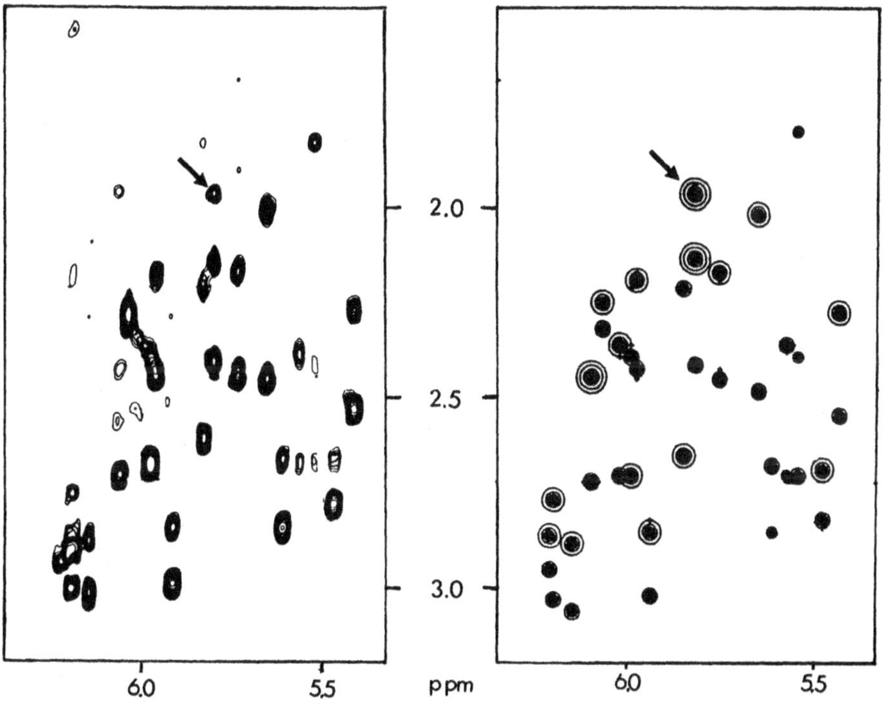

FIGURE 9. Part of the 500 MHz ^{1}H NMR NOESY spectrum (ω_1 = 1.5 to 3.2, ω_2 = 5.3 to 6.4 ppm) exhibiting connectivities between H1' and H2'/H2'' resonances. The connectivity indicated by an arrow is the exceptional cross peak between the methyl resonance of base L4 and the H1' resonance of base L3 (see text). Cross peaks in the left box are part of the experimental NOESY spectrum; cross peaks in the right box are part of a theoretical NOESY spectrum simulated on the basis of the energy minimized structure presented in Fig. 8.

and the H1' sugar proton of L3, is particularly interesting. As was mentioned above, in the experimental spectrum a cross peak is observed for this set of proton spins and apparently the energy minimized structure also neatly reproduces this aspect of the loop folding.

In summary, we conclude that our results are in good agreement with the principles of loop folding described elsewhere (11,12). We interprete the formation of the T-T pair in the loop as a local optimization of the hairpin structure within the conformational space delineated by our loop folding principle.

REFERENCES

1. Heerschap, A., Haasnoot, C.A.G. and Hilbers, C.W.: Nucleic Acids Res. 10, 6981-7000 (1982).
2. Heerschap, A., Haasnoot, C.A.G. and Hilbers, C.W.: Nucleic Acids Res. 11, 4483-4500 (1983).
3. Heerschap, A., Haasnoot, C.A.G. and Hilbers, C.W.: Nucleic Acids Res. 11, 4501-4520 (1983).
4. Roy, S. and Redfield, A.G.: Nucleic Acids Res. 9, 7073-7083 (1981).
5. Roy, S. and Redfield, A.G.: Biochemistry 22, 1386-1390 (1983).
6. Hare, D.R. and Reid, B.R.: Biochemistry 21, 5129-5135 (1982).
7. Reid, D.R., Salisbury, S.A., Bellard, S., Zhakked, Z. and Williams, D.H.: Biochemistry 22, 2019-2025 (1983).
8. Clore, G.M. and Gronenborn, A.M.: FEBS Letters 175, 117-123 (1984).
9. Clore, G.M. and Gronenborn, A.M.: Biomol. Stereodynamics (Eds. R.H. Sarma and M.H. Sarma) IV p.139-155 (1986).
10. Haasnoot, C.A.G., de Bruin, S.H., Berendsen, R.G., Janssen, H.G.J.M., Binnendijk, T.J.J., Hilbers, C.W., van der Marel, G.A. and van Boom, J.H.: J. Biomol. Struct. Dyns. 1, 115-129 (1983).
11. Haasnoot, C.A.G., de Bruin, S.H., Hilbers, C.W., van der Marel, G.A and van Boom, J.H.: J. Biosci. 8, 767-780 (1985).
12. Hilbers, C.W., Haasnoot, C.A.G., de Bruin, S.H., Joordens, J.J.M. van der Marel, G.A. and van Boom, J.H.: Biochimie 67, 685-695 (1985).
13. Haasnoot, C.A.G., Hilbers, C.W., van der Marel, G.A., van Boom J.H., Singh, U.C., Pattabiraman, N. and Kollman, P.A.: J. Biomol. Struct. Dyns. 3, 843-857 (1986).
14. Haasnoot, C.A.G., Westerink, H.P., van der Marel, G.A. and van Boom J.H.: Biomol. Struct. Dyns. 1; 131-149 (1983).
15. Scheek, R.M., Russo, N., Boelens, R., Kaptein, R. and van Boom J.H.: J. Am. Chem. Soc. 105, 2914-2916 (1983).
16. Hare, D.R., Wemmer, D.E., Chou, S.-H., Drobny, G. and Reid, B.R.: J Mol. Biol. 171, 319-336 (1983).
17. Arnott, S. and Hukins, D.W.L.: Biochem. Biophys. Res. Comm. 47 1504-1509 (1972).
18. Haasnoot, C.A.G. and Hilbers, C.W.: Biopolymers 22, 1259-126 (1983).
19. Sanchez, V., Redfield, A.G., Johnston, P.D. and Tropp, J.: Proc Natl. Acad. Sci. USA 77, 5659-5662 (1980).
20. Gralla, J. and Crothers, D.M.: J. Mol. Biol. 78, 301-319 (1973).
21. Uhlenbeck, O.C., Borer, P.N., Dengler, B. and Tinoco, I.: J. Mol Biol. 73, 483-496 (1973).
22. Altona, C.: Recl. Trav. Chim. Pays-Bas 101, 413-433 (1982).
23. Rinkel, L.J. and Altona, C.: J. Biomol. Struct. Dyns. (in th press).

24. Weiner, S.J., Kollman, PA., Case, D.A., Chandra Singh, U., Ghio, C. Alagona, G., Profeta, S. and Weiner, P.: J. Am. Chem. Soc. 106, 765-784 (1984).
25. Singh, U.C., Weiner, S.J. and Kollman, P.A.: Proc. Natl. Acad. Sci USA 82, 755-759 (1985).

STRUCTURE AND FUNCTION IN NUCLEIC ACIDS: MUTAGENESIS

IGNACIO TINOCO, JR., STEVEN WOLK, FRANCES ARNOLD AND FAREED ABOUL-ELA

Chemistry Department, and Laboratory of Chemical Biodynamics, University of California, Berkeley, CA 94720

1. INTRODUCTION

The biological functions of nucleic acids depend on their conformations; we will discuss one example: the mechanism of mutation in DNA. A mutation occurs when a base pair in DNA is substituted for another one, or when one or more base pairs are added or deleted. A base pair substitution can change one amino acid into another, it can change an amino acid into a stop signal which aborts protein synthesis, it can affect the control of replication, transcription or translation, or it can be a silent mutation which has no apparent effect. One or two base pair additions or deletions can lead to frame-shift mutations. Because the reading frame is altered, a frame-shift mutation does not change just one amino acid. It will change all the amino acids coded by the base sequence which follows the frame-shift. There are many causes of mutations and many steps in the mechanism of mutagenesis. We will concentrate on the steps which, through an error in replication of one strand, produce a heteroduplex which then is replicated to form a mutant duplex. We have synthesized oligonucleotides with base mismatches (1,2) to assess the thermodynamic stability of heteroduplexes and to determine their conformations. The structure and stability of the heteroduplex will affect the probability that an error in replicating a strand occurs, and the probability that the error is repaired before correct replication of this mutant strand occurs. This is an example of a biological event--a mutation, which we want to explain in terms of DNA structure.

MUTAGENESIS

We define a mutation as a change in the sequence of base pairs in a DNA, and we will concentrate on changes involving one or two adjacent base pairs. There are many physical, chemical and biological agents which ultimately lead to a mutation, but all the agents which lead to a mutation involve the following basic steps.

1. Formation of a heteroduplex. An incorrect nucleotide (carrying a non-complementary base) is placed in one strand of a DNA duplex. This nucleotide may be incorporated during normal replication of the duplex; it may be added during repair, or during recombination. In these cases the incorrect nucleotide is a naturally occurring nucleotide; it is just in the

wrong location on one of the strands. Another class of incorrect nucleotides can be introduced into one strand by chemical reaction. The bases involved can be nearly normal. Examples are uracil (from deaminated cytosine), hypoxanthine (from deaminated adenine), xanthine (from deaminated guanine), 5-methyl cytosine and 6-methyl adenine (from naturally occurring methylases). Or the modified bases can be very significantly altered by reaction with strong mutagens such as acetylaminofluorene, alkylating agents and so forth.

2. Formation of a mutant duplex. The incorrect strand is copied to produce a new double strand which can be further replicated to provide many copies of the mutant duplex. The mutation is complete. Before the incorrect strand is copied there is opportunity for repair to occur. In fact for major modifications of the bases (or for depurination, strand breaks and other damage) repair nearly always occurs before DNA polymerases act to copy the strand. In this context repair means to form a strand which contains nucleotides that can be used as a template by a DNA polymerase to make a new strand. It does not necessarily mean that the nucleotides introduced are the correct ones. Repair DNA polymerases have less fidelity than the DNA replication complex of polymerase and other proteins, so "repair" often leads to mutation. The various steps involved in mutagenesis are illustrated in Figure 1. The examples shown lead to a transversion mutation (Pu•Py→Py•Pu) and to a frame-shift mutation (addition of one nucleotide). A transition mutation (Pu•Py→Pu•Py) is not illustrated.

Figure 1 omits all the enzymes and other proteins involved in each step shown. Replication is done by a large set of proteins which separate the

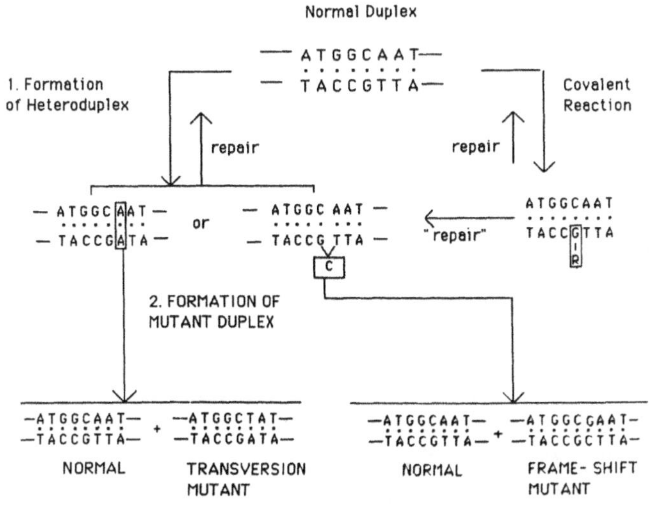

Figure 1. A diagram of the reactions important in mutation. The steps involved are selection and adding of the wrong nucleotide, proofreading, and repair of heteroduplexes

strands, introduce nucleotides complementary to each strand, check that the newly synthesized strands are correct and remove incorrect nucleotides (3). Thus DNA polymerases, nucleases and single strand binding proteins are involved. Recombination (the exchange of segments of duplex between two DNA molecules) uses a similar, but different set of proteins (4). Repair of covalently modified DNA, or of base mismatches in heteroduplexes, involves a very large number of enzymes which can recognize the error, remove the wrong nucleotides and attempt to replace them with right nucleotides (5). As mentioned earlier the repair enzymes are much more error prone.

NUCLEOTIDE MISINCORPORATION

We will focus our attention on the steps involved in the selection of the correct nucleotide to incorporate into the growing DNA strand. A simple mechanism for the binding of a nucleoside triphosphate to the polymerase-DNA complex, and for the subsequent chemical steps (characterized by k_{cat}) is:

$$\text{Enz} \cdot \text{DNA}_n + \text{pppN} \underset{k_{-1}}{\overset{k_1}{\rightleftharpoons}} \text{Enz} \cdot \text{DNA}_n \cdot \text{pppN} \xrightarrow{k_{cat}} \text{Enz} \cdot \text{DNA}_{n+1} + \text{pp}$$

The relative rates of reaction for two competing substrates is given by Fersht (6). The rate of an enzyme catalyzed reaction in terms of <u>free</u> enzyme, E, concentration is:

$$\text{rate} = \frac{k_{cat}}{K_m}[E][S]$$

The relative rate is then (with [A] and [B] equal to the concentrations of competing substrates)

$$\frac{A_{rate}}{B_{rate}} = \frac{(k_{cat}/K_m)_A[A]}{(k_{cat}/K_m)_B[B]}$$

The same chemical steps are involved in the covalent attachment of the complementary or non-complementary nucleotide. A 3' hydroxyl is attached to the α phosphate of a nucleoside triphosphate, and pyrophosphate is released. It is thus reasonable to make the approximation that k_{cat} does not depend on whether the correct or incorrect nucleoside triphosphate is being incorporated (7). The selection of the correct substrate is attributed to preferential binding; once any substrate is bound the chemistry is the same. For equal concentrations of substrates

$$\frac{A_{rate}}{B_{rate}} = \frac{K_{mB}}{K_{mA}}$$

This is illustrated in the free energy transition-state diagram given in Figure 2. An incorrect trinucleoside phosphate raises the free energy of binding, ΔG°, it does not change the transition free energy, $\Delta G^{\neq}$. If k_{cat} (and $\Delta G^{\neq}$) does depend on the substrate, the selectivity should be even greater than stated above. The non-complementary nucleotide binds more poorly and also reacts slower. In any case it is clear that substrate binding has a large effect on the probability of incorporating the wrong nucleotide.

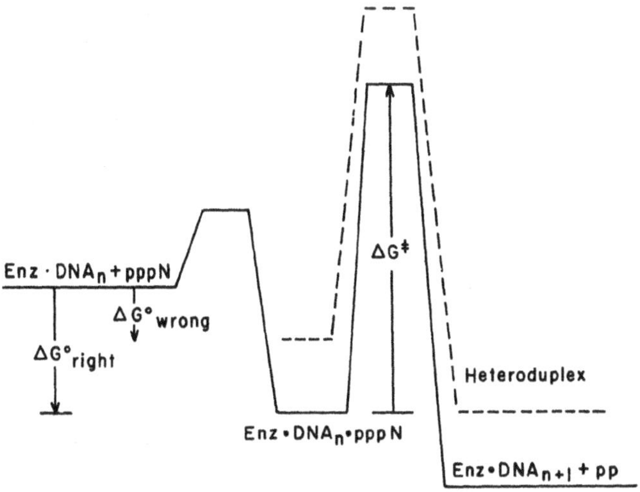

Figure 2. A transition state diagram for adding a nucleotide to a daughter strand in DNA synthesis. The probability of an error depends on the equilibrium constant for binding a nucleoside triphosphate (Michaelis constant, K_m).

DNA polymerases have 3'-exonuclease activity (3). They not only add nucleotides, they also remove 5'-nucleotides from the newly synthesized strand. This exonuclease activity is an important part of the proofreading which helps provide fidelity of replication. Figure 3 illustrates how a wrong nucleotide is more likely than a complementary nucleotide to be removed by exonuclease activity. The end of the newly synthesized strand is in equilibrium between a base-paired double strand form and a dissociated single strand end. The rate of exonuclease hydrolysis depends on the equilibrium constant for dissociation. A non-complementary nucleotide will have a more negative free energy of dissociation to single strands and thus a higher rate of removal.

Experimentally, the range of error frequencies is 5×10^{-3} to 10^{-7}; this includes the effects of both nucleotide binding and catalytic selectivity, and proofreading (8). If each of the two steps has similar error frequencies, the error range per step is 0.07 to 3 x 10^{-4}.

Exonuclease proofreading

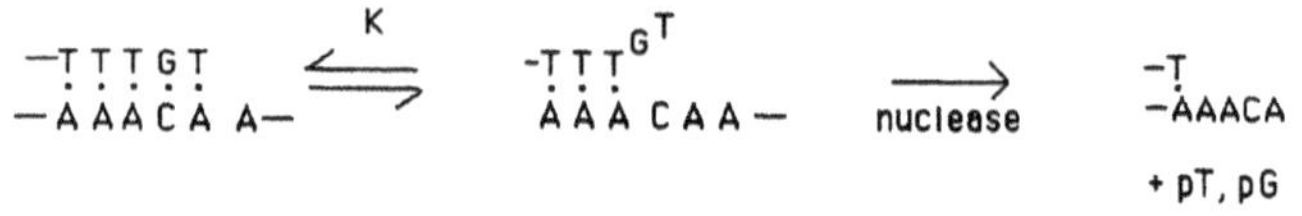

$$\frac{\text{rate A}}{\text{rate B}} \cong \frac{K_A}{K_B} = e^{-(\Delta G^\circ_A - \Delta G^\circ_B)/RT}$$

Figure 3. DNA polymerase has exonucleoase activity which removes newly added nucleotides from the daughter strand. The rate of exonuclease activity depends on the equilibrium constant, K, and free energy, ΔG^o, for double strand dissociation.

THERMODYNAMICS

The thermodynamics of misincorporating a nucleotide are independent of the enzymes that catalyze the reactions. Catalysts affect rates, not equilibria. It is also true that in general thermodynamics can not explain rates. However, in the previous section relative rates of misincorporation have been related to the thermodynamics of binding complementary vs. non-complementary nucleoside triphosphates, and to double strand-single strand equilibria.

We have begun a systematic study of the thermodynamics of nucleotide misincorporation. Deoxyoligonucleotide duplexes which are models for DNA heteroduplexes are synthesized, and their single strand-double strand equilibria are measured. From the equilibrium constants and from the temperature dependence of the equilibrium constants, the thermodynamic parameters can be obtained. We define K for the reaction between two single strands.

$$S_A + S_B \rightleftharpoons D$$

$$K = \frac{[D]}{[S_A][S_B]}$$

The standard thermodynamic parameters for forming the double strand are:

$$\Delta G^o = -RT \ln K$$

$$\Delta H^o = \frac{-R\partial(\ln K)}{\partial(1/T)}$$

$$\Delta S^o = \frac{\Delta H^o - \Delta G^o}{T}$$

Here T is the absolute temperature and R is the gas constant (1.987 cal deg^{-1}). For this simple analysis to be valid, it is necessary that the reaction be all-or-none as written. A test of this two-state assumption is to use different methods of calculating or measuring ΔH^o. If the assumption is valid the different methods must give the same result. There are two ways to take the derivative of ln K--at constant concentration of strands, or at constant fraction of double strand. For concentrations of S_A and S_B each equal to c/2, and a fraction, f, of double strands at equilibrium,

$$K = \frac{2f}{(1-f)^2 c}$$

At constant f

$$\frac{\partial(\ln K)}{\partial(1/T)_f} = \frac{-\partial(\ln c)}{\partial(1/T)_f}$$

and at constant c

$$\frac{\partial(\ln K)}{\partial(1/T)_c} = \frac{(1+f)}{f(1-f)} \frac{\partial f}{\partial(1/T)_c}$$

Usually the derivative at constant f is taken at f = 1/2, thus ln c is plotted vs. (1/T) at f = 1/2. A better way to treat the data at constant concentration is to fit K as a function of temperature to

$$K = e^{-\Delta H^o/RT} \; e^{\Delta S^o/R}$$

ΔH^o can also be obtained directly by scanning calorimetry (9). This can provide an independent check of ΔH^o values obtained by the van't Hoff equation. Experimentally, the two-state assumption seems to be valid for oligonucleotides containing ten base pairs or less. As the length of helix increases, partially base-paired structures become significant and a multi-state analysis is necessary. Partially base-paired structures may also occur for sequences which are stable in one area, but unstable in another (oligonucleotides with C•G base pairs plus a mismatch).

We have measured the thermodynamic stability of a series of heteroduplexes of the type $CA_3XA_3G + CT_3YT_3G$ in 1M NaCl, pH 7 (2). Values of ΔH°, ΔS° (assumed independent of temperature) and ΔG° as a function of temperature were obtained. The data are summarized in Figures 4-7 as the increase in free energy at 25°C of each heteroduplex with a base-base mismatch relative to the Watson-Crick duplex. These data only apply to mismatches with A•T base pairs as neighbors, so studies with G•C neighbors must be done before generalizations can be made. However, if the stability of the heteroduplex has a significant effect on the probability of incorporating the wrong nucleotide, Figure 4 shows that the sequence

A-[T]-A
• • •
T-[A]-T

should have a higher rate of misincorporation than the three other possible base pairs in the box as shown in Figures 5-7. The free energy increase of a non-Watson-Crick pair at that position is the smallest of the sequences studied. The figures also imply the relative rates of adding each wrong

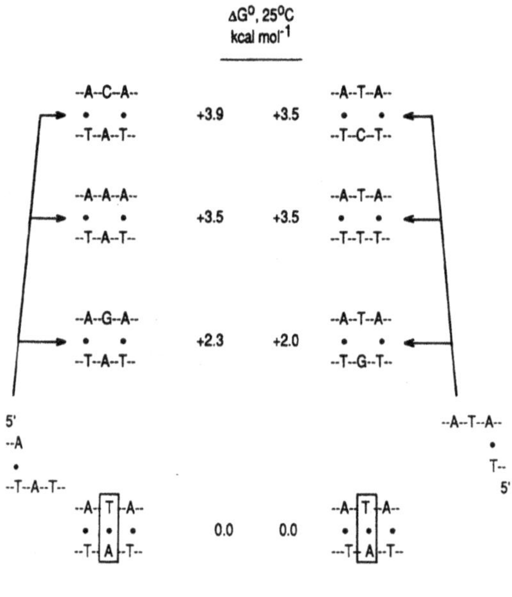

Figure 4. Standard free energies at 25°C for a base-base mismatch surrounded by A•T base pairs.

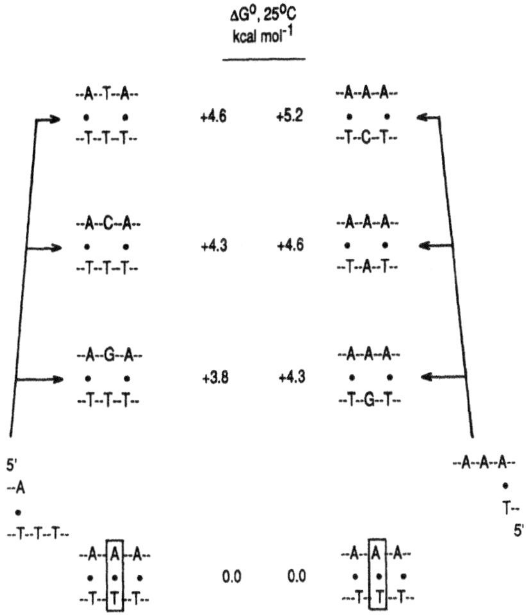

Figure 5. Standard free energies at 25°C for a base-base mismatch surrounded by A•T base pairs.

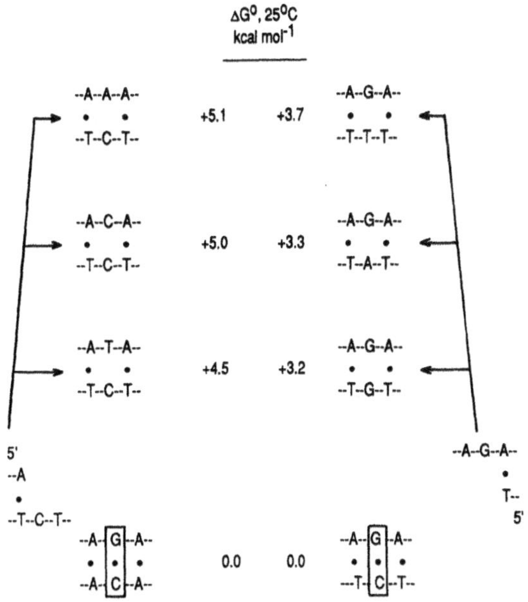

Figure 6. Standard free energies at 25°C for a base-base mismatch surrounded by A•T base pairs.

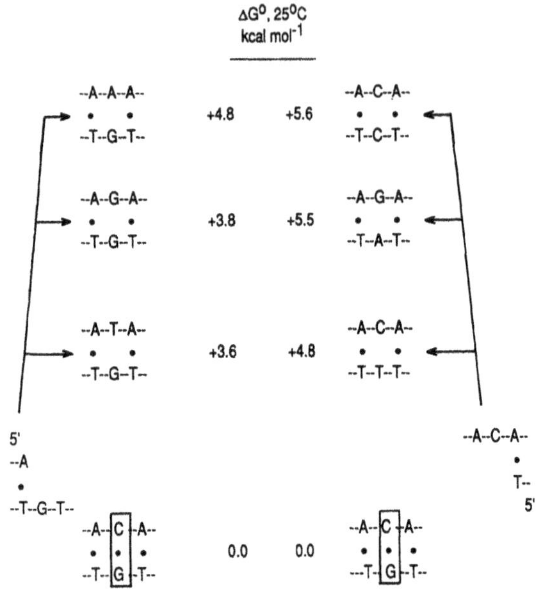

Figure 7. Standard free energies at 25°C for a base-base mismatch surrounded by A•T base pairs.

nucleotide. We note that the range of probabilities of nucleotide misincorporation is from 3 x 10^{-2} (G•T) to 10^{-4} (C•C). For a given sequence there can be a selectivity of one wrong nucleotide for another of a factor of 10. From Figures 4-7 we infer the order of incorporation of the wrong nucleotide given in Table 1.

Table 1. Probability of Incorporation of Each Non-complementary Nucleotide (Based on Double Strand Stability)

Parent Strand		Daughter Strand Mismatch*
A:	A-A-A	G > A > C
	T-A-T	G > A > C
C:	A-C-A	T > C, A
	T-C-T	T > C, A
G:	A-G-A	G, A > T
	T-G-T	T, G > A
T:	A-T-A	G > C, T
	T-T-T	G > C > T

*The mismatch is at the center nucleotide of the parent strand.

We have measured only three duplexes of the type C_3XG_3 + C_3YG_3, therefore further work is needed. However, the data show that A•A and T•T mismatches with C•G neighbors behave differently than the duplexes seen in Figures 4,5. In Figure 8 we see that with C•G neighbors, T•T mismatches are more stable than A•A mismatches by 0.7 kcal mol^{-1} for

ΔG^o at 25°C. With A•T neighbors they have the same stability.

More sequences must be measured to assess better the effect of nearest-neighbors. We have only measured 19 of the 136 duplexes of the type -W-X-W- + -C-Y-C-, where W•C are Watson-Crick complements and X,Y are any of the bases. Solvent effects may also be important. The data we have been discussing were obtained (2) in 1M NaCl, pH7. In 30mM $MgCl_2$, 150 mM KCl at pH 7, which may better approximate physiological

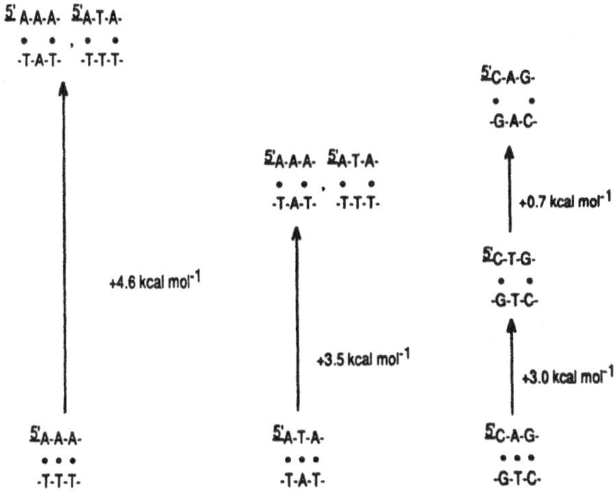

Figure 8. Comparison between free energies of a mismatch surrounded by A•T and by G•C base pairs.

conditions, three duplexes without mismatches were measured. The duplexes are less stable, but their order of stability was not changed. The standard free energy of formation from the single strands at 25°C was 85% of $\Delta G^o(25°C)$ for each duplex in 1M NaCl. We do not know if changing the solvent composition will change the relative order of stability of heteroduplexes.

For purposes of understanding mutation, there is some question about the most appropriate solvent to measure duplex stability. Petruska et al. (10) think that the DNA polymerase excludes water from the binding site, thus free energies obtained in aqueous solution are not directly relevant. It would be very useful to have data for enzymatic K_m values for non-complementary nucleoside triphosphates, and for double strand-single strand equilibria for enzyme-bound nucleic acids to begin to answer these questions.

CONFORMATION

Knowledge of the conformations of non-Watson-Crick base-base mismatches will help in understanding the thermodynamics and the probability of mutation. The mismatches have been studied by NMR (11-13) by X-ray diffraction (14-16), and by theoretical calculations (17). We will summarize NMR results obtained in our laboratory on bulges (1) and mismatches.

NUCLEAR MAGNETIC RESONANCE

The most useful NMR phenomenon to apply to conformation studies is the nuclear Overhauser effect (NOE). We will limit the discussion to proton-proton NOE's, although the effect is quite general. In the NOE experiment one irradiates the sample at a frequency of a particular proton resonance in the NMR spectrum and then measures changes in the intensities of peaks at other frequencies. The intensity change, or NOE, is a result of magnetic dipole coupling between the nuclei. Transitions between the spin energy levels (spin up or down) of a given nucleus are induced by local magnetic fields oscillating at the resonance frequency of that nucleus. Thus the NOE depends on the amplitude and frequency of the local magnetic field produced by neighboring nuclei. The effect is proportional to the square of the dipole-dipole interaction and depends on the rotational and internal motion of the molecule.

For short irradiation times, the NOE is proportional to the cross relaxation rate between the irradiated proton, proton 1, and the observed proton 2. The cross relaxation rate, σ_{12}, can be used to obtain the distance between the two protons.

$$\sigma_{12}\,(\mathrm{sec}^{-1}) = \frac{\hbar^2\gamma^4}{10r_{12}^{\;6}}\,[-J(0) + J(2\omega)]$$

γ is the proton gyromagnetic ratio, and h is Planck's constant divided by 2π. J(0) and J(2ω) are spectral density terms which characterize the frequency dependence of the magnetic field at nucleus 2. J(0) identifies the magnetic field at zero frequency, whereas J(2ω) characterizes the field at twice the resonance frequency of the proton.

The form of the spectral density term depends on the model used to describe the motional modes of the molecule. The simplest model, and the only one that we will discuss is that for a rigid, isotropic molecule--a rigid sphere. For this case the spectral density at any frequency is

$$J(\omega) = \frac{\tau_c}{1+\omega^2\tau_c^{\;2}}$$

where τ_c is the rotational correlation time, which is proportional to the reciprocal of the rotational diffusion coefficient. Thus, if τ_c is known, we can calculate J(ω) and obtain r_{12} from the NOE. In fact if the molecule is rigid, then $J(\omega)$ is the same for all proton pairs, and relative NOE's can be used to obtain estimates of interproton distances. The NOE for two protons of known separation, such as H5-H6 of cytosine, is measured and compared with the NOE for two protons of unknown separation.

$$\frac{NOE_{12}}{NOE_{56}} = \frac{\sigma_{12}}{\sigma_{56}} = \frac{r_{56}^{6}}{r_{12}^{6}}$$

In spite of the many assumptions and approximations involved in this approach, NOE measurements provide the best method for obtaining short distances (<5 Å) from NMR data. In the preceding discussion we assumed only two spins interacting in a rigid molecule. In general we must consider the interaction among all the nuclei, and we should acknowledge the flexibility of the molecule. A flexible molecule has different spectral density functions for different pairs of nuclei. Furthermore, the distance r_{12} is no longer constant. On the experimental side the quantitative measurement of σ_{ij} can be difficult. In a 1D NOE experiment one peak is saturated; this removes it from the spectrum and causes changes in the intensities of interacting protons through the NOE. The difference spectrum--normal spectrum minus the spectrum with one proton saturated--will contain only the NOE changes and the irradiated peak. The magnitude of the NOE, defined as the area of a peak in the difference spectrum divided by the area of the saturated peak, depends on the length of time that the peak is saturated, t.

$$NOE = \frac{\sigma}{\rho}\,(1 - e^{-\rho t})$$

σ is once again the rate of cross relaxation from the saturated proton, and ρ is the spin-lattice relaxation rate, a result of interactions with all other nuclei in the system. At short irradiation times, the NOE is dominated by the cross relaxation from the proton closest in space. At longer times direct interactions with more distant nuclei and indirect magnetization transfer pathways become increasingly important. In order to interpret the measured NOE quantitatively in terms of interproton separations, it is best to measure NOE's at various irradiation times. The initial slope of the curve of the NOE versus irradiation time is then used to obtain the σ_{ij}.

In a 2D NOESY experiment (19,20) analogous data are obtained. In a 2D experiment, however, NOE couplings between all pairs of protons are obtained in a single experiment. The volumes of off-diagonal peaks are measures of the strengths of the NOE interactions. As in the 1D experiments, the NOE's in 2D experiments must be measured as a function of the mixing time in order to obtain σ_{ij} values.

Here NOE measurements will be used only to get qualitative information about the distances between certain protons. Essentially we will answer the question of whether two protons are closer than about 4Å, or not.

This answer is sufficient to establish the gross features of the conformations of some intermediates in mutagnesis.

BULGED BASES

For a frame-shift mutation to occur a heteroduplex with an extra nucleotide on one strand (a bulge) must be formed. Is the base of this nucleotide intercalated within the helix, or is it outside the helix? This question was answered for cytosine in $CA_3CA_3G{\cdot}CT_6G$ using NOE's for the imino protons (1). Figure 9 shows NOE difference spectra at 500 MHz for

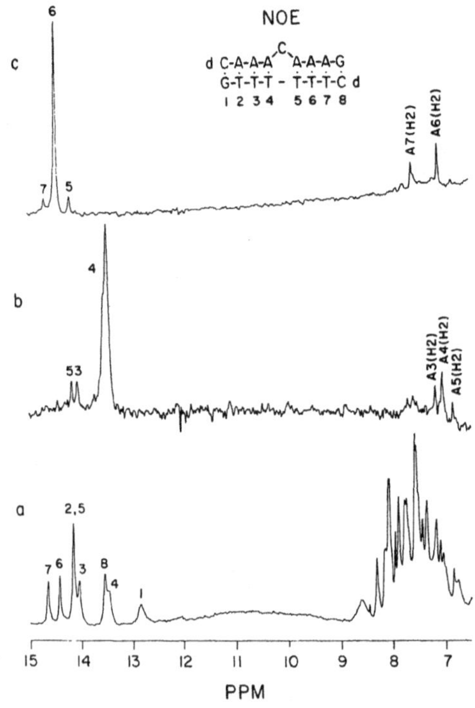

Figure 9. NMR spectra of 1 mM $CA_3CA_3G{\cdot}CT_6G$ in H_2O, pH 7, 0.2 M NaCl, 1°C. Spectrum (a) shows the eight imino protons. The difference NOE spectra after 500 ms saturation (b, c) indicate that the extra cytosine is outside of the helix.

the imino region of the spectrum of the heteroduplex in H_2O. Spectrum (a) shows the eight imino protons for the eight base pairs. In spectrum (b) proton 4 is saturated and NOE intensity changes are seen at protons 3 and 5. The NOE's are similar for peaks 3 and 5, therefore the distances between protons 3 and 4, and between protons 4 and 5 must be similar. Thus cytosine must be out of the helix. Spectrum (c) shows similar NOE intensity changes at protons 5 and 7 when proton 6 is saturated. Clearly the distance between the base pairs on each side of the extra nucleotide is about the same as the distance between normally stacked base pairs. Figure 10 shows a possible conformation. We do not know how commonly this type of structure will occur. Based on chemical shift data, the extra adenine in the sequence CGCAGAATTCGCG seems to be in the helix (11). The equilibrium between base intercalation and an extrahelical position may well depend on the stacking properties of the bases involved. As adenine is the strongest stacking base, this hypothesis is consistent with the two experimental results.

Figure 10. Picture of a model for the extrahelical cytosine in a duplex with an extra cytidylic residue.

BASE-BASE MISMATCHES

We have used NMR measurements to study A•A mismatches and T•T mismatches in C_3AG_3•C_3AG_3 and C_3TG_3•C_3TG_3. The two strands in each duplex are found to be equivalent; the NMR spectrum corresponds to seven nucleotides in both the single strands and the duplex. This means either that the strands in the duplex are in identical conformations, or that there is rapid exchange (shorter than msec) among different conformations. Our NMR studies further show that the mismatched bases are within the helix. In B-form geometry with all bases in an anti conformation and all sugars 2' endo (see Fig. 11), there is a well established NOE connectivity path from H8 of adenine or guanine to H1' of its own nucleotide and to H1' of the sugar on its 5' side. There is a similar connectivity involving the H6 of cytosine or thymine. This means that in B-form geometry each base (except the 5' terminal one) is NOE linked to two H1' protons, on its own nucleotide and one on the preceding nucleotide. In other conformations, such as A-form, this connectivity is changed. A 2D-NOESY spectrum for $C_3AG_3 \cdot C_3AG_3$ showing the aromatic--H1' crosspeaks (Figure 12) shows this B-form connectivity from H6 on cytosine 1 through each nucleotide to H1' on guanosine 7. Therefore the average conformation must have the adenines inside the helix

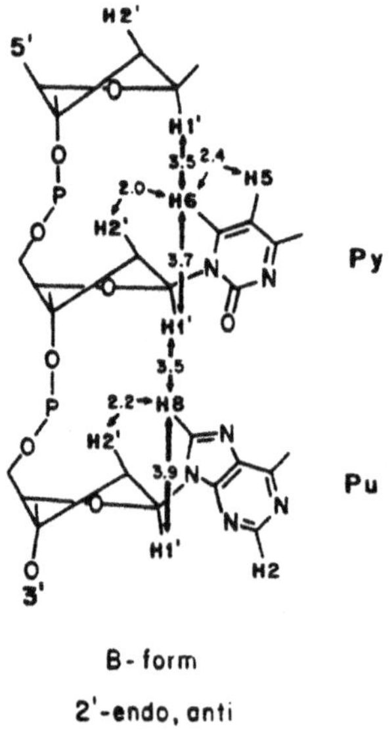

Figure 11. Short distances between protons in a B-form duplex. Strong NOE's are expected between these protons.

with the adenine H8 within ~4Å of the H1' of the preceding cytosine. Further evidence for this conformation with intercalated adenines comes from NOE measurements of imino protons for the duplex in H_2O. In water three imino protons are seen near 13.0 ppm. When the G5 imino proton is saturated (this proton is in base pair 3 and 5), an NOE is seen to the 2 position of adenine (A4H2). This NOE requires that the average conformation has the adenines within the helix, with the adenine H2 within ~4Å of the next imino proton. Chemical shift data as a function of temperature are consistent with this picture. When a Watson-Crick duplex melts there is a downfield shift of adenine H2 because of the decreased stacking in the single strands relative to the double strands. We found about a 0.2 ppm downfield shift for A4H2 on melting the duplex. In spite of the A•A mismatch the adenine is more stacked in the duplex than in the single strands.

All the NMR data show that the adenine bases in the A•A mismatch are stacked inside the helix, and that this stacking does not significantly perturb the C•G base pairs on both sides. The C_3AG_3•C_3AG_3 duplex, including the A•A, essentially has the conformation of a B-DNA helix. NMR studies of

```
CCCTGGG    GCGATTCGC
•••••••    •••••••••
GGGTCCC    CGCTTAGCG
```

indicate that a T•T mismatch is also inside the helix and does not seriously perturb the B-form geometry. The two strands in each of the T•T duplexes are equivalent as was found for the A•A mismatch. We conclude that these

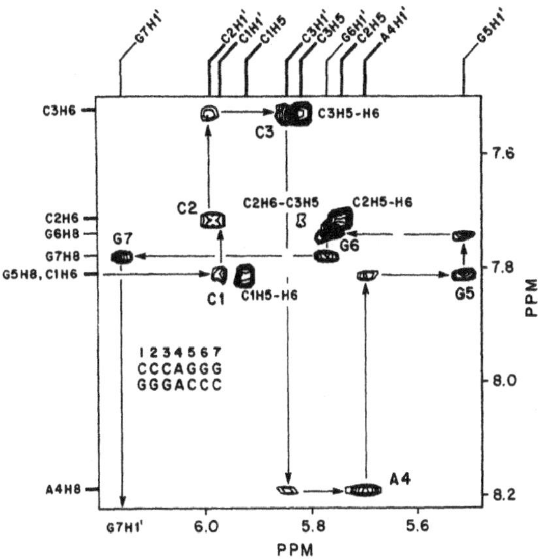

Figure 12. A two-dimensional NOESY spectrum for 5mM C_3AG_3•C_3AG_3 in H_20, pH 7, 0.2 M NaCl, 20°C; the region involving aromatic protons and 1' protons of deoxyribose is shown. The B-form NOE connectivites indicate that the A•A mismatch does not disrupt the B-form geometry.

two mismatches do not make large changes in the double helix, although they must change the details of the conformation in their neighborhood. We do not know if hydrogen bonding is an important part of the mismatch structures; one possibility for each bonded structure is compared with a Watson-Crick base pair in Figure 13. Only one of the two equivalent structures for each mismatch pair is shown; for example the A•A pair is shown as a rapid equilibrium between the amino group of the left adenine bonded to the N1 ring nitrogen of the right adenine and vice versa.

$$A-NH_2 \cdots N1A \rightleftharpoons A1N \cdots H_2N-A$$

We expected to find conformations for these deoxynucleotides corresponding to B-family structures (2'endo sugars), and we did. However, Kennard's group (21) has found that dG_4C_4 crystallizes in A-form geometry with 3' endo sugars. Either the conformations of DNA duplexes are very dependent on sequence, or crystal forces can shift the conformation from B-form to A-form.

BASE MISINCORPORATION AND REPAIR

In order to relate structural and thermodynamic data on mismatches and bulges to mutagenesis, it is necessary to have quantitative data on probabilities of base misincorporation and repair in biological systems. Studies of the fidelity of DNA polymerase in vitro have been done (8,22), but data for quantitative comparison with our sequences are not available. The group of Radman (23,24) has recently published data on the repair of

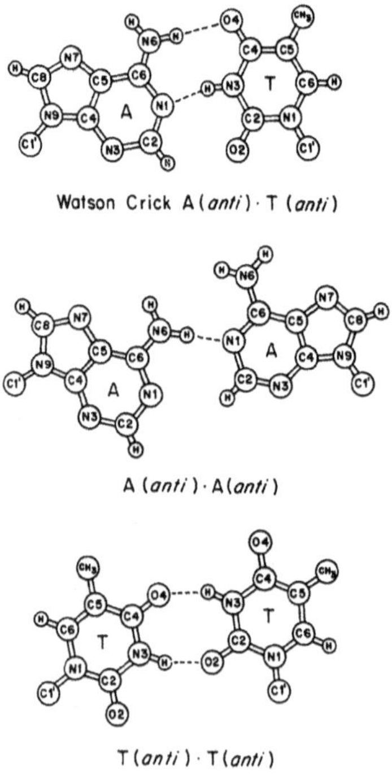

Figure 13. Possible hydrogen bonded structures for A•A and T•T mismatches. Each mismatch is in rapid equilibrium so as to have both bases in equivalent positions.

DNA heteroduplexes in E. coli. Separated strands of bacteriophage λ DNA were annealed to form duplexes with base-base mismatches (23) or single-base bulges (24). The E. coli were transfected with the λ heteroduplex, and the phage progeny were measured to determine the efficiency of repair. Their data on repair of base-base mismatches can be compared directly with our thermodynamic data for the same nearest-neighbor sequences. The order of efficiency of repair is

```
 TGT     TTT     TAT     TAT     TAT     TCT
 ...  >  ...  >  ...  >  ...  >  ...  >  ...
 ATA     ATA     ACA     AAA     AGA     ATA
94.4%   93.1%   89.7%   87.8%   27.3%   20.0%
```

There is no correlation with thermodynamics of double strand dissociation. The thermodynamically most stable mismatches are G•T (94.4% repaired) and G•A (27.3% repaired). The least stable are C•A (89.7% repaired) and C•T (20% repaired). The repair enzyme is definitely not recognizing a dissociated double helical region (an open state) to initiate repair. The E. coli repair system apparently senses some aspect of the conformation of the double helix at the mismatch.

Further studies on the detailed structures at mismatches will be necessary to help understand the repair data. There are crystal structures for the following mismatches.

```
G·T: GGTC  (14)  ;  TTG (14)
     ....           ...
     CTGG           AGC

G·A: TAG  (15)
     ...
     AGC

C·A: CAA  (16)
     ...
     GCT
```

There are two surprises; the G•A mismatch has the adenine syn and the C•A mismatch has the adenine protonated at physiological pH's. NMR data on a G•A mismatch in the sequences

```
-G A-  (12)  ;  -G A G-  (13)
 . .             . . .
-A G-           -C G C-
```

found both G and A anti. It will be extremely useful to obtain biological, structural and thermodynamic data on the same sequences. Until this is done we can not decide if the conformations are very sensitive to sequence, or if the conformations are different between solutions and crystals.

The studies on thermodynamics and conformations of base-base mismatches described here are just the beginning. They suggest that physical studies may eventually provide a rationale for understanding the rates of mutation. We want to know (1) the effect of each step on the mutation rate, (2) which sequences are most susceptible to mutation and what change in sequence will likely occur, (3) how does the environment affect the mutation rate. To be able to understand mutation it is important that biological and physical studies be done on the same molecules under the same conditions.

REFERENCES

1. K. M. Morden, Y. G. Chu, F. H. Martin and I. Tinoco, Jr., Biochemistry 22, 5557-5563 (1983).
2. F. Aboul-ela, D. Koh, I. Tinoco, Jr., and F. H. Martin, Nucleic Acids Research, 13, 4811-4824 (1985).
3. A. Kornberg, DNA Replication, W. H. Freeman, San Francisco, 1980.
4. F. W. Stahl, Genetic Recombination, W. H. Freeman, San Francisco, 1979.
5. E. C. Friedberg, DNA Repair, W. H. Freeman, San Francisco, 1985.
6. A. Fersht, Enzyme Structure and Mechanism, 2nd edition, W. H. Freeman, San Francisco, 1985, p. 112.
7. J. J. Hopfield, Proc.Nat'l. Acad. Sci., U. S. A. 71, 4135-4139 (1974).
8. L. A. Loeb and T. A. Kunkel, Ann. Rev. Biochem. 52 429-457 (1982).
9. K. J. Breslauer, R. Frank, H. Blocker and L. A. Marky, Proc. Nat'l. Acad. Sci., U. S. A., 83, 3746-3750 (1986).
10. J. Petruska, L. C. Sowers and M. F. Goodman, Proc.Nat'l. Acad. Sci., U. S. A. 83, 1559-1562 (1986).
11. D. J. Patel, A. S. Kozlowski, L. A. Marky, J. A. Rice, C. Broka, J. Dallas, K. Itakura and K. J. Breslauer, Biochemistry 21, 437-444 (1982).
12. L. Kan, S. Chandrasegaran, S. M. Pulford and P. S. Miller, Proc. Natl. Acad. Sci.,U. S. A. 80, 4263-4265 (1983).
13. D. J. Patel, S. A. Kozlowski, S. Ikuta and K. Itakura, Biochemistry 23, 3207-3226 (1984).
14. G. Kneale, T. Brown and O. Kennard, J. Mol. Biol.186, 805-814 (1985).
15. O. Kennard, J. Biol. Struct. Dynamics 3, 205-226 (1985).
16. W. N. Hunter, T. Brown, N. N. Anand and O. Kennard, Nature 320, 552-555 (1986).
17. J. W. Keepers, P. Schmidt, T. L. James and P. A. Kollman, Biopolymers 23, 2901-2929 (1984).
18. I. Solomon, Phys. Rev. 99, 559-565 (1955).
19. S. Macura and R. R. Ernst, Mol. Physics 41, 95-117 (1980).
20. A. Bax and L. Lerner, Science 232, 960-967 (1986).
21. M. McCall, T. Brown and O. Kennard, J. Mol. Biol. 183, 385-396

(1985).
22. S. M. Watanabe and M. F. Goodman, Proc. Nat'l. Acad. Sci., U. S. A. 79, 6429-6433 (1982).
23. C. Dohet, R. Wagner and M. Radman, Proc. Nat'l. Acad. Sci., U. S. A. 82, 503-505 (1985).
24. C. Dohet, R. Wagner and M. Radman, Proc. Nat'l. Acad. Sci., U. S. A. 83, 3395-3397 (1986).

The Structure and Physical Chemistry of Cruciform Structures in Supercoiled DNA

David M J Lilley

1. DNA Structure and DNA supercoiling

Our perception of DNA structure has undergone marked changes in the last few years. This has arisen from an enhanced awareness of the variation in local structure which can occur in DNA, resulting in a structure with considerable local micro-heterogeneity which is sequence-dependent. This has been revealed in the crystal by means of near-atomic resolution crystallographic studies of oligonucleotides of defined sequence (Dickerson *et al* 1983; Dickerson, 1983), and by studies upon DNA in solution, principally using nucleases (Lomonosoff *et al* 1981; Drew & Travers, 1984). Two factors combine together to promote structural variation in DNA. These are base sequence and negative supercoiling. Supercoiling provides free energy which may be employed to stabilise otherwise unstable structural polymorphs which are permitted by specific local sequence arrangements.

Why does supercoiling stabilise novel DNA structures? DNA supercoiling is a very common phenomenon (Vinograd *et al* 1965) in circular or constrained loops of DNA, and arises from the double-stranded structure of the DNA molecule. If a linear DNA molecule is cyclised, the two single-strands of DNA become linked circles, characterised by a linking number Lk. If this is done without any net rotation about the helix axis, the resulting circular molecule is torsionally relaxed, and we define this linking number as Lk^o. However, if before ligation we unwind the molecule by n turns of duplex, or n x 360°, then the resulting circular molecule has a linking deficiency.

$$Lk = Lk^o - n$$

We define the linking difference of any circular DNA molecule ΔLk by

$$\Delta Lk = Lk - Lk^o$$

Because the molecule is *underwound*, the sign of ΔLk is negative, and we speak of *negative* supercoiling. DNA molecules differing only in Lk are topological isomers, or topoisomers. As the linkage changes from the most relaxed state we must, by definition, put free energy into the system. In fact, the free energy of DNA supercoiling (ΔGs) is quadratically related to the linking difference (Depew & Wang, 1975; Pulleyblank *et al* ,1985)

$$\Delta G_s = f.(\Delta Lk^2)$$

These energies are very significant. A 5kb plasmid extracted from *Escherichia coli* has a linkage reduction equivalent to about 1 turn removed in 17. We define a specific linking difference, or superhelix density, σ, to express this

$$\sigma = \Delta Lk/Lk^o$$

and the plasmid taken from the bacterial cell typically has a superhelix density of

$$\sigma = -0.06$$

This corresponds to a free energy of about 100kcal mole^{-1} in the above example.

How does this affect the structure of the supercoiled molecule? The linking difference is partitioned between changes in helical twist (ΔTw) and writhing the duplex axis in three dimensions (Wr), according to Fuller (1971)

$$\Delta Lk = \Delta Tw + Wr$$

Since ΔLk is negative for natural DNA molecules, formation of a local structure which results in a reduction in helical twist, ie. a negative ΔTw, will partially compensate for the linking deficiency. This is thermodynamically equivalent to fixing this linkage reduction in the reference state, and the extent to which this lowers the free energy of supercoiling may be offset against the free energy of formation of the unwound structure. To take an example, if local structural perturbation in pBR322 results in a ΔTw of -1, this will release nearly 10 kcal mole^{-1} of supercoiling free energy at σ = -0.06.

What kind of structures may be formed which confer a local negative ΔTw? This might be a local melting, such as occurs during open complex formation during transcription initiation (Walter *et al*, 1967), or perhaps a B to A transition which results in a net underwinding. More dramatic structural variants may be stabilised. Alternating C-G and related sequences may form the left-handed Z conformation (Wang *et al*, 1979). These allow the largest known ΔTw to occur, and Z-DNA is stabilised by negative supercoiling (Peck *et al*, 1982; Singleton, *et al*, 1982). Still more striking rearrangements can occur. If the sequence possesses a local two-fold axis of symmetry, referred to as an inverted repeat, then intra-strand base-pairing is possible, to form hairpin loops, see Figure 1. The structure which

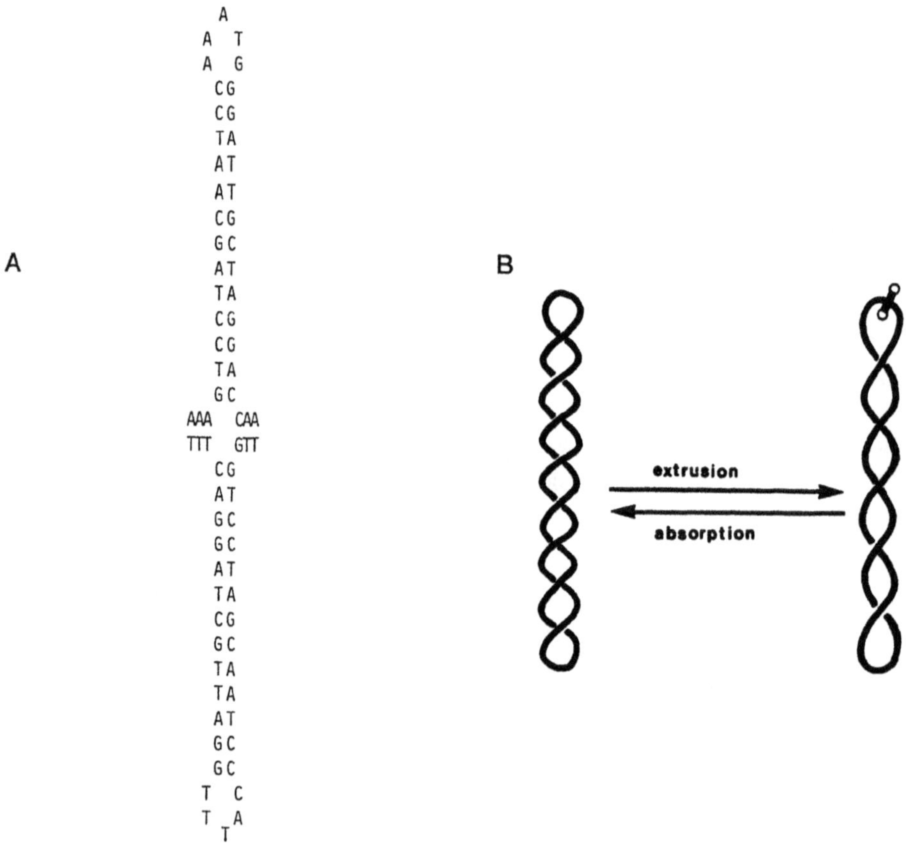

Figure 1. Cruciform formation by inverted repeat sequences in supercoiled DNA molecules.

A. An inverted repeat drawn in its cruciform (hairpin) conformation. The sequence shown is the ColE1 cruciform (Lilley, 1980)
B. Schematic to illustrate the relaxation of supercoiling which accompanies cruciform formation. Note the smaller number of writhing turns in the extruded state.

forms, usually called a cruciform structure, is topologically equivalent to melting the participating sequence, ie a negative ΔTw. Once again, stabilisation of cruciform structures has been experimentally observed for negatively supercoiled molecules (Gellert *et al*, 1979, Lilley, 1980; Panayotatos & Wells, 1981). Other structures such as slippage by direct repeats (Hentschel, 1982; Trifonov, *et al*, 1985) and a novel poly-purine conformation (not yet fully defined structurally) (Cantor & Efstradiadis, 1984; Lyamichev *et al*, 1985; Pulleyblank *et al*, 1985) are also possible, but in the remainder of this chapter I will concentrate on the cruciform structure.

2. Cruciform Structures

The existence of cruciform structures was predicted many years ago (Platt, 1955; Gierer, 1966), but experimental verification required an appreciation of the importance of DNA supercoiling. Two contrasting and complementary methods may be used to reveal any such structure. These are gel electrophoretic methods which reveal the topological changes in global conformation which accompany the transition, and enzymic or chemical probing methods, which attempt to recognise some chemically abnormal aspect of the perturbed conformation relative to the background of normal DNA. Gellert *et al* (1979) generated giant cruciforms *in vitro* by gyration of head to head dimers of pBR322, and revealed the formation of cruciforms by gel electrophoresis and electron microscopy. The first demonstration of the existence of cruciforms in naturally supercoiled plasmids taken from bacteria was made simultaneously by Lilley (1980) and Panyotatos & Wells (1981) who used SI nuclease to reveal the single-stranded loop in a cruciform present in ColEI and pVH51, as well as other cruciforms in pBR322 and ϕX174RF.

We have subsequently developed a battery of enzyme and chemical probes of cruciform structure, summarised in Table 1, some of which are likely to be useful for the study of other structures. In principle there are two aspects of the cruciform which may be susceptible to such attack. These are the unpaired loop, and the four-way junction. The majority of probes attack the loop region by virtue of single-strand specificity or selectivity. These are either single-strand specific nucleases, such as the prototypical SI nuclease, or chemicals. We may also employ restriction enzymes as probes, which *fail* to cleave the DNA when it is the non-standard conformation (Mizuuchi *et al*, 1982; Lilley & Hallam, 1983). In general, greater structural specificity is

Probe	Target	Reaction	Reference
S1 nuclease	Loop	Single-strand-specific cleavage (pH 4.6)	Lilley, 1980 Panayotatos & Wells, 1981
Micrococcal nuclease	Loop	Single-strand-specific cleavage (pH 7.4)	Lilley, 1983a
Bal31 nuclease	Loop	Single-strand-specific cleavage (pH 8.1)	Lilley & Hallam, 1984
Bromoacet-aldehyde	Loop	Ethano adduct formation at A and C	Lilley, 1983b
Osmium tetroxide	Loop	*cis* ester formation at T and C	Lilley & Palecek 1984
Glyoxal	Loop	Ethano adduct formation at G	Lilley, 1986; Gough, 1986
Bisulphite	Loop	Deamination of C	Gough *et al*, 1986
Diethyl pyrocarbonate	Loop	Carbethoxylation of A > G	Furlong & Lilley, 1986
T4 endo-nuclease	Junction	Resolution cleavage	Lilley & Kemper, 1984

Table 1. Summary of the probes employed and developed in this laboratory for the study of cruciform structures.

likely to be found with probes which recognise the four-way junction. T4 endonuclease VII (Kemper & Garabett, 1981) is just such an enzyme, which cleaves supercoil-stabilised cruciforms with high specificity (Mizuuchi *et al*, 1982; Lilley & Kemper, 1984). Similar enzymes from other sources are now becoming available (Symington & Kolodner, 1985; West & Korner, 1985). Our experience in recent years has taught us that whenever possible the maximum amount of data, both probing and topological, should be obtained when a new structure is studied, in order to avoid pitfalls and arrive at the correct solution.

In addition to having many probes of cruciform structure, we now have a great many sequences for study. These may be natural, such as those of ColEI, pBR322, ϕX174 (Lilley, 1980) and alternating $(A\text{-}T)_{34}$ (Greaves *et al*, 1985), but we are not limited by the sequences with which nature provides us. Some years ago we developed a synthesis and cloning approach (Lilley & Markham, 1983), as a result of which we can make construct inverted repeats of any desired sequence. Systematic application of such procedures is currently being used in the dissection of sequence-dependent properties of cruciform formation.

3. Cruciform structures - Questions and Problems

In order to understand the properties of cruciform structures, just as with any novel DNA structure, we must ask questions of two types.

1. The molecular geometry of the cruciform.

There are two elements of the cruciform which are not normally found in DNA. These are the unpaired loops, and the four-way junction. The latter is of particular biological interest since it is formally equivalent to the Holliday junction of homologous genetic recombination (Holliday, 1964, Meselson and Radding, 1975).

2. The physical chemistry of cruciform formation.

This divides into two aspects, thermodynamic and kinetic. The former asks how much free energy is required to form a stable cruciform. The latter addresses the question of the mechanism by which the cruciform becomes extruded from regular DNA. Questions of this nature must be answered before any possible biological role can be assessed.

4. Molecular Geometry of the Cruciform

One of our major approaches in the analysis of cruciform geometry has been to study accessibility to enzyme and chemical probes. There are a number of chemical species which react with DNA in a single-strand-selective manner, and by studying their reactivity at individual nucleotides, estimates of single-strand character may be made. Recently we have employed two small-molecule probes, bisulphite (Gough *et al*, 1986) and diethyl pyrocarbonate (Furlong & Lilley, 1986). A series of cruciforms formed by perfect inverted repeats were probed using bisulphite, in order to deduce optimal loop size. In the case of an inverted repeat with a central assymetric region, the ColEI sequence for example, the size of the loop is determined by the sequence. For a perfect inverted repeat the symmetry extends to the centre, and hence a number of base pairs will open to form a loop which optimises the free energy of the structure. We observed maximal bisulphite-catalysed cytosine deamination in the central four bases, significant but lower modification at the next base either side, and negligible reaction beyond this. We concluded that the optimal loop size is four bases, but that the next base pair, ie that at the top of the stem, is loosely associated. Hilbers and coworkers (Hasnoot *et al*, 1986) have estimated an optimal loop size of between 4 and 6 bases in physical studies of synthetic DNA hairpins.

The ColEI cruciform was probed using diethyl pyrocarbonate (DEPC) to estimate adenosine reactivity, and the result is shown in Figure 2. Two points are apparent from inspection of this autoradiograph. First, the only adenine bases to be significantly modified are present in the loops of the cruciform. Second, not all loop adenines are modified with equal intensity. In the AAATG loop there is a distinct reduction of reactivity in the 5' direction, implying a reduced accessibility of N7A on the 5' side of the loop. We suggest that the adenine bases may stack, thus continuing the helical geometry on this side of the loop, and thereby filling space in the interior of the loop. Model building shows this to be stereochemically reasonable, as presented in Figure 3. Loop stacking has previously been proposed in a hairpin loop composed of thymine bases (Hasnoot *et al*, 1986).

Very little chemical modification has been observed at bases in the four-way junction of any of our cruciforms. In Figure 2 for example, whilst A residues in the loops are extensively carbethoxylated by DEPC, there is only background modification at any adenine in or around the four-way junctions. Cytosine deamination in pIRbph13, which was almost 60% in the loops, revealed only two events from 80 screened of cytosine bases in the four-way junction (Gough *et al*, 1986). We conclude that the junctions exhibit very little single-strand character,

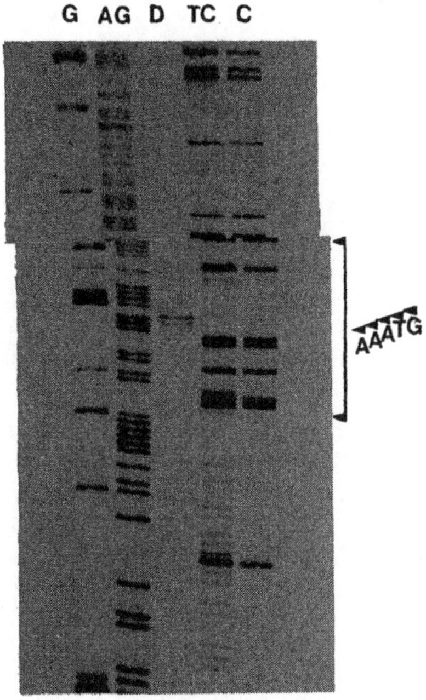

Figure 2. Diethyl pyrocarbonate modification of adenosine bases in the loop of the ColE1 cruciform.

Supercoiled pColIR315 was reacted with DEPC, and the carbethoxylated bases chemically excised with hot piperidine. The DNA (radioactively labelled at the *Eco*RI site) was electrophoresed on a sequencing gel (track D) flanked by Maxam-Gilbert sequence markers. The extent of the entire inverted repeat is indicated by the vertical line, and the AAATG loop is shown.

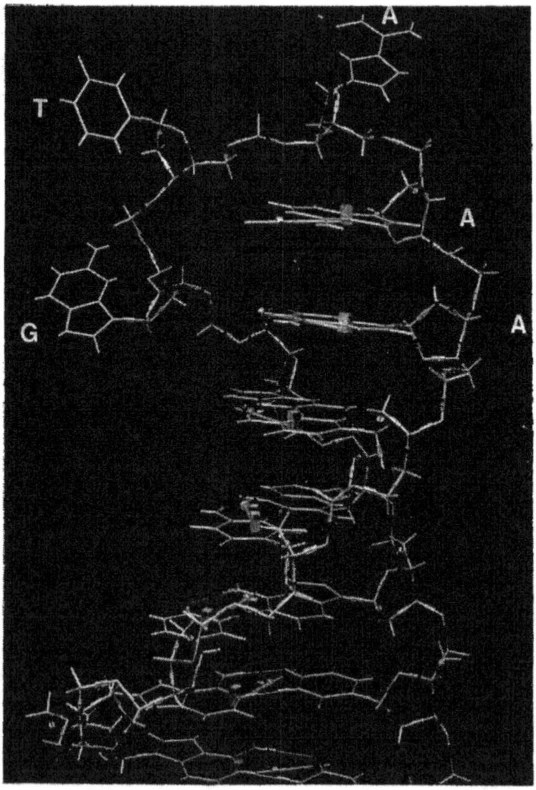

Figure 3. A model for the ColE1 cruciform loop.

A skeletal model for the ColE1 cruciform loop was constructed based upon chemical modification (DEPC) and nuclease cleavage (S1 nuclease) data, in which the 5' **AA**ATG 3' adenosine bases are stacked on top of the cruciform stem.

and are most probably fully base paired.

In addition to examining the base-pairing in the junction, we may also ask how this is disposed in three dimensions. Is it really flat as most frequently drawn? To attempt to answer this point we constructed pseudo-cruciform fragments which are stable in linear DNA molecules (Gough & Lilley, 1985). The principle behind this construction is simple - the equivalent DNA fragments which have different inverted repeat sequences cloned into a given site are heteroduplexed. Hybridisation is perfect everywhere except the inverted repeats, which are not homologous. To minimise their free energy these then form stem-loop structures, and the overall structure is stereochemically equivalent to a cruciform. In practise, an extra cloning step into single-strand phage is employed, resulting in the construction of a unique heteroduplex product. These pseudo-cruciform structures are cleaved by both SI nuclease and T4 endonuclease VII.

A striking feature of these constructs is their anomalously slow migration in polyacrylamide gels. For example, a 440bp pseudo-cruciform fragment migrates with an apparent size of 660bp in a 5% polyacrylamide gel. The extent of this anomaly depends on the relative position the pseudo-cruciform occupies on the fragment. Retardation is maximal when the pseudo-cruciform is central, and reduces as it moves towards either end. This behaviour is strongly reminiscent of that of the bent kinetoplast DNA fragments first observed by Wu & Crothers (1984). If we propose that the junction is not flat but is bent at an angle, tetrahedral for example, then the trajectory of the long arms of the fragment bends at the pseudo-cruciform, resulting in an L-shaped molecule. This is likely to experience greater resistance to passage through the gel pores, the more so as the location of the bend is centralised. In one respect, our fragments do behave differently from the sequence-directed bent kinetoplast fragments. Whilst their anomalous migration is lost with increasing temperature, the pseudo-cruciform fragments exhibit temperature-independent migration to a first approximation (S Diekmann & D M J Lilley, unpublished data). This is taken to confirm that whilst both fragments are bent, they achieve this by different mechanisms.

We may summarise our current ideas about cruciform structure.

1. The optimal size of the DNA loop is four bases. The pairing of the first base pair in the stem is probably rather weak.
2. The loop is probably structured, whereby bases are stacked on one side.

3. The four-way junction is fully base-paired.
4. The four-way junction is bent.

Whilst these data do not yet fully define the structure of a cruciform, they severely limit the possibilities, and we are currently using molecular graphics techniques to explore this further.

5. Thermodynamics of Cruciform Formation

Cruciforms are unstable relative to their unextruded states in torsionally unstressed DNA molecules, but may become stabilised in supercoiled DNA due to the local negative twist change which accompanies their formation. Thus there is a critical level of supercoiling for any given cruciform, at which the extruded and unextruded states are of equal stability. If this can be measured experimentally, then we may calculate the free energy of formation of the cruciform.

We have performed this exercise for the ColE1 cruciform (Lilley & Hallam, 1984). This was subcloned into pColIR515, a 2270 bp plasmid containing the 440bp *Eco*R1 to *Bam*HI ColE1 fragment, and individual topoisomers were prepared. These were separately tested for site specific reaction with S1 nuclease, Bal31 nuclease, T4 endonuclease VII and osmium tetroxide. The results for each probe were identical - there was a sharp threshold at which the cruciform became detectable, at a linking difference of about -13 (when corrected for temperature dependent unwinding of DNA). If these same topoisomers were electrophoresed in a composite polyacrylamide - agarose gel, a transition to a retarded, ie, extruded, species was observed at exactly the same linking difference. By measuring the extent of the band shift, the local twist change upon extrusion was estimated as

$$\Delta Tw = -3.2$$

This is close to the value expected for a B to cruciform transition by a 31 bp inverted repeat. We may calculate the free energy of formation from these data. The free energy of supercoiling (ΔGs) of a topoisomer is given by

$$\Delta G_S = \frac{1050RT}{N} . \Delta Lk^2$$

(Depew & Wang, 1975; Pulleyblank *et al*, 1975; Horowitz & Wang, 1983) where R is the gas constant, T the absolute temperature and N the plasmid size in bp. The free energy of cruciform formation is therefore given by

$$\Delta G_C = \frac{1050RT}{N} \cdot [\Delta Lk_{crit}^2 - (\Delta Lk_{crit} - \Delta Tw)^2]$$

$$= 18 \text{ kcal mole}^{-1}$$

for the ColE1 cruciform, at 37°C.

This value is typical of several which have been observed for other sequences (Courey & Wang, 1983; Gellert *et al*, 1983) although we have measured rather smaller values for some alternating sequences with two-fold symmetry, *vide infra*. Further studies on the sequence-dependence of cruciform stability are in hand.

6. Alternating purine-pyrimidine sequences - the B-Z-cruciform equilibrium.

Alternating purine-pyrimidine sequences have generated considerable interest in recent years by virtue of their adoption of the left-handed Z conformation (Wang *et al*, 1979), which is favoured in negatively supercoiled DNA molecules for the same topological reasons as cruciform structures. Some alternating sequences may theoretically adopt either Z or cruciform geometry. This applies to both (G-C)n and (A-T)n sequences. Z-DNA formation has been demonstrated for (G-C)n sequences by topological band shift methods (Peck *et al*, 1982; Singleton *et al*, 1982) and by antibody binding (Nordheim *et al*, 1982). No evidence for cruciform formation has yet been found. In contrast, we have studied a series of (A-T)n sequences in supercoiled plasmids. In particular we have examined an $(A\text{-}T)_{34}$ sequence isolated from an intron of a *Xenopus* globin gene in considerable detail (Greaves *et al*, 1985). Some of the data are shown in Figure 4. These show that the structure is cleaved near the ends of the $(A\text{-}T)_{34}$ tract by T4 endonuclease VII, and at its centre by SI nuclease. These results strongly indicate cruciform formation by the $(A\text{-}T)_{34}$ sequence. Two dimensional gel electrophoresis of this molecule reveals a striking discontinuity corresponding to a ΔTw of -5.8. This is approximately the value expected for a cruciform, but

A

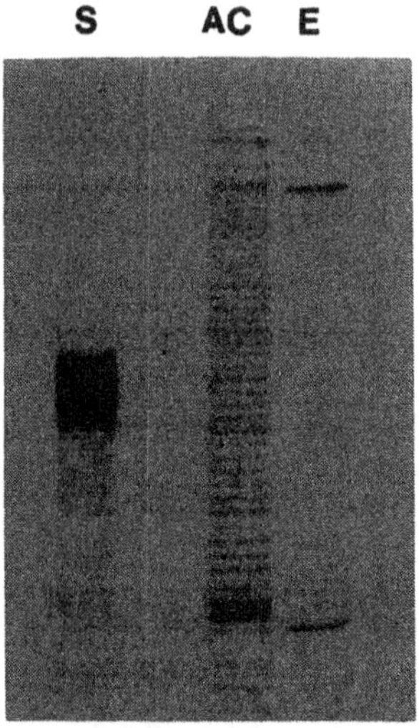

C

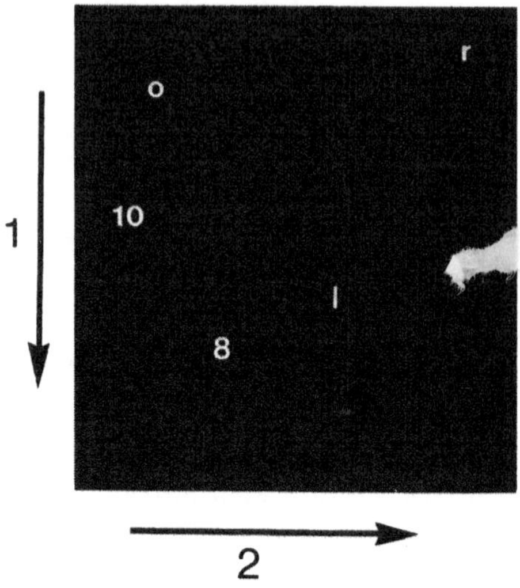

B

Figure 4. Cruciform formation by an $(A\text{-}T)_{34}$ sequence in supercoiled pXG540.

A. S1 nuclease (S) and T4 endonuclease VII (E) cleavage of $(A\text{-}T)_{34}$ at single-nucleotide resolution on a sequencing gel. AC is a Maxam-Gilbert A>C sequencing ladder.
B. S1 nuclease and T4 endonuclease VII cleavage positions at the $(A\text{-}T)_{34}$ sequence, derived from analysis of (A).
C. Two-dimensional gel electrophoresis of pXG540 topoisomers. A mixture of topoisomers was electrophoresed from a single well in an agarose gel (dimension1). The gel was soaked in 1.5 µg/ml chloroquine for several hours, rotated 90° and electrophoresed once more, in the presence of chloroquine (dimension 2). The 2D gel shows a discontinuity at a linking difference of approximately -9, which is flanked by the topoisomers labelled 8 and 10. r denotes topoisomers which are virtually relaxed in the first dimension, while o and l are open circular and linear species respectively.

much smaller than that predicted from a B-Z transition (ΔTw = -12.2). Furthermore, no second transition is seen, up to σ = -0.06. These results indicate the formation of a cruciform by the $(A\text{-}T)_{34}$ sequence, with a free energy of formation of 13kcal $mole^{-1}$, and interestingly, no kinetic barrier. We have also studied two other *Xenopus* globin (A-T)n sequences, with n = 23 and 37, and both form cruciform structures. Synthetic (A-T)n sequences have been cloned, and also adopt cruciform geometry (Haniford & Pulleyblank, 1986; Panyutin *et al*, 1986).

Why do (G-C)n inserts form Z-DNA under torsional stress, while (A-T)n always adopt cruciform geometry? The answer mainly lies in the energetic cost of flipping a base pair from the B to Z conformations, which varies with the nature of the base pair involved (Peck & Wang 1983; Vologodskii & Frank-Kamenetskii 1984; Ellison *et al*, 1985). We may summarise the energetic 'balance-sheet' for the two possible alternative conformations.

1. Cruciform. There is a fixed free energy of formation which is not length-dependent, although some sequence-dependence occurs. The free energy return is provided by the local twist change which is approximately n/10.5, where n is the total number of base pairs in the inverted repeat.

2. Z-DNA. There are two components to the free energy of formation of Z-DNA. There is the fixed cost of two B-Z junctions, plus a length-dependent cost of flipping each base pair. The latter is very sequence-dependent, being about 0.3 kcal/mole bp for CG, rising to over 1 kcal/mole bp for AT. The free energy return is again provided by the local twist change,which is approximately n/10.5 + n/12 for Z-DNA, where n is the number of base pairs in alternation.

On the basis of these simple considerations we may compute phase diagrams for B-Z-cruciform equilibria for alternating sequences as a function of superhelicity and length of alternation, shown in Figure 5. These calculations successfully predict that for most insert lengths and superhelix densities (G-C)n sequences will adopt Z-DNA and (A-T)n will form cruciform structures. What of a sequence of intermediate character? We have studied two alternating sequences $(C\text{-}A\text{-}T\text{-}G)_{10}$ (Naylor, Lilley & Van de Sande, 1986) and $(G\text{-}T\text{-}A\text{-}C)_{10}$ (L Naylor, D M J Lilley & J H Van de Sande, unpublished data) in a supercoiled 2.1 kb plasmid. Both sequences unequivocally adopt cruciform geometry, judged by nuclease cleavage and 2D gel electrophoresis. Repeating the calculation of the phase diagram using the measured free energy of

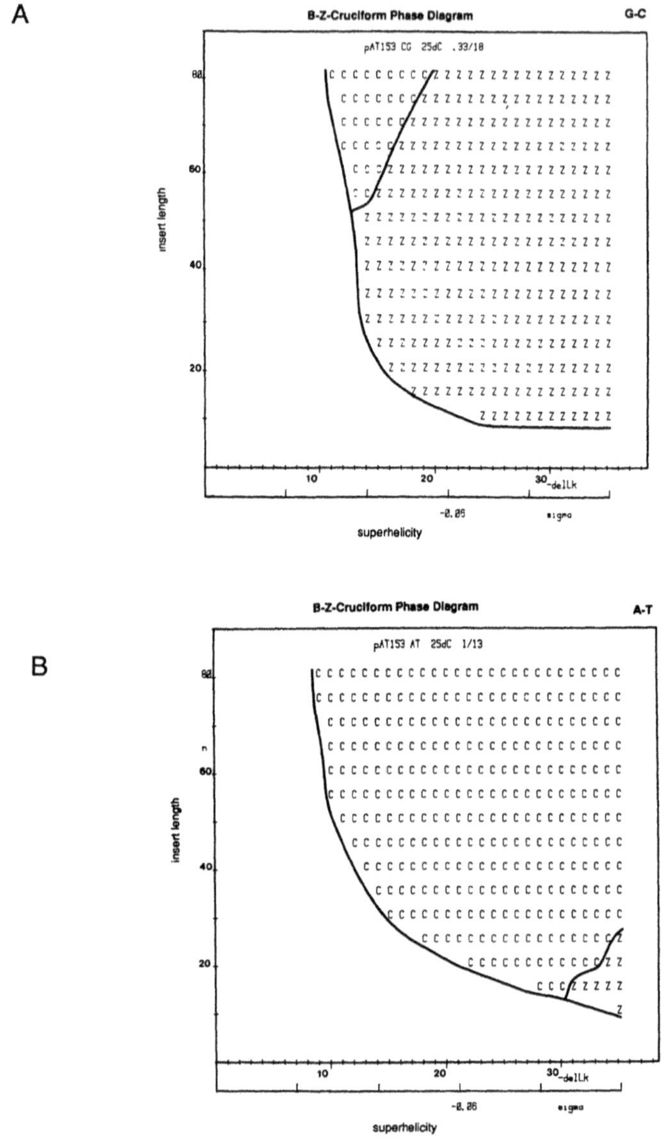

Figure 5. Computer-generated phase diagrams of B-Z-cruciform equilibria for (G-C)n and (A-T)n sequences.

The plots show the most stable conformation for a given length insert at a particular superhelix density. The conformations are indicated B-DNA (blank), cruciform (C) and Z-DNA (Z).

A. (G-C)n inserts. B-Z junction 5kcal $mole^{-1}$junction, 330 cal $mole^{-1}$bp Z and 18kcal $mole^{-1}$plasmid with cruciform.

B. (A-T)n inserts. 1 kcal $mole^{-1}$bp Z , 13 kcal $mole^{-1}$ plasmid with cruciform.

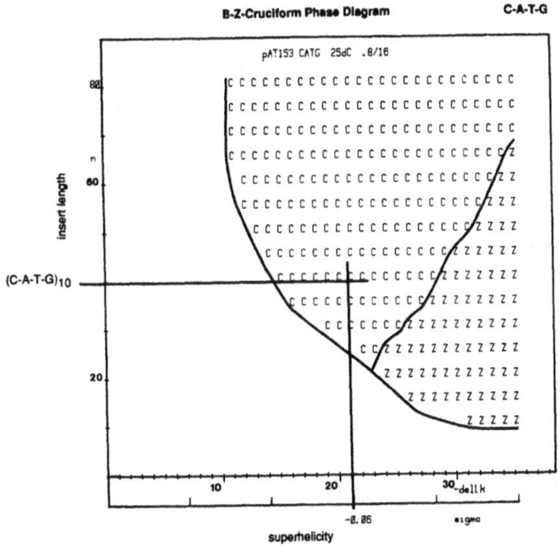

Figure 6. Computer generated phase diagram for B-Z-cruciform equilibria for (C-A-T-G)n sequences.

Parameters employed were 800 cal $mole^{-1}$ bp Z and 16 kcal $mole^{-1}$ plasmid with cruciform. The intersection of the drawn lines indicates the position for a $(C\text{-}A\text{-}T\text{-}G)_{10}$ sequence at a superhelix density of -0.06.

cruciform formation for $(C\text{-}A\text{-}T\text{-}G)_{10}$, and an assumed free energy of formation for Z-DNA of 800 cal/mole bp, we obtain the plot, presented in Figure 6. Once again the calculation correctly predicts cruciform formation for the 40 bp insert at $\sigma = -0.06$, but it also shows that the formation of Z DNA is more probable at some values of length and superhelicity. Indeed, although these diagrams can only be regarded as a qualitative guide to the likely behaviour, it shows that $(C\text{-}A\text{-}T\text{-}G)_n$ is probably not too far removed from a triple point. Of course, it must be emphasised that these thermodynamic calculations provide no kinetic information concerning the ease of interconversion of these structural forms. Nevertheless, this approach provides a useful framework for thinking about these equilibria, and should be applicable to a number of biologically interesting sequences.

7. Cruciform Kinetics - the Extrusion Mechanism

How are cruciforms extruded from regular unperturbed DNA? Gellert and colleagues (Mizuuchi *et al*, 1982) first showed that this reaction can be far from facile, and may be subject to substantial kinetic barriers. We made a detailed comparison of the extrusion kinetics of two plasmids (Lilley, 1986). pColIR315, containing the ColEI cruciform, and pIRbke8 which contains a synthetic inverted repeat, the bke cruciform. To our surprise, we found that these two molecules exhibited strikingly different kinetic properties, which are summarised in Figure 7. The kinetics are described by the following properties.

1. *pColIR315*. Extrusion is maximal at very low ionic strength, and severely suppressed as it is raised, with almost complete suppression by 75 mM NaCl. The temperature dependence of this reaction is very pronounced, with an Arrhenius activation energy (Ea) of about 180 kcal mole^{-1}. We term this behaviour C-type kinetics.
2. *pIRbke8*. No extrusion occurs in the absence of cation, and extrusion rate shows an optimum at 50mM NaCl or 100μm $MgCl_2$. There is a more modest temperature dependence, corresponding to an Ea of about 40 kcal mole^{-1}. We term this kinetic behaviour S-type.

The properties are so contrasting that we believe they probably reflect two difference mechanistic pathways for the extrusion process. Two extreme mechanisms may be proposed *ab initio* (Lilley & Markham, 1983), which are summarised in Figure 8. In the upper pathway, the transition state contains a considerable extent of

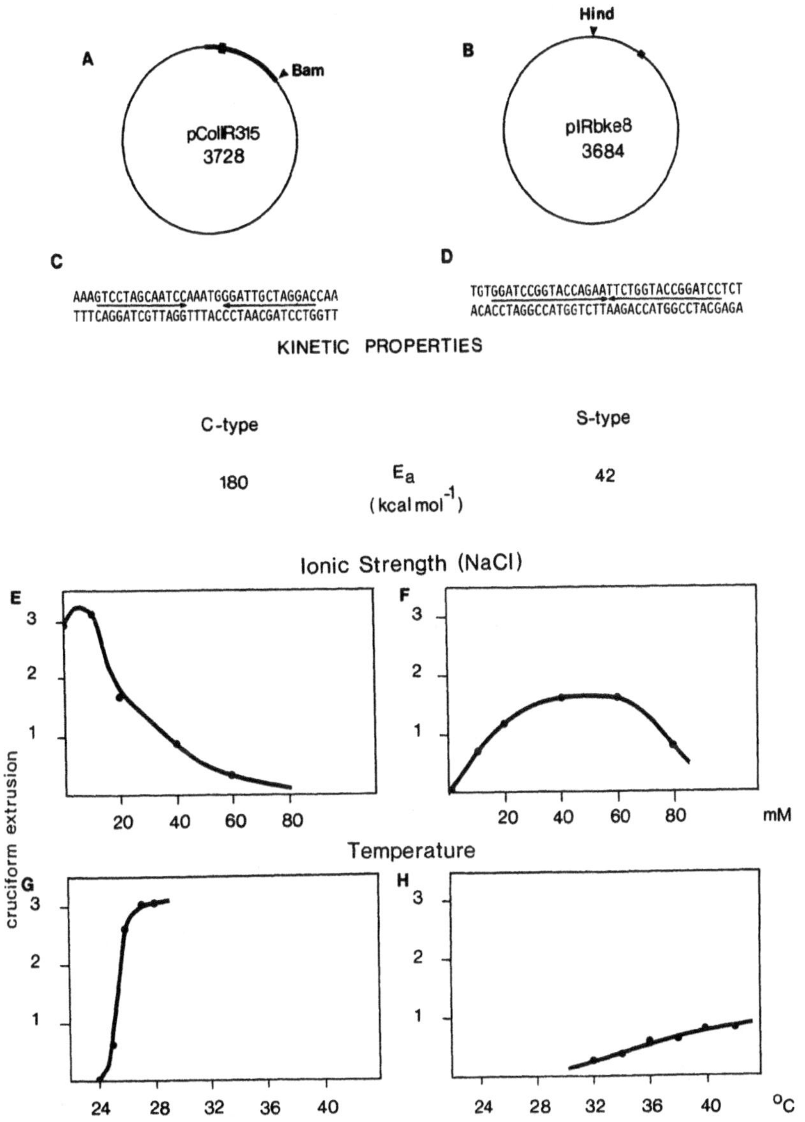

Figure 7. Summary of the kinetic properties of cruciform extrusion by pColIR315 and pIRbke8.

A. Map of pColIR315 showing the extent of ColE1 DNA (thicker) and the inverted repeat (box).
B. Map of pIRbke8. The inverted repeat is shown by the box.
C and D. Sequences of the inverted repeats of pColIR315 and pIRbke8 respectively. Symmetrically related nucleotides are indicated by the arrows.
E and F. Extrusion of pColIR315 and pIRekb8 respectively as a function of sodium chloride concentration.
G and H. Extrusion of pColIR315 and pIRbke8 respectively as a function of temperature.

unpairing and unwinding, and the formation of the fully extruded cruciform is a single step 'snap-back' event. In the lower pathway the initial disruption is much more limited, and is followed by the formation of a relatively small proto-cruciform. Sequential transfer of base pairing allows the extrusion to be completed by branch migration. These two mechanisms account for the observed kinetic properties of C-type and S-type kinetics rather well. However, they should still be regarded only as a working hypothesis, although more rigorous testing of the model is in hand.

S-type kinetics appear to be more common than C-type, since most sequences studied have properties akin to those of pIRbke8 (Gellert *et al*, 1983; Courey & Wang, 1983; Sinden & Pettijohn, 1984). Thus it is the ColEI cruciform of pColIR315 which must be scrutinised for unusual features which might help to explain its kinetic character. If the sequences of the two inverted repeats are compared, no clearcut differences are discernible. However, in one respect the ColEI sequence is quite remarkable. If the sequences which flank the ColEI inverted repeat for 100bp or more each side are examined, it is soon apparent that these are very unusual in one respect. They are about 80% or more A + T. Could this have some role in the selection of kinetic pathway?

We explored this possibility in the following series of experiments (K M Sullivan & D M J Lilley, submitted for publication).

1. We deleted the ColEI cruciform from pColIR315 and replaced it with the xke inverted repeat, which is closely similar to the bke inverted repeat. By this means we constructed a plasmid in which a bke-like cruciform resides in a context of the A + T rich ColEI sequences. The new plasmid, pIRxke/col, exhibits maximal cruciform extrusion at OmM NaCl, with an Ea of 215 kcal $mole^{-1}$. These properties are typically C-type.

2. We also performed the reverse of this experiment, by cloning oligonucleotides to generate a ColEI-like inverted repeat in the location of pAT153 which the bke sequence occupies in pIRbke8. The resulting plasmid, pIRCol/vec, shows no extrusion at 0mM NaCl, maximal extrusion at 50mM NaCl with an Ea of 50 kcal $mole^{-1}$. These are typical S-type kinetics.

These two results indicate that the kinetic class, ie mechanism, is determined by the flanking sequence, and that the inverted repeat sequence itself is of secondary importance. Further experiments have revealed the following properties of these flanking sequences.

1. The A + T rich ColEI DNA acts as C-type inducing sequence on inverted repeats in *cis*.

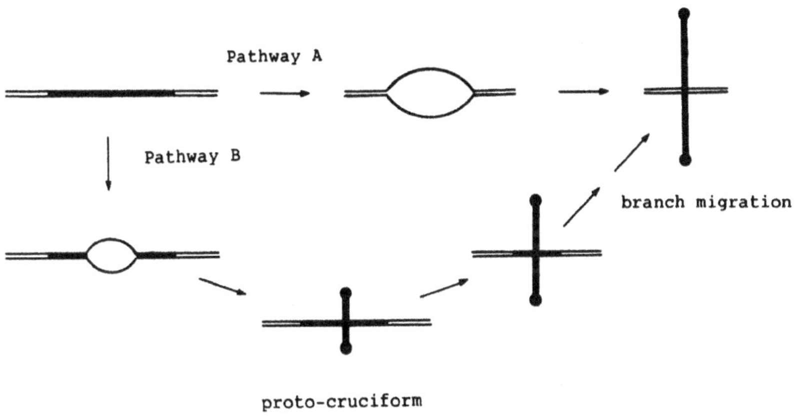

Figure 8. Two mechanistic models for cruciform extrusion.

The upper pathway (A) represents an unwinding of the entire inverted repeat and a single-step cruciform formation. In the lower pathway (B), the initial disruption is more localised, forming the proto-cruciform, which branch migrates to the final state. In our current thinking, C- and S-type kinetics are probably a consequence of paths A and B respectively.

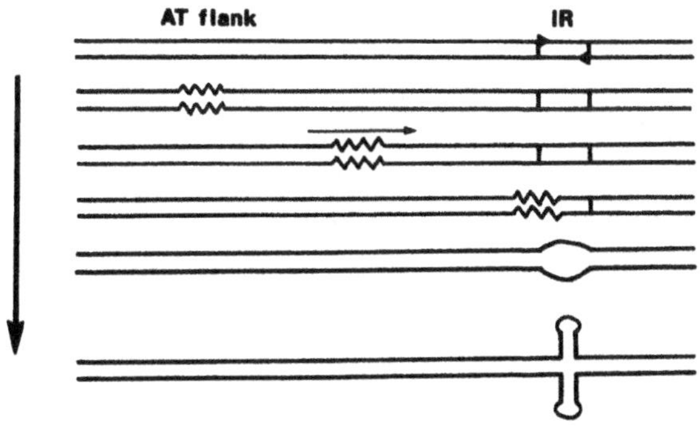

Figure 9. Proposed mechanism by which A+T rich flanking sequences may confer C-type extrusion kinetics on an ajoining inverted repeat.

A pre-melted conformation arises in the A+T rich region, which is mobile. This may migrate into the inverted repeat, where it triggers C-type cruciform extrusion.

2. Only a single such sequence is necessary. Either the right or left hand side flanking sequences function equally.
3. When an inverted repeat has one C-type inducing sequence, and one S-type, then the dominant sequence is determined by the ionic strength.
4. Polarity is unimportant. The left hand side 100bp ColEI sequence can confer C-type character on inverted repeats placed at either end.
5. The effects may be modulated over considerable distance, at least 100 bp.

What is the origin of these effects? Since either side of the ColEI sequence can function equally well in inducing C-type kinetics, it seems probable that the effect arises from the abnormal base composition in general, rather than any specific local sequence. DNA rich in A and T is both structurally polymorphic (Klug *et al*, 1979; Arnott & Selsing, 1974; Rhodes & Klug, 1981; Shindo, 1981; Mahendrasingham *et al*, 1983; Greaves *et al*, 1985) and dynamic (Patel *et al*, 1985a & b). It is possible that thermally induced conformational fluctuations may arise, and that these are mobile in one dimension. Such soliton excitations have been proposed previously for DNA molecules (Englander *et al* 1980, Sobell *et al*, 1982) and might be regarded as a locally pre-melted conformation. Upon diffusing into an inverted repeat this may induce cruciform extrusion *via* the C-type mechanistic pathway. This idea is illustrated diagramatically in Figure 9.

We may point to two pieces of additional evidence in support of these ideas. First, the ColEI flanking sequences are chemically reactive towards bromoacetaldehyde, glyoxal and bisulphite, indicating geometry perturbed from the normal B structure. Second, 2μM distamycin, which stabilises A+T rich DNA structure by binding in the minor groove, suppresses C-type cruciform extrusion but not S-type extrusion. These data provide evidence for the existence of pre-melted conformations, and show that by stabilising this DNA we may inhibit this extrusion pathway.

These data have a significance which extends far beyond the details of cruciform extrusion. They are a dramatic example of how conformational 'information' may travel along DNA, resulting in an influence on a structural transition at a remote location. Such effects have been observed previously in the study of the melting of block polymers, and have been termed 'telestability' (Burd *et al*, 1975). This sort of behaviour is of considerable biological significance. The initial events in cruciform formation and transcription are rather similar, and the importance of contextual effects on the latter are obvious. *cis* acting regulatory elements control the rate of initiation of transcription at promoters, an important example being the eukaryotic enhancer elements (Banerji *et al*, 1981; Benoist & Chambon, 1981; Gruss *et al*, 1981).

Clearly cruciform extrusion is a complex process. Nevertheless, we are making rapid progress currently, and we believe we will ultimately understand this in considerable detail. In the process we learn a great deal of general applicability to the structure and dynamics of DNA.

8. Acknowledgements

I wish to thank all my coworkers in these studies - Karen Sullivan, Gerald Gough, James McClellan, Judy Furlong, Laurence Hallam, Alastair Murchie and Anamitra Bhattacharyya (Dundee), Louise Naylor and Hans Van de Sande (Calgary), David Greaves and Roger Patient (Kings College, London), Emil Palecek (Brno) and Boerries Kemper (Koeln). I thank the Medical Research Council, the Cancer Research Campaign and the Royal Society for financial support.

References

Arnott, S & Selsing, E (1974) J.Molec.Biol. 88, 509-521.
Banerji, J, Rusconi, S & Schaffner, W (1981) Cell 27, 299-308.
Benoist, C & Chambon, P (1981) Nature 290, 304-310.
Burd, J F, Wartell, R M, Dodgson, J B & Wells, R D (1975) J. Biol. Chem. 250, 5109-5113.
Cantor, C R & Efstratiadis, A (1984) Nucleic Acids Res. 12, 8059-8072.
Courey, A J & Wang, J C (1983) Cell 33, 817-829.
Depew, R E & Wang, J C (1975) Proc. Natl. Acad. Sci. USA 72, 4275-4279
Dickerson, R E (1983) J. Molec. Biol. 166, 419-441.
Dickerson, R E, Drew, H R, Conner, B N, Wing, R W, Fratini, A V & Kopka, M L (1982) Science 216, 475-485.
Drew, H R & Travers, A A (1984) Cell 37, 491-502.
Ellison, M J, Kelleher, R J, Wang, A H-J, Habener, J F & Rich, A (1985) Proc. Natl. Acad. Sci. USA 82, 8320-8324.
Englander, S W, Kallenbach, N R, Heeger, A J, Krumhansl, J A & Litwin, S (1980) Proc. Natl. Acad. Sci. USA 77, 7222 - 7226.
Fuller, F B (1971) Proc. Natl. Acad. Sci. USA 68, 815-819.
Furlong, J C & Lilley, D M J (1986) Nucleic Acids Res. 14, 3995 - 4007.
Gellert, M, Mizuuchi, K, O'Dea, M H, Ohmori & Tomizawa, J (1979) Cold

Spring Harbor Symp. Quant. Biol. 43, 35-40.
Gellert, M, O'Dea, M H & Mizuuchi, K (1983) Proc. Natl. Acad. Sci. USA 80, 5545-5549.
Gierer, A (1966) Nature 212, 1480-1481.
Gough, G W (1986) Ph.D. Thesis, University of Dundee
Gough, G W & Lilley, D M J (1985) Nature 313, 154-156
Gough, G W, Sullivan, K M & Lilley, D M J (1986) EMBO J. 5, 191-196.
Greaves, D R, Patient, R K & Lilley, D M J (1985) J. Molec. Biol. 185, 461-478.
Gruss, P, Dhar, R & Khoury, G (1981) Proc. Natl. Acad. Sci. USA 78, 943-947.
Haniford, D B & Pulleyblank, D E (1985) Nucleic Acids Res. 13, 4343-4363.
Hasnoot, C A G, Hilbers, C W, van der Marel, G A, van Boom, J H, Singh, U C, Pattabiraman, N & Kollman, P A (1986) J. Biomolec. Struc. & Dynamics 3, 843 - 857.
Hentschel, C C (1982) Nature 295, 714-716.
Holliday, R (1964) Genet. Res. 5, 282-304.
Horowitz, D A & Wang, J C (1984) J. Molec. Biol. 173, 75-91.
Kemper, B & Garabett, M (1981) Eur. J. Biochem. 115, 123-131.
Klug, A, Jack, A, Viswamitra, M A, Kennard, O, Shakked, Z & Steitz, T A (1979) J, Molec. Biol. 131, 669-680.
Lilley, D M J (1980) Proc, Natl. Acad. Sci. USA 77, 6468-6472.
Lilley, D M J (1983a) Cold Spring Harbor Symp. Quant. Biol. 47, 101-112.
Lilley, D M J (1983b) Nucleic Acids Res. 11, 3097-3112.
Lilley, D M J (1985) Nucleic Acids Res. 13, 1443-1465.
Lilley, D M J (1986) IARC Special Report on Nucleic Acid Cyclic Adducts and Carcinogenesis, No. 70, 33-39.
Lilley, D M J & Hallam, L R (1983) J. Biomolec. Struc. & Dynamics 1, 169-182.
Lilley, D M J & Hallam, L R (1984) J. Molec. Biol. 180, 179-200.
Lilley, D M J & Kemper, B (1984) Cell 36, 413-422.
Lilley, D M J & Markham, A F (1983) EMBO J. 2, 527-533.
Lilley, D M J & Palecek, E (1984) EMBO J 3, 1187-1192.
Lomonossoff, G P, Butler, P J G & Klug, A (1981) J. Molec. Biol. 149, 745-760.
Lyamichev,V I, Mirkin, S M & Frank-Kamenetskii, M D (1985) J. Biomolec. Struc. & Dynamics 3, 327 - 338.
Mahendrasingham, A, Rhodes, N J, Goodwin, D C, Nave, C, Pigram, W J, Fuller, W,, Brahms, J & Vergne, J (1983) Nature 301, 535-537.
Meselson, M S & Radding, C M (1975) Proc. Natl. Acad. Sci. USA 72,

358-361.
Mizuuchi, K, Kemper, B, Hays, J & Weisberg, R A (1982a) Cell 29, 357-365.
Mizuuchi, K, Mizuuchi, M & Gellert, M (1982b) J. Molec. Biol. 156, 229-243.
Naylor, L H, Lilley, D M J & Van de Sande, J H (1986) Submitted.
Nordheim, A, Lafer, E M, Peck, L J, Wang, J C, Stollar, B D & Rich, A (1982) Cell 31, 309-318.
Panayotatos, N & Wells, R D (1981) Nature 289, 466-470.
Panyutin, I, Lyamichev, V & Mirkin, S (1095) J. Biomolec. Structure & Dynamics 2, 1221 - 1234.
Patel, D J, Kozlowski, S A, Hare, D R, Reid, B, Ikuta, S, Lander, N & Itakura, K (1985a) Biochemistry 24, 926-935.
Patel, D J, Kozlowski, S A, Weiss, M, Bhatt, R (1985b) Biochemistry 24, 936-944.
Peck, L J, Nordheim, A, Rich, A & Wang, J C (1982) Proc. Natl. Acad. Sci. USA 79, 4560-4564.
Peck, L J & Wang, J C (1983) Proc. Natl. Acad. Sci. USA 80, 6206-6210.
Platt, J R (1955) Proc. Natl. Acad. Sci. USA 41, 181-183.
Pulleyblank, D E, Shure, M, Tang, D, Vinograd, J & Vosberg, H-P (1975) Proc. Natl. Acad. Sci. USA 72, 4280-4284.
Pulleyblank, D E, Haniford, D B & Morgan, A R (1985) Cell 42, 271-280.
Rhodes, D & Klug, A (1981) Nature 292, 378-380.
Shakked, Z, Rabinovich, D, Kennard, O, Cruse, W B T, Salisbury, S A & Viswamitra, M A (1983) J. Molec. Biol. 166, 183-201.
Shindo, H (1981) Eur. J. Biochem. 120, 309-312.
Sinden, R R & Pettijohn, D E (1984) J. Biol. Chem. 259, 6593-6600.
Singleton, C K, Klysik, J, Stirdivant, S M & Wells, R D (1982) Nature 299, 312-316.
Sobell, H M, Sakore, T DD, Jain, S C, Banerjee, A, Bhandary, K K, Reddy, B S & Lozansky, E D (1983) Cold Spring Harbor Symp. Quant. Biol. 47, 293-314.
Sullivan, K M & Lilley, D M J (1986) Submitted.
Symington, L & Kolodner, R (1985) Proc. Natl. Acad. Sci. USA 82, 7247-7251.
Trifonov, E N, Konopka, A K & Jovin, T M (1985) FEBS Lett. 185, 197-202.
Vinograd, J, Lebowitz, J, Radloff, R, Watson, R & Laipis, P (1965) Proc. Natl. Acad. Sci. USA 53, 1104-1111.
Vologodskii, A V & Frank-Kamenetskii, M D (1984) J. Biomolec. Structure & Dynamics 1, 1325-1333.

Walter, G, Zilling, W, Palm, P & Fuchs, E (1967) Eur. J. Biochem. 3, 194-201.
Wang, A H-J, Quigley, G J, Kolpak, F J, Crawford, J L, van Boom, J H, van der Marel, G & Rich, A (1979) Nature 282, 680-686.
West, S C & Korner, A (1985) Proc. Natl. Acad. Sci. USA 82, 6445-6449.
Wu, H-M & Crothers, D M (1984) Nature 308, 509-513.

TOROIDAL DNA CONDENSATES; ASPECTS OF FORMATION, STRUCTURES, DYNAMICS AND BIOLOGICAL IMPLICATIONS

Kenneth A. Marx

Department of Chemistry
University of Lowell
Lowell, Mass. 01854 USA

ABSTRACT

DNA, of varying secondary structure, condenses from solution both intramolecularly and intermolecularly under a variety of conditions to form a range of structures including rods, spheres, toruses, lariats and amorphous banded aggregates. Toroidal DNA condensates, formed by intramolecular collapse with the naturally occuring trivalent cation spermidine, are of great interest as self-assembling structures since they represent convenient in vitro model systems for DNA condensation in certain bacteriophage and viruses. In this report we review the DNA condensate structures observed by an array of physical techniques, but concentrate our focus on torus DNA tertiary structure and certain aspects of their formation, stability, dynamics and relevance to the problem of DNA packaging in bacteriophage and viruses.

INTRODUCTION

DNA is capable of existing in a wide variety of condensed tertiary and quaternary structures in vivo and in vitro. In cells of higher organisms DNA exists condensed as a left-handed superhelix wound about cationic histone protein octamers in the repeating nucleosomal arrays that comprise the basic chromatin fibre(1). Higher order folding of the nucleosomal array provides a level of quaternary structure that is not yet fully understood. In certain bacteriophage and viruses, available evidence suggests that packaged DNA may exist as superhelical segments hexagonally packing to form a macroscopic torus shaped object(2). These two examples provide some indication of the scope of DNA tertiary and quaternary structure to be found and understood in biological systems.

To provide a basis for explaining the biological structures and attendant functions of DNA it has proved valuable over the years to investigate condensed DNA tertiary and quaternary structures formed in vitro under a variety of conditions. Investigations of these structures were made all the more timely in the mid-to-late seventies by the then surprising demonstrations of DNA flexibility and structural polymorphism at both tertiary(nucleosome) and secondary(left-handed Z DNA) structure levels(1,3).

Historically, DNA intermolecular condensation into macroscopic fibres has been exploited since the earliest days of structural biochemical investigation to determine DNA secondary structure by x-ray diffraction methods. Gel and then fibre formation occurs at very high DNA concentrations in the presence of monovalent cations.

More recently, intramolecular DNA condensation has been shown to occur at low DNA concentration under a wide variety of solution and cation conditions(4-32). The range of tertiary and quaternary structures formed and the details of their microscopic structure and dynamics will provide valuable insights into the thermodynamically accessible states of DNA in different biological systems.

DNA condensate systems have revealed a wide variety of structures. Constant length rods, toruses, spheres, lariat shaped complexes, and banded aggregates are all structures that have been observed by transmission electron microscopy(EM) at low DNA collapse concentrations under varying solution conditions and EM preparation protocols(4,8-11,13,17,22,23, 25-33). We have focused our efforts on studying DNA tertiary structure organization in hydrated spermidine(spd)-condensed toruses, in vitro model systems for DNA packaging in certain bacteriophage and viruses(23). In this review we describe the current understanding of DNA tertiary structure organization in toruses, then broaden the discussion of toruses to include aspects of their formation, dynamics and biological implications.

MATERIALS AND METHODS

Detailed methodologies for the micrococcal nuclease digestion, sedimentation assay(16,18) and freeze fracture, deep-etch EM experimental work have been previously published(17,23,33) and will not be described again here except occasionally in the text where particular emphasis is desired. All samples studied (except those for calf thymus EM) were prepared by mixing DNA and spermidine stock solutions to the following final concentrations: DNA 5 μg/ml, 1 mM spermidine(spd), 1 mM NaCl, 10 mM Tris pH 7.0. Calf thymus DNA for EM visualization only was condensed at 0.2 mM (spd). Spermidine homologues used in the pH dependent sedimentation studies were either purchased from Sigma Chemicals [(spd3,2), (spd3,3) and (spd); numbers refer to the two methylene bridge carbon chain lengths in the homologues relative to spermidine having 3,4 chain lengths] or were generously provided by Dr. S. Bunce[(spd3,5) and (spd4,4)] and used without further purification.

RESULTS

Structural studies of spermidine-condensed DNA preparations have been performed by a number of largely physical techniques. In the Discussion section we relate the results of other investigators to our structural results, presented here. Our experiments rely on two major approaches, one biochemical, the other EM.

Reversible condensation and ion competition. The biochemical experiments utilize a sedimentation assay for the presence of condensed DNA, following either: enzymatic digestion of the DNA, pH change or addition of competitor counterions. Collapse of the DNA from its extended solution conformation to a compact tertiary structure results in a much higher sedimentation coefficient and therefore velocity under fixed sedimentation conditions. The collapse transitions occur over a very narrow concentration range of condensing counterion(18). If one considers a simple reversible two state collapse transition, the spd midpoints of such transitions plotted against the competitor ion concentration provide convenient phase diagrams for comparisons such as shown in Figure 1 where Na+ is the competitor. Here the condensation direction midpoints curve lies

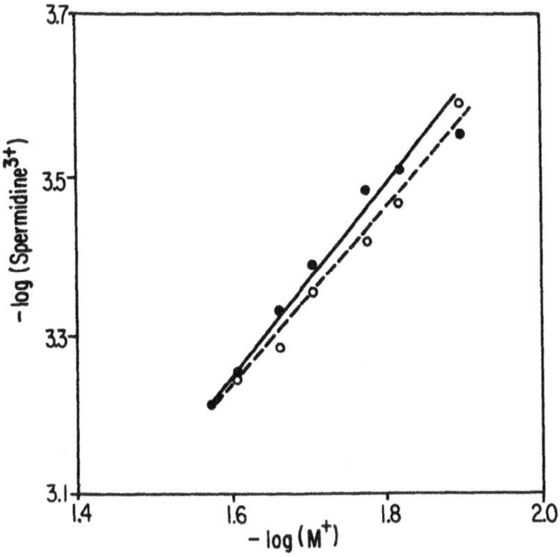

Figure 1. Phase diagram of condensation-decondensation transition of spd-calf thymus DNA with added Na+. O cond. transition; ● decond. transition.

lower than that of the decondensation direction midpoints curve. This hysteresis in the forward and reverse reaction midpoints suggests slow kinetics to equilibrium in the condensation direction but rapid kinetics during decondensation. The decondensation data have been shown to be consistent with Manning's idea of DNA collapse by counterion condensation, extended to two counterions, when a critical DNA charge fraction is neutralized(18,23).

We have performed experiments in which various competitor divalent cations were compared for their ability to destabilize the DNA condensates. These data, in Figure 2, show that specific ion effects exist in the counterion condensation systems, a not too surprising result seen by others(7,8,10,11,14,15,18,20-23) but not predicted by the ion point charge model of Manning(34). This competitor effect order: Mg2+ > Ca2+ > putrescine2+, may be explained electrostatically since this order describes increasing effective ionic radii and lower effective surface charge density.

Torus structure probed by micrococcal nuclease. We have utilized micrococcal nuclease as a probe for the tertiary

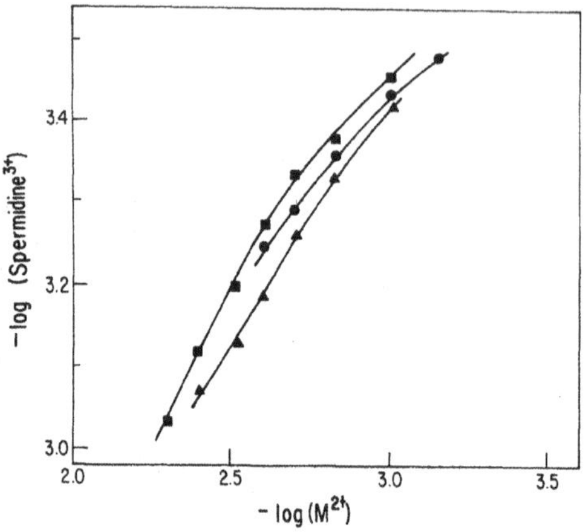

Figure 2. Phase diagram for condensation direction transition o spd-calf thymus DNA with added divalent competitor ions: ▲ Mg2+, ● Ca2+, ■ putrescine2+.

structure organization of spd-condensed DNA preparations. Micrococcal nuclease is able to digest condensed DNA to very small fragment lengths. It makes double stranded cleavages at a

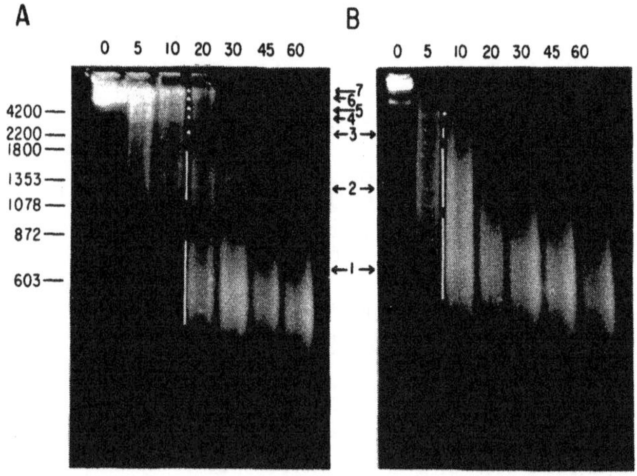

Figure 3. Agarose gel electrophoresis of spd-ϕx-174 DNAs digested with micrococcal nuclease: A. RFII nicked, circular form; B. linear form. Fragment size calibrations(bp) are indicated (left) and lines and dots indicate all bands present.

rate up to thirty-fold slower in lambda DNA condensates compared to uncondensed lambda DNA(35). Gel electrophoretic profiles of the DNA fragment distribution with time of digestion are shown in Figure 3 for spd-ϕX-174 DNA. Here it is clear that digestion proceeds, not with a continuous DNA fragment size distribution, but rather with one in which broad bands of discrete fragment size appear. We have shown discrete bands for ϕx-174 nicked circular and linear DNAs(16,23), calf thymus DNA(18) and lambda DNA(35).

In all populations the band centers progress to shorter fragment size with time due to the subsequent and continuous endonucleolytic cleavage of newly generated DNA fragments. To accurately determine the band sizes initially cleaved in the DNA condensates we plotted the band centers against time and extrapolated to zero digestion time. These band size data are compiled in Table 1.

Table 1. Zero Digestion Time Extrapolated DNA Band Sizes

DNA	Band size and size ratio					
	1	1/1	2	2/1	3	3/1
ϕX-174						
linear	830±120	1.0	1470±250	1.8	-	-
circular	780±80	1.0	1580±150	2.0	2450±250	3.1
calf	760±87	1.0	1355±135	1.8	-	-
lambda	1003±115	1.0	1972±190	2.0	3100±350	3.1

Circumferentially wrapped DNA torus cleavage model. The most striking feature of these data are that relative to the lowest molecular weight band 1 for each digested DNA, each of the higher band sizes fall into an arithmetic series.

How may the arithmetic DNA fragment series be rationalized? This requires a model for DNA tertiary structure organization within toruses, the predominant condensed DNA structures visualized in these preparations(see Discussion). Since our electron microscopic observations presented later in this section provide clear cut evidence for toruses organized by continuous circumferential wrapping of B form DNA we developed the Figure 4 model for micrococcal nuclease cleavage of DNA toruses(16,23). Monomer is produced by the 1,2 cleavage sequence; dimer is produced by the 1,2' cleavage sequence where 2' can be a nearest neighbor strand cleavage event in a three dimensional torus. Higher members in the arithmetic sequence of fragment sizes are produced by cleavage at 1 followed by cleavage of all other nearest neighbor strands in this initial cleavage region in the hexagonally packed three dimensional torus.

Two pieces of evidence support the torus cleavage model. The first is the actual circumference distribution measured in the individual torus populations, presented later in this section. The second is that for the monodisperse ϕx-174 DNA condensate, seven bands were observed, which represents seven circumferential DNA wraps or a 7(780)=5460bp DNA size according to our model. The actual DNA size(5386 bp), agrees well with that required by the model.

Our model for micrococcal nuclease cleavage of DNA is based on a torus organized by a continuous circumferential wrapping of DNA. While our EM observations support this type of DNA organization it represents a nonunique organization within toruses capable of producing an arithmetic DNA band series. In Figure 5 we describe two general mechanisms of torus formation based on circumferential wrapping. One has unidirectional strands; the other has alternating strand directions.

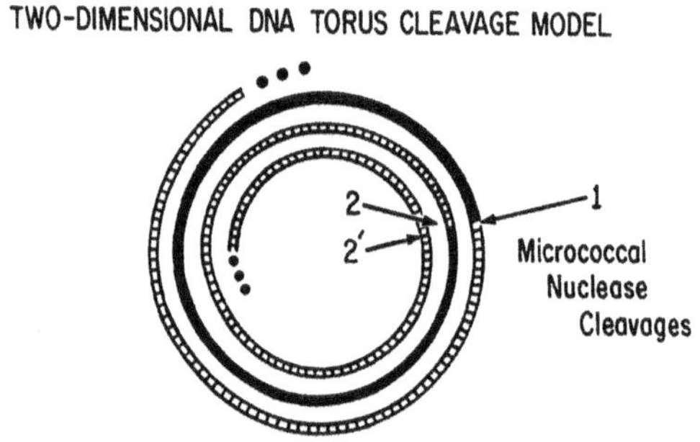

Figure 4. Two dimensional model of spd-DNA torus cleavage by micrococcal nuclease: 1. Slow kinetic cleavage of hexagonal close packed surface DNA; 2. Rapid kinetic cleavage of nearest neighbor DNA strands in opened region of first cleavage.

Fortunately, our nuclease cleavage data for ϕx-174 DNA (circular) allow us to determine that the Unidirectional Model 1 operates in solution. For simple topological reasons only even numbers of strand wraps (n) could form Model 2 toruses(Table inset), resulting in the same even number of micrococcal nuclease bands in the arithmetic series. Model 1 toruses can be comprised of any integer number of bands, giving rise to the same integer numbers of arithmetic series bands in the gel electrophoresis experiment. Since our result for ϕx-174 showed a 7 band arithmetic series and 7 bands accounted for all of the possible DNA in this molecule, 7(780)=5460 bp compared to 5386 bp known size, Model 1 alone can account for the solution results.

Freeze-etch electron microscopy of toruses. We have utilized freeze-etch transmission EM to visualize DNA toruses from all the DNA preparations studied by micrococcal nuclease digestion.

The freeze-etch preparation technique has clear advantages over other EM preparation techniques(23,33). Of primary importance to biological sample preservation is the fact that quick frozen hydrated samples are visualized without any distortions due to disruptive surface tension forces during drying, added solvents, or large pH or ionic strength changes during staining steps, undesirable features of other preparation protocols.

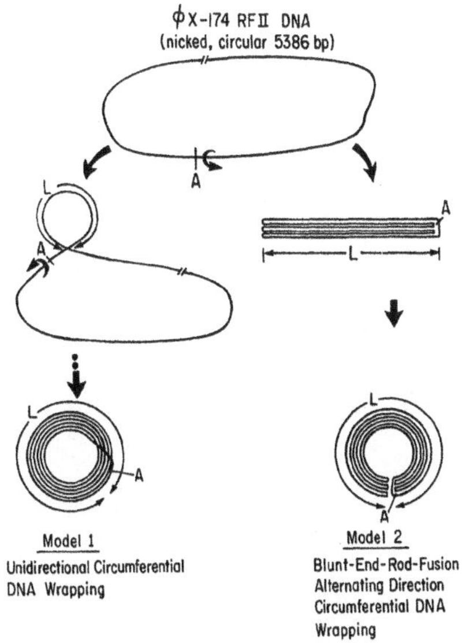

Torus Model Parameters

n	L (5386/n bp)	Model 1	Model 2
6	898	+	+
7	770	+	-
8	670	+	+

Figure 5. Formation mechanisms and topology of two circumferentially wrapped toruses. Table inset shows model compatibility with integer n wrapped, average wrap length L torus.

Our studies demonstrated that toruses exist in the hydrated, solution-like environment of the freeze-etch experiment. Toruses were the only regular condensed DNA structure noted in all of the preparations. A group of representative spd-calf thymus DNA toruses are shown in Figure 6.

Although the toruses vary somewhat in size they appear continuous around their entire circumference and all possess continuous, unidirectional Pt/C(9Å) decorated surface fibres. We believe these to be circumferentially wrapped B form DNA helices since the fibre widths average 30Å in regions close to coincident with the shadow direction. Moreover, we have visualized the B secondary structure major and minor groove features in selected regions of these spd-DNA fibres using freeze-etch EM(23). This demonstration highlights the utility and ultimate resolving power of the freeze-etch technique(10Å) for macromolecular surface structure.

For all DNA condensate preparations we have measured torus

population dimensions(31,32,36). Our most extensive collection of torus measurements was obtained for the ϕx-174 DNA (circular) population(30,36). Inner and outer circumference measurements for all toruses are highly correlated. In Figure 7 is shown data for inner circumference values plotted against the torus ring thickness. It is very clear that two populations of torus dimensions exist. There is a diffuse group of large dimension and a small group of smaller dimension. The lines indicate dimensions calculated from models of monomer DNA packed toruses(see inset). The smaller dimension group of toruses are

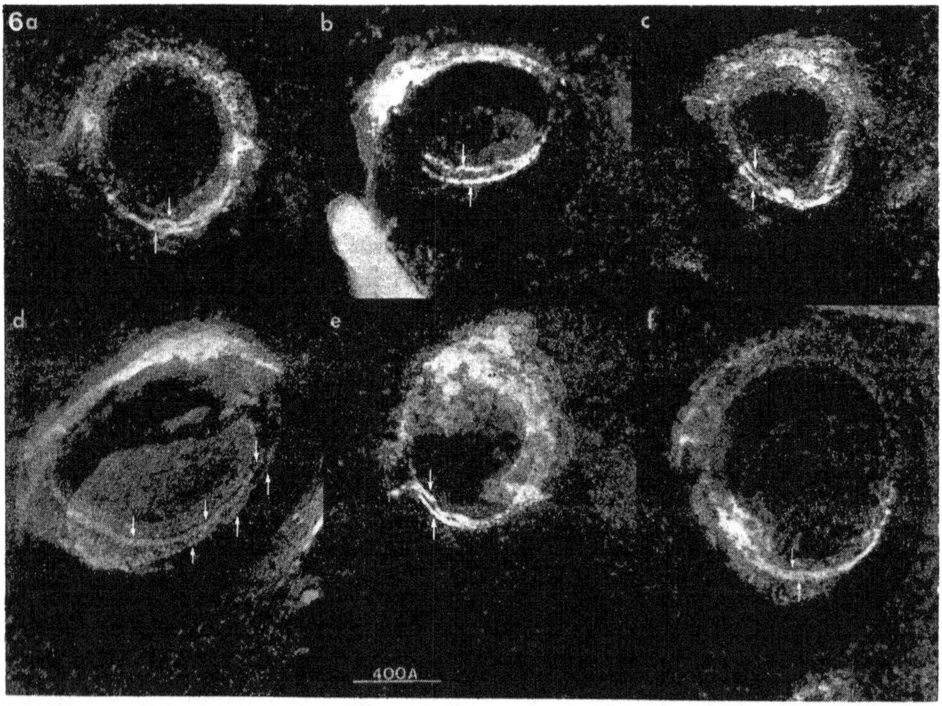

Figure 6. Spd-calf thymus DNA toruses visualized by freeze-etch electron microscopy. Single direction shadowing at 45° with Pt/C(9Å). Arrows indicate continuous B DNA helices.

best fit by the MHP model(minimum hexagonal packing) which maintains hexagonal packing of nearest neighbor strands. Toruses of larger dimension are clearly multimeric DNA structures since their dimensions exceed the limit MLD(minimum limiting disk) model which is clearly an energetically unfavorable collapsed structure since it maximizes solution-helix interactions. That the smaller dimension toruses possess the dimensions appropriate to the monomer DNA models has been demonstrated in a number of cases by the freeze-etch torus replica being tilted through large angles to the end-on view(27,30,33,36) as is shown for one

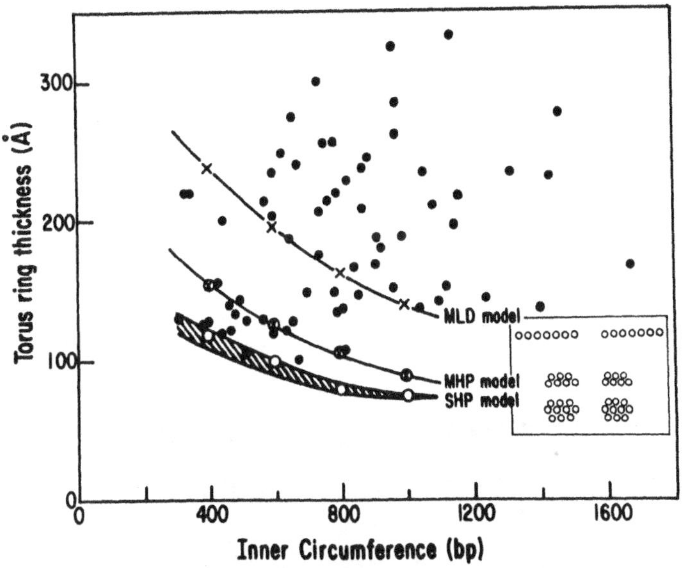

Figure 7. Torus ring thickness vs. inner circumference (bp of circumf. B DNA) plot of ϕx-174 (circ.) toruses. Thickness is measured across torus annulus from 2-D EM projection.

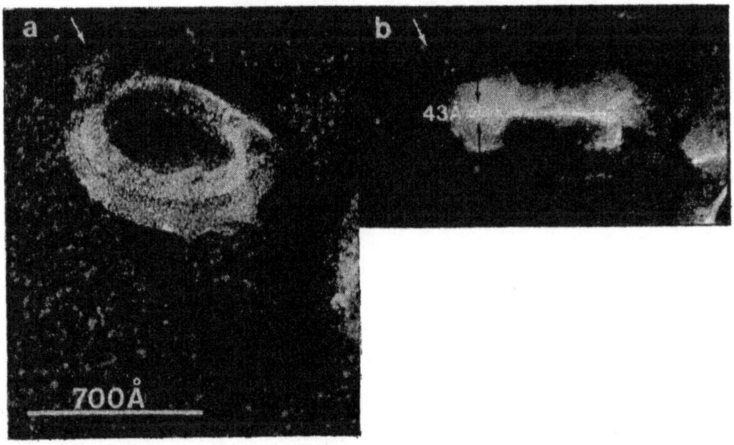

Figure 8. Freeze-etch replica of spd-ϕx-174 DNA (linear) torus a); tilted to end-on view b). Arrows indicate shadow direction.

ϕx-174 torus in Figure 8. Here, a 43Å end-on measurement is shown, about that expected for minimum hexagonal packing of a singlè DNA molecule into a thin disk like torus.

Besides corroborating our interpretations of torus dimensions in terms of the models discussed in Figure 7 the value of extensively tilting replicas to end-on and other views, allows us to confirm details of DNA writhe in the toruses as well as directly confirm stereometer measurements made from stereopairs of micrographs(33). Such stereometer measurements on a representative torus are shown in Figure 9. Here two circumferentially wrapping DNA helices have been measured and their relative elevations in the EM beam axis direction are indicated. From these and other stereometer measurements as well as visualizing extensively tilted toruses we have concluded that there is far from perfect hexagonal packing of DNA helices in these objects(17,23,33,36).

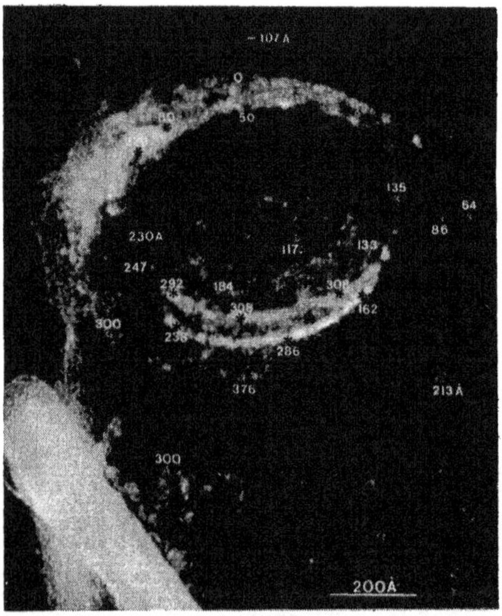

Figure 9. Stereoscopic measurements of DNA helices in spd-calf thymus DNA torus made from a 10° tilt stereomicrograph pair.

DISCUSSION

Reversibility and specific ion effects. A number of trivalent and tetravalent cations, both inorganic and organic polyamines, have been shown to produce a DNA condensed phase at low DNA concentration in purely aqueous solution in vitro(4-33). Even divalent cations such as Mg2+ and putrescine2+ induce condensation under the reduced water activity conditions of mixed alcohol:water solvents(7,14,15). Our choice of the trivalent spermidine cation for DNA condensation studies was based upon the fact that polyamines occur as major ionic constituents complexed with DNA or RNA in certain virus and bacteriophage heads(2,37-40).

Using a sedimentation assay we measured the DNA tertiary structure transition equilibrium starting from both condensation and decondensation directions. In the presence of varying monovalent and divalent competitor ion concentrations(Fig. 1 and 2) we have shown a hysteresis in the forward and reverse reaction midpoint positions(18,23). The hysteresis in equilibrium reaction midpoint positions may be interpreted in terms of rapid decondensation kinetics but slow condensation kinetics resulting from intermolecular aggregation effects. A similar hysteresis was noted from qualitative light scattering kinetic measurements in two other systems(14,15,19) and a number of investigators have documented the DNA concentration sensitivity of light scattering intensity at concentrations up to five-fold lower than that used in the author's laboratory(7,11,14,15,19,21,22). Moreover, in agreement with these data, Porschke(41) has recently measured the stopped flow light scattering kinetics of DNA condensation and finds good evidence for a slow intermolecular process(100s time constant) as well as a rapid intramolecular condensation. Decondensation was found to be rapid.

DNA condensation requires phosphate charge neutralization. Manning's theory of counterion condensation(34) predicts that counterions will condense on the polyelectrolyte to neutralize a significant charge fraction even at quite low cation concentrations. First Bloomfield(7) and then a number of other investigators have used primarily light scattering methodology to measure condensation and then have calculated the critical DNA charge fraction neutralized at collapse. They have established that DNA collapse into condensed tertiary structures occurs, in most cases, with 88-90% charge neutralization by any combination of counterions for DNAs in aqueous/ mixed solvents and in mixed valence counterion systems(8,14,15,19,21,22,41). Our data(Fig. 1; 18,23), obtained using a less sensitive sedimentation assay, is in general agreement with this charge neutralization value, as is the data obtained by measuring macroscopic DNA gel stability(12).

However, ion specific effects on DNA condensation have been noted by a large body of investigators. We have shown differential divalent competitor cation effects on the DNA tertiary structure phase transition(Fig. 2; 18,23), with the smaller cations(higher charge density) being the more effective competitors. A number of light scattering studies have demonstrated specific ion effects(7,8,14,15,20,21,22). These are manifest in the different critical concentrations of multivalent cations required for DNA condensation in identical buffer systems and as variations in total DNA charge neutralization level at collapse for different condensing counterions. Also, Widom and Baldwin(19) have noted some exceptions to the general rule that 88-90% charge neutralization produces collapse. Baase et.al.(42) report in their spd-T7 DNA system that at 90% DNA charge neutralization spd is well below the concentration required for DNA collapse. Benbasat(22) has studied a series of DNAs of varying charge density including the ϕW-14 bacteriophage DNA, containing the hypermodified, positively charged(+2), base alpha-putrescinylthymine, and shown that condensation occurs with hexamminecobalt(3+) at 87% charge neutralization for most of the DNAs but at 89% with spermidine for the same DNAs. The wild type ϕW-14 DNA, possessing maximum reduced linear charge density, behaves somewhat differently with DNA charge neutralization values 2% lower than the above values for each respective condensing agent at collapse.

Polyamines as specific ions and methylene bridging effects. The success of the Manning theory is somewhat surprising given its appealing simplicity. It does not include specific ion effects in its description of counterion condensation by purely electrostatic interactions(34). In fact, ions are treated as point charges. More recently, LeBret and Zimm(43,44) have reported counterion size and charge effects on the mixed counterion radial concentration distribution about uniformly charged DNA cylinders generated by Monte Carlo computer simulations. The considerable body of experimental evidence demonstrating specific ion effects suggests that nonelectrostatic cation interactions with DNA play a significant factor in the collapse process and may effect the structure and dynamics of the condensed DNA. The naturally occuring polyamines are biological cations that serve metabolic regulatory functions as well as interact with nucleic acids in all cells, viruses and bacteriophage. Their structure, amino and imino groups positively charged at pH values below 9 and separated by methylene carbon bridging groups of varying length, provides for possible H-bonding and van der Waals interactions as well as purely electrostatic interactions with DNA.

In addition to the specific ion effects noted above for polyamine-DNA condensation transitions, a number of investigators have specifically examined the effect of spermidine homologues of varying methylene carbon bridge lengths on the properties of spd-DNA complexes. Thomas and Bloomfield(45) have studied the Tm of spd-DNA complexes where spd homologues had one methylene carbon spacer length varied from 2-8. They found elevated Tm values with small Tm differences for homologues 4-8. Homologues 3 and particularly 2 showed significantly lowered Tm values. This is consistent with the idea that spd(3,4) can stabilize the B secondary structure by spanning the DNA minor groove. A model for spd-DNA incorporating: electrostatic interactions, van der Waals interactions of 4 carbon methylene spacers with DNA base edge atoms and H-bonding of amino and imino groups with DNA phosphate O atoms has been proposed(46). According to this model decreasing methylene spacer length would be expected to lower the stabilizing energy of homologues spanning the DNA minor groove.

Schellman and Parthasarathy(47) have described X-ray diffraction experiments on DNA gels formed with spd homologues. They showed that spd homologues of longer methylene spacer length produced greater DNA interhelical spacings, presumably due to DNA nearest neighbors in the hexagonal lattice cross linked and stabilized by the amino and imino charges of the spd homologues separated by increasing methylene spacer lengths. This direct demonstration of polyamine specific ion effects on a structural parameter of the DNA condensate argues that interactions other than purely electrostatic influence the condensation process. A study by Baase et.al.(42) emphasizes the same aspect of polyamine collapse of DNA. In their study of the T7 DNA precollapse region of cation critical concentration they concluded that spd cross linking promoted the transition to the compact DNA state. In an earlier study Allison et.al.(11) studied DNA condensation of ϕ-29 DNA with spd homologues and found differences in the degree of intermolecular aggregation produced. Longer methylene spacer length spd homologues produced considerably greater aggregation, again presumably through cross linking DNA strands.

We have performed experiments in which the pH dependent sedimenting fraction of spd homologue-DNA condensates was monitored. In Figure 10 are plotted phase diagrams of the added competitor cation concentration versus midpoint pH values

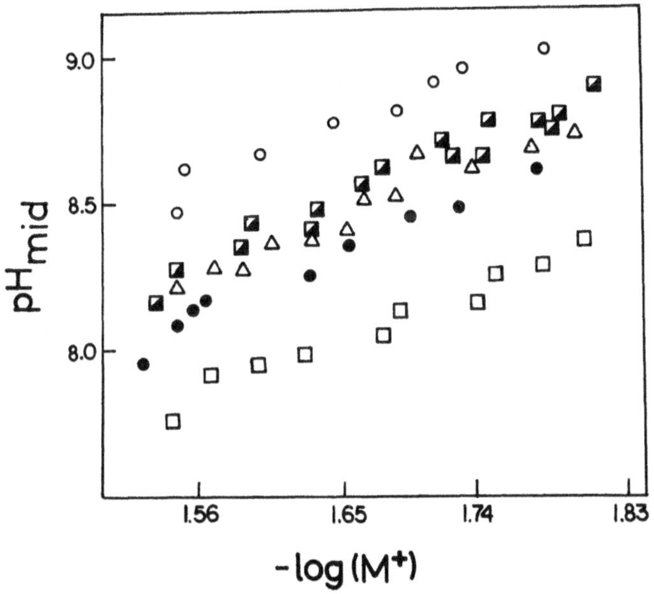

Figure 10. Phase diagram of added monovalent cation versus pH midpoints of pH dependent sedimentation transitions of spd homologue-DNA condensates: (□) spd 3,2; (●) spd 3,3; (△) spd 3,4; (◪) spd 4,4; (○) spd 3,5.

determined from the sharp transitions measured for the loss of fraction spd homologue DNA sedimenting with increasing pH. It is clear that specific polyamine cations affect the transition differently. We interpret the sharp transitions in sedimenting DNA condensate fraction as an aggregation -deaggregation transition. The pH midpoint position is due to spd homologue loss of cross linking ability with titration of the homologue from 3+ to 2+. The aggregation-deaggregation interpretation is consistent with many observations of DNA aggregation by spermidine at the collapse concentration and DNA concentration we have used(7,8,11,14,15,19,21,45). We observed no pH transition when hexamminecobalt(3+) is the condensing counterion, again in agreement with the studies mentioned above and specific spd homologue cross linking ability.

The homologue pH midpoint values and the order of the homologue effects is roughly consistent with the magnitudes and known order of homologue pKa values with increasing methylene spacer length(48). Some investigators(11) have argued that counterion condensed polyamines ought to have pKa values shifted upward by close to 4 units based on the measurements of Wilson and Jones(49) of the pKa of the ethidium dye intercalated into DNA. The rationale is that the large electrostatic potential of the DNA groove will cause the upward shift. While we agree with this interpretation for counterion condensed polyamines lying in the DNA grooves, it may not be the case for the spermidine homologues in our study. Here, one may expect the cross linking amino and imino groups of the homologues to lie outside of the DNA grooves. Calculations of the electrostatic potential around B form DNA have shown the deepest potentials within the grooves, due to all DNA functional groups on the bases and sugars, not just charged phosphates(50). Therefore, pKa upshifts for cross linking homologues outside of the DNA grooves may be considerably less than 4 units, in line with the pH transition midpoints we have observed.

An alternate explanation for our pH dependent spd homologue values may lie in the data of Lin et.al.(51) on dynamic light scattering of spd-DNA condensates as a function of pH. They measured changes in the DNA internal motion parameters in the pH range 8.5 to 10.5, a range over which no significant titration of pure DNA functional groups occurs. With spd binding they postulate that a basic spd ligand facilitates local DNA denaturation by stabilizing the titration of one or more base ring protons near the spd binding site to form a pH titrable joint in the DNA. The creation of joints in this pH region could explain our DNA tertiary structure transition data if the spd homologue pH effect order could be rationalized with a specific model for homologue binding.

There are many examples of the specificity of polyamines in different biological systems. Thomas and Bloomfield(52) have shown differential effects on the B to Z DNA secondary structure transition for two different monoacetylated spd derivatives, metabolic species found in living cells. Krasnow and Cozzarelli(53) provide evidence that DNA condensation and aggregation by spermidine to a collapsed tertiary structure is the form in which the catenation-decatenation reactions occur, catalyzed by topoisomerase enzymes. They feel that the condensed DNA tertiary state, stabilized by polyamines, is important as the substrate in site specific DNA recombination such as in bacteriophage lambda. Morris and Lockshon(54) have demonstrated that reduced spermidine levels in spermidine auxotrophs of E. coli. result in altered plasmid DNA supercoiling. They have proposed a theory based on altered chromosomal supercoiling to explain the macromolecular synthetic effects in these cells. Moreover, they have shown that different polyamine homologues have different effects on cellular synthetic rates in these cells, suggesting structural specificity for polyamines in specific roles in the cellular machinery(55). One such role may be stabilizing a particular tRNA structure in the ribosome(56). Polyamines are known to specifically affect the protein translation rate in cells(57). And Cohen's group has shown that spermidine is a specific structural component of the RNA within certain RNA viruses(40,58). Recent reviews(38,39) of polyamine biochemical effects in cells catalogue their many functions.

Enzymatic cleavage and circumferentially wrapped toruses. We have shown by electron microscopy that toruses are unidirectional circumferentially wrapped DNA structures(13,17, 23,33,36). Our studies of the arithmetic series of DNA fragment sizes produced by micrococcal nuclease cleavage of spd-condensed calf thymus, ϕx-174 and lambda DNAs corroborates this model(16,18,23,31,32,35). Since it has been demonstrated by CD spectroscopy that condensed DNAs possess the B secondary structure(5) and since our cleavage model requires the monomer fragment to wrap once about the torus, the monomer fragment sizes from digestion studies of all the DNAs must lie between the respective inner and outer circumference values we measured

Table 2. DNA Torus Parameters from Nuclease and EM

DNA	MW (bp)	monomer DNA fragment (bp)	EM Parameters Inner (bp)	Outer (bp)	Ring thickness (Å)	End-on (Å)
ϕx-174						
lin	5386	830±120	539±165	858±75	169±15	30-43
circ	5386	780±80	766±69	1147±81	186±14	-
calf	20000	760±87	541±60	825±65	143±18	(60)
lambda	48000	1003±115	682±123	1086±120	200±17	60-210

by EM for each torus population. This can be seen to be the case in Table 2 for all the DNAs we measured.

It may be noted that the torus dimensions of the ϕx-174 circ. DNA population are considerably higher than the linear while the monomer fragment size is uncharacteristically close to the inner circumference dimension. Although this phenomenon is not fully understood a number of points may be made to rationalize these data. To begin with, this DNA population is topologically distinct from the other three in Table 2. It is circular while the others are linear. This difference must result in certain topological constraints being placed upon collapse pathways in the circular population that are not the case with the linear DNAs. We have already noted that the ϕx-174 DNA torus dimension distribution(Fig. 7) clearly shows the existence of a small monomer DNA torus population in equilibrium with a population of toruses comprised of more than one condensed ϕx-174 DNA molecule. This is not the case with the larger DNAs and therefore, aggregation helps to explain the large torus dimensions noted for the circ. ϕx-174 DNA sample.

We have already discussed the fact that the micrococcal nuclease digestion data for spd-condensed ϕx-174 DNA allows us to discriminate on topological grounds between unidirectional and alternating direction circumferential wrapping models for DNA collapse(16,23). At least in solution the digestion data only support continuously wrapped DNA toruses, perhaps condensing by a general mechanism of the type shown in Figure 5.

Using a variety of both specific restriction enzymes and nonspecific nucleases(DNAase I, DNAase II) we found that spd-DNA condensates could be digested. Hae III was able to cleave at all of its normal restriction sites in spd-condensed ϕx-174 DNA samples. Different DNA condensates were cleaved by nonspecific nucleases at different rates(35), undoubtedly due to the condensate size differences we have already noted in Table 2. However, micrococcal nuclease was the only enzyme capable of sensing the torus structure in such a way as to give rise to the distinctive gel banding patterns.

Our torus cleavage model(Fig. 4) has received no direct experimental confirmation. Indirectly, one can argue that hexagonally packed interior DNA strands will be digested at a slower rate than exterior ones. Our digestion experiments with spd-condensed lambda DNA show that this DNA is digested some 30 fold slower than uncondensed lambda while under identical conditions the relative cleavage rates for condensed and uncondensed calf thymus DNA were only 3.6 fold different(18,35). Since monomer lambda DNA toruses contain 2.4 fold as much DNA per torus as do the calf thymus they will contain a larger fraction of hexagonally packed interior strands than will calf thymus toruses(Table 2). Therefore, the large digestion rate difference, due to the slower kinetics of digesting the larger interior strand fraction of the lambda torus supports our model's qualitative kinetic description giving rise to arithmetic series fragments by regional cleavage. Along the same line, in studies of chromatin digestion rates by micrococcal nuclease, it has been shown that DNA packaged with cationic histone octamer protein in chromatin is digested some 6 fold slower than DNA in solution(59).

Hydrated torus dimensions. We have discussed in detail the superiority of freeze-etch TEM using low Pt/C(10Å) metal deposition levels for the visualization of the native structure of any hydrated macromolecular system at high resolution(23,33).

Application of the single direction shadowing freeze-etch protocol to spd-DNA condensates formed from ϕx-174, calf thymus and lambda DNA allowed us to visualize toruses and demonstrate conclusively that these objects are organized by continuous circumferential wrapping of B DNA helices. Moreover, by extensive tilting of the torus replica and performing high precision stereoscopic measurements($\pm$5-10Å, 95% error) of torus surface fibres(17,23,27,30,32,33,36,60) we have demonstrated

Table 3. Torus Dimensions

		circumference (bp)		
group	DNA	inner	outer	technique
Benbasat(22)	T4 170000 bp	325	-	EM
Chattoraj. et.al.(4)	T7 38700 bp	266 356	738 696	EM grid stain EM sol'n stain
Widom & Baldwin(19)	lambda 48000 bp	277 370 296	891 913 794	light scat. fits " EM grid stain fit
Marx & Ruben(32,36)	lambda 48000 bp	682	1086	EM freeze-etch
Allison et.al.(11)	ϕ-29 17700 bp	-	1710	EM grid stain
Marx & Ruben(36)	ϕx-174 5386 bp	766	1147	EM freeze-etch
Thomas & Bloomfield(52)	dA-dT:dA-dT B to X	92-139	554 924	EM grid stain
Thomas & Bloomfield(26)	dG-m5dC:dG-m5dC Z	185-280		EM grid stain
Castleman et.al.(25)	dG-dC:dG-dC Z	115-185	-	EM grid stain & shadowing

that, while a high degree of order exists in the DNA packing, there is far from perfect hexagonal packing in the circumferential strand lattice of the torus.

The torus population dimensions measured for four DNAs have been presented in Table 2. All of our torus dimension distributions show rather abrupt lower limits in inner circumference at about 350 bp, with few toruses having values below that. These data are in excellent agreement with a thermodynamic stability theory developed by Manning(61) that predicts a critical maximum curvature for DNA in toruses corresponding to a minimum inner torus circumference of 315 bp B DNA. This agreement is not fortuitous since we have discussed previously(23) how other EM preparation protocols utilize steps involving dehydration, the application of solvents and uncertainties of substrate surface. These are circumvented by the freeze-etch technique where rapid freezing produces a solution-like state in which the torus objects remain hydrated on the ice surface during etching and replication steps.

Literature values for torus dimensions vary considerably(Table 3.). In most cases the ranges seen can be rationalized on the basis of DNA size, concentration dependent intermolecular condensation and differing EM preparation protocols. Benbasat(22) has measured T4 DNA toruses and found good agreement with our freeze-etch measured data and Manning's theoretical limit. Chattoraj. et.al.(4) demonstrate the variability possible in EM dimensions for toruses using two slightly different staining protocols. Widom & Baldwin(19), using light scattering of condensed lambda DNA, calculated fits to equivalent hydrodynamic toruses of inner circumference considerably lower than those we measured, although outer circumference values were in fair agreement. These relatively low inner circumference values point up what a number of investigators have noted, that standard EM preparation protocols seem to result in smaller dimensions than those measured by other physical techniques, including the freeze-fracture, deep etch technique.

Allison et.al.(11) measured large torus dimensions from collapse of a relatively small DNA. They ascribe the large dimensions to intermolecular condensation. This agrees with our measured torus dimension distribution for ϕx-174 DNA toruses(Fig. 7) where we can discriminate a large population of toruses formed from intermolecular DNA condensation. Post and Zimm(14) have formulated an extension of the Flory-Huggins theory of polymer-solvent interaction to describe DNA collapse and its relationship to aggregation that confirms our interpretations for DNAs of small size.

Thomas and Bloomfield(52), studying toruses formed from poly(dA-dT:dA-dT), found considerably smaller EM dimensions than their light scattering measurements indicated. Two other studies of toruses formed from Z secondary structure DNA polymers, poly(dG-m5dC:dG-m5dC)(26) and poly(dG-dC:dG-dC)(25), measured very small torus dimensions. In the first study this discrepancy was ascribed to EM preparation artifacts based on their comparison of the EM data with light scattering measurements of the structures.

Other counterion condensed DNA structures. DNA has been shown to condense from solution at high concentration in the presence of high salt and polymeric compounds such as polyethylene glycol, to form structures that have been playfully labeled "shish kabobs"(62). At lower DNA concentrations other investigators have shown that toruses form under similar solution conditions(63,64). Laemmli(65), using nuclease digestion, has detected a repeating DNA structural unit in these condensates based on an organization other than circumferential DNA wrapping of the type we detected in spd-DNA condensates. We have performed experiments micrococcal nuclease digesting PEG condensed DNA, but have had no success in detecting DNA structural units of any kind(35). One characteristic shared by these structures is a CD spectrum of ten fold higher intensity than that of spd or other counterion condensed DNA toruses. These condensed structures have been termed psi. Both psi(+) and psi(-) spectra have been shown to be produced by different DNAs(66) condensed with different agents: PEG/salt; certain transition metal ions(66); polylysine(67-69); histone H1(70-72), etc. These unusual CD effects have recently been shown theoretically to result from a long range chiral order of right or left handed pitch in the aggregated DNA(73-75). The psi state has the characteristics of a cholesteric phase while the toruses we have studied have the order characteristics of a nematic phase.

In counterion condensed DNA preparations at low DNA concentrations, structures other than toruses have been observed. These include spheres(4,8,10,11,14,15,52), rods(10, 52), toroidally bent rods(10) and occasional lariat shaped toruses(4). Since spheres and toruses seem to coexist in many of the preparations studied, they may represent equilibrium solution structures of similar energy. We have not observed spherical shaped DNA structures using the freeze-fracture, deep-etch methodology.

It has been proposed that toruses are formed by the circumferential bending of folded fibre rods(10;as in Fig. 5 Model 2). As we have discussed our solution nuclease structural studies and freeze-fracture, deep-etch EM data all argue strongly against this model. In fact we have never observed a rod or folded rod structure in our EM preparations. Rods observed in the above study(10) were produced using a 95% ethanol immersion step in the EM preparation protocol. It has been shown by Zehfus and Johnson(76,77) that DNA stacking interactions are significantly weakened under these conditions. They demonstrated, using UV, CD and infrared spectroscopy, that the collapsed DNA state contained little base stacking. Therefore, we feel that ethanol dehydration and perhaps other EM dehydration protocols may produce thermodynamically stable structures that bear little resemblance to those that exist in solution.

A ring structure has been shown to form when circular superhelical DNA is complexed with a tripeptide, dansyl hydrazide trivaline(78). The ring structures are proposed to contain DNA helices in close contact due to DNA-peptide and peptide-peptide interactions. Complexes formed from linear DNAs show thick banded fibres that may contain superhelically wrapped DNA. Torus shaped complexes of unknown tertiary structure are formed using relaxed circular DNA(79).

It has been shown(80) that DNA condensates are decondensed by intercalating dyes such as ethidium(+) and propidium(2+) by a cation competition mechanism. However, in this same study it was shown that intercalation and the DNA collapsed state are not incompatible since methidium-spermine(4+) was found to intercalate and condense the DNA simultaneously. Recently, it has been demonstrated that DNAs and RNAs are condensed by the intercalating aromatic cations mitoxantrone and DHAQ(81). Moreover, condensation of DNA and RNA has been observed to occur in situ in cells at the physiological concentrations of these same cations, some of which play pharmacologic roles as mutagens and antitumour agents at physiological concentrations(82).

Coupled 2°-3° structure. Many examples exist of B DNA tertiary structures in which DNA is bent. These include nucleosomes(1), gyrasomes(83) and possibly the Cro protein-DNA complex(84). In the case of nucleosomes it has been suggested that certain base sequences promote bending of DNA around the histone octamer core(85). Certain small DNAs, in the absence of proteins, exhibit anomalous physical properties such as electrophoretic mobility(86). In the case of the kinetoplast trypanosome DNA, this can be attributed to DNA sequence specific bending in solution(87). An explanation for this behavior may lie in altered DNA secondary structure regions appropriately phased to the helix repeat so that their local curvatures are coupled producing a characteristic long range tertiary structure-macroscopic curvature(88).

We have already discussed the long range ordered chiral tertiary structure of condensed psi states. At least in some cases the psi structure formed seems to be correlated with particular DNA secondary structures(66). And in a study of a DNA fragment containing a B and Z secondary structure junction a kink was measured at the junction which changed the helix axis direction of the two regions(89). Therefore, it is not surprising that secondary structure variations appear to give rise to characteristic tertiary structures.

Thomas and Bloomfield(90) have shown secondary structure as well as cation specific effects on tertiary structure for poly(dA-dT:dA-dT). This polynucleotide undergoes a B to X secondary structure transition(by CD) with increasing concentration of condensing trivalent counterion. These authors suggest a sequence of events in which B DNA, condensed by cations, undergoes further tertiary structure changes (aggregation) while converting to the X form. Their EM data show that rods, spheres and aggregates result with hexammine-cobalt(3+) condensation while spd produces toruses at high concentration. These results are different from random sequence DNA behavior where no secondary structure change occurs either before, during or after the tertiary structure transition. Data from another study of the same investigators(26) demonstrated that the B to Z secondary structure transition in poly(dG-m5dC:dG-m5dC) was coupled exactly to the tertiary structure transition. Here EM pointed up an intermediate rod like form which corresponded to equimolar B and Z secondary structure in the DNA(CD spectra). When the DNA is fully converted to Z form a dense population of very small toruses was observed by EM. These investigators specifically noted that their torus EM dimensions were likely to be artifactual since they were considerably lower than dimensions calculated from light scattering.

Castleman et.al.(25) have also described a coupled secondary, tertiary structure transition in poly(dG-dC:dG-dC). They observed the appearance of large banded fibre structures which they ascribed to extensive intermolecular aggregation during the conversion to Z DNA. By converting back to the B form the banded aggregates dissolved into toruses and interwound DNA structures. They propose that banded aggregates may be hollow, resulting from lateral aggregation of stacks of individual toruses.

Finally, Benbasat(22) has discussed how ϕw-14 DNA may be in an altered secondary structure that results from its lowered linear charge density[alpha-putrescinylthymine(2+) covalent modification]. In this study light scattering and EM measurements showed maximally condensed toruses with large central holes and exceptionally tightly wrapped strands.

Covalent transition metal ion modification of DNA appears also to result in tertiary structure transitions. Osterberg et.al.(91) have shown an altered tertiary structure for Cr^{3+}-DNA by EM. And, in Ru^{3+}-DNA, Clarke et.al.(92) have seen evidence of condensation phenomena.

Studies of the mechanism of torus formation. We have indicated schematically in Fig. 5 Model 1 that the mechanism of torus formation in solution is likely to be a multistep event. Some studies have been performed recently that obtain information on this sequence of events. In an earlier study with ϕ-29 DNA(11) and now more recently with T7 DNA(42) it has been shown by combined linear dichroism and Zimm-Crothers viscometry that below the critical spd collapse concentration(30 μM) there are measureable changes in the DNA hydrodynamic parameters. Up to 10 μM spd the DNA persistence length decreases in a way consistent with counterion condensation theory. By 20 μM spd the persistence length decrease cannot be accounted for by theory and may result, as the authors suggest, from the DNA conformation no longer being a random-coil or Gaussian chain. At 20 μM spd the light scattering interpretation of DNA structure was that of an inhomogeneous suspension of partially collapsed intermediate structures. These authors suggest that counterion crosslinked intermediates may be necessary to promote the intramolecular collapse transition.

In this regard, Schellman and Parthasarathy's(47) measured variation in x-ray interhelix spacing with spd homologue chain length increase and this author's results with pH dependent spd homologue specific aggregation effects lend support to the crosslinking idea. Also in direct support are the data of Widom and Baldwin(19) on diffusion coefficients measured throughout the collapse transition region. These investigators claim that the DNA condensation process is not a simple two-state reaction.

The most detailed description of the DNA collapse mechanism has recently been provided by Porschke(41,93,94). This study used stopped flow and field jump techniques with light scattering detection to investigate the fast kinetics of DNA condensation and provide a picture of the dynamics of the collapsed state. It was demonstrated that two processes occur. One is a fast process on the msec time scale in which DNA condenses intramolecularly with a characteristic induction time due to counterion binding to achieve a critical DNA binding fraction before the onset of the condensation kinetics. The second is a slow process(100 sec time constant) due to intermolecular condensation. In the field jump experiments condensed DNA toruses exhibit complex dynamic behavior with a broad range of time constants ranging from 25 μs to 2 ms. These studies have shown that DNA possesses a high degree of flexibility, with very fast bending motions. Below the critical counterion concentration the gradual decrease of measured DNA dichroism suggests that the helix is being bent gradually into an ordered superhelical structure before the transition to the collapsed state.

Structural-energetic units in DNA condensates. Recently, some measurements have been made and a model proposed for possible structural-energetic units in DNA condensates. These are presented in Table 4 for comparative purposes.

Table 4. Structural/Energetic Units in DNA condensates.

Investigator	Ave. Length	Model/Measurement
Manning(95)	40 bp	size of axially compressed straight DNA segment between kinks
Porschke(41)	60 bp	energetic cooperativity size of smoothly bent DNA in toruses
Marx & Reynolds(35)	85 bp	length of thermodynamically unstable DNA fragments in nucl. digested DNA condensates
Griese, Serwer & Horowitz(98)	83 bp	DNA length between 1.2% high affinity ethidium sites in T7 DNA within bacteriophage

Manning has developed a model for breathing and bending fluctuations in DNA in which an open base pair kink is coupled to a slight axial compression of the helix of average straight segment length 40 bp(95). With directionally correlated kink angles this may serve as a mechanism to allow DNA to condense into toruses.

In the studies by Porschke discussed above he measured a 60 bp cooperativity size for the condensation-decondensation equilibrium(41). Independent electric dichroism measurements by Porschke's group of small DNA restriction fragment flexibility(96) showed that smooth bending can account for the observed relaxation times of 0.1 to 1 μs in the chain length range between 100 and 200 bp. Smooth bending of fragments 700-1000 bp, appropriate to toroidal condensation by circumferential wrapping, would possess time constants in agreement with the range(25 μs and up) he measured(41) for DNA condensates. Furthermore, using the curvature represented by the average bending amplitudes measured for the short DNA restriction fragments, Porschke has calculated a complete circular circumference to be about 740 bp(97). This value is in excellent agreement with the measured circumferential wrapping size for all our DNA samples measured by both micrococcal nuclease cleavage and EM(Table 2).

Our studies of spd-DNA torus structures by high resolution EM indicate that for the most part smoothly bent DNA helices appear to lie in the hexagonal strand matrix. However, we cannot rigorously exclude kinked DNA since the freeze-etch technique prepares replicas of the objects at low temperatures (-73° C) by covering them with Pt metal(10-15Å). We have investigated the thermodynamic length stability of DNA fragments in the condensates. Previous experiments on torus formation, using small DNA fragments, suggested that below 400 bp fragments were incapable of forming toruses, perhaps for kinetic reasons(8). Therefore, we performed the reverse experiment starting with high MW spd-DNA toruses. Using gel electrophoresis we followed the DNA fragment length distribution released into the supernatant of a sedimented DNA condensate preparation which was being digested by micrococcal nuclease. With longer digestion times an increasing percentage of the total DNA was released into the supernatant. However, the DNA fragments released were found to be about 85 bp in average size at all times in digestion, although an exonucleolytic shortening of released fragment size was noticed at longer digestion times. This number is in good agreement with Porschke's(41) calculated 60 bp energetic cooperativity size in the DNA condensates. It is interesting that the fast bending relaxation measured by Dieckmann et.al.(96) is present only in DNA fragments 95 bp and greater. Therefore, our measured unstable DNA length of 85 bp may be related to its thermodynamic inability to remain stably bent.

Finally, in an interesting study, Griese et.al.(98) have measured the binding of ethidium dye to DNA in intact T7 bacteriophage and some T7 deletion mutants. They find evidence for two DNA intercalation site classes at low ethidium binding ratios. One class is comprised of the bulk of the DNA sites while the second class is comprised of 1.2% of the DNA sites and possesses a significantly higher ethidium affinity constant. These investigators suggest that the class of high affinity sites represent kinks introduced into the packaged T7 DNA on binding the first ethidium molecules. On average, 83 bp exists between these high affinity binding sites in the condensed T7 bacteriophage DNA, assuming a random site distribution. The exact distribution of these high affinity sites in the T7 bacteriophage DNA is not known. This correlation in DNA size with the others measured in Table 4 is striking, although possibly fortuitous. It may be suggesting a thermodynamically preferred DNA structural unit in the highly condensed DNA within the bacteriophage head.

Molecular stabilization forces in condensed DNA. This section will present a brief account of some of the molecular ideas proposed to explain DNA condensate stability. Oosawa(100) proposed a model for DNA condensation in which mobile multivalent cations can provide a net interhelical attractive force due to net transient dipoles produced along the helix axis. Experimental evidence has been published that is in disagreement with the predictions of the theory(101).

Schellman and Parthasarathy(47) have noted the charge distribution analogy between counterion condensed DNA and stable ionic crystals. Bloomfield et.al.(99) have assumed that cations simply reduce unfavorable DNA segment-segment interactions. Their calculations find a net repulsive force between DNA segments that they postulate is overcome by London dispersion forces between the DNA segments.

As part of the electrostatic picture of condensate stability one must consider the many examples of specific ion effects we have reviewed. Especially important is the DNA interhelical x-ray spacing measured for polyamine homologues(47). This provides direct evidence for the importance of crosslinking helices in ways which maximize close contact between ions of different sign. And it provides some confirmation of the ion site binding model proposed by Widom and Baldwin(8). They postulated that the entire attractive force for condensation was provided by cation site binding to particular phosphates on adjacent helices with the loss of water from the hydration shells of the phosphates and cations.

Two important considerations of condensate stability discussed in (47) are species entropy and condensate surface energy. To begin with, the loss of counterion entropy in the condensates is relatively unimportant since counterion condensation already neutralizes a major fraction of the DNA charge in the random coil form. DNA conformational entropy must be very important and its loss may be balanced by energetic changes due to solvent reordering. Condensate surface energy may be an important component of stability because of the high surface to volume ratio of toruses. This may account for why condensate aggregation has been observed for smaller DNAs(11, Fig. 7 multimer toruses) and may have something to do with limits measured on aggregation extent in certain systems(8,11,14,15,).

Recently, measurements of repulsive hydration forces have been made on DNA gels(102). By measuring the force per unit length versus x-ray interhelical spacing of parallel DNA helices a strong repulsive force was detected when the helix surfaces were 15Å apart. The repulsive force grew exponentially with 2.5-3.5Å characteristic distance making it incompatible with electrostatic double layer theory. These investigators feel that the hydration force represents the work of removing polarized water on facing molecular surfaces. Specific ion effects on water polarization can be incorporated into their model as modulations of the hydration force while the characteristic decay distance is a fundamental property of the water itself and does not vary with cation type as their experiments indicate.

Implications for DNA structure in bacteriophage. As we alluded to in the Introduction certain double stranded DNA bacteriophage[lambda(4,103,104); T4,T7(105,106)] and viruses [herpes simplex(107)] are thought to have their DNA organized into torus shaped structures. And polyamines, including spd, are naturally occuring cations in the bacteriophage head(37-39). Therefore, spd-DNA toruses represent an in vitro model system for DNA organization in these biological pathogens.

We studied spd-DNA torus populations from a number of different DNAs, including lambda. These have been presented already in Table 2. The lambda DNA torus is of interest since it may be compared to the results from physical studies of lambda DNA bacteriophage packaging. In Figure 11 is shown the dimension distribution for spd-lambda DNA toruses.

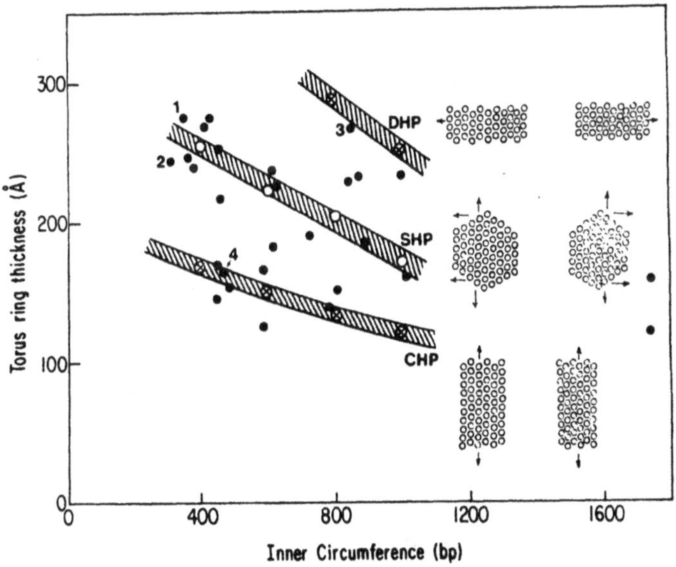

Figure 11. Torus ring thickness vs. inner circumference plot of lambda DNA toruses. Thickness is measured across torus annulus from 2-D EM projection.

There is a broad range of torus dimensions but the range may be fit by assuming intramolecular lambda DNA collapse according to the three idealized tertiary structure packing models(in cross section) shown as insets next to their indicated dimensions. By extensive replica tilting and direct end-on view measurements of a few of these toruses we were able to confirm the validity of these shape models and concluded that the toruses are intramolecularly condensed(36).

Some specific conclusions concerning the progressivity of DNA self-assembly have been drawn from the data in Table 2 and the shape information in Figs. 7,8 and 11(31,32,36). The ϕx-174 DNA toruses have been shown to be thin, relatively wide disk shaped objects. This can be understood on energetic grounds. DNA must self-assemble into a torus with, at minimum, hexagonal packing to avoid helix-solution interactions. However, due to unfavorable bending energy, it will minimize helix curvature of successively condensing strands by assembling strands outward forming a thin, wider disk. For a longer length DNA, like calf thymus, toruses grow thicker but not substantially wider. For the largest DNA studied, lambda, the toruses are wider and substantially thicker, with a range of observed tertiary/quaternary structure shapes represented.

Our studies of the pattern of DNA strand self-assembly in toruses lead us to some conclusions that are in fundamental agreement with recent studies on the organization of lambda DNA in the bacteriophage head. While our measured lambda DNA torus dimensions are considerably larger(1176Å outer diameter) than those of packaged bacteriophage[590Å outer diameter(103)], this is understandable in light of the energetic cost of DNA bending and the fact that lambda phage requires energy input for its DNA packaging(2). This difference in the two condensed structures is reflected in the measured densities of the structures. A variety of bacteriophage have been shown to have higher B DNA density 0.5 ± 0.02 g/cm3(2), reflecting tighter strand packing, than that measured for DNA toruses, 0.43 g/cm3(108). The ϕW-14 bacteriophage, containing the divalent cation modified base discussed earlier, may assume a normal bacteriophage DNA packing density if the DNA alters secondary structure(22).

We have observed great variability in the tertiary structure packing of DNA strands in the lambda DNA toruses. This agrees with the idea that the process of DNA "crystallization" is not solely what determines the nearest neighbor order of DNA strands in the lambda bacteriophage head. In a careful x-ray diffraction study Earnshaw and Harrison(103) have obtained evidence for coherent diffraction by DNA in a hexagonal lattice near the opposite interior surfaces of the phage particle. From their study of lambda deletion mutants they conclude that the DNA layers closest to the protein shell are highly ordered by the shell and that this order is progressively lost going towards the phage interior.

Both our general observations of DNA strands and our specific stereoscopic measurements of local DNA tertiary structure describe the lack of perfect hexagonal packing order in toruses(17,23,30,33). These data support the idea of the falling off of lambda DNA packing order farther from the protein shell. Also in support are recent biochemical crosslinking studies indicating a nearly complete lack of ordering at the level of DNA sequence-protein shell interaction and DNA sequence nearest neighbor interaction in the lambda bacteriophage head(109,110). Therefore, although the protein shell may locally order DNA helices it must be organizing different regions of the lambda genome in individual phage. This is in the spirit of flow dichroism studies of DNA organization in T4 bacteriophage(111, 112) which concluded that there may be no unique orientation of the DNA spool axis in the bacteriophage.

It should be mentioned that in a study of DNA-protein crosslinked lambda bacteriophage lysed for EM visualization, evidence was obtained for DNA supercoiled structures(113). Moreover, these investigators seemed able to distinguish between two models for gross DNA strand packing. They clearly favored the one in which DNA packaged as stacks of disks rather than as sets of concentric shells. Our DNA size series study of the progression of torus assembly supports this interpretation.

While spd has been shown to not be an absolute requirement(39) for bacteriophage packaging [spd(-) mutants can package, but with lowered efficiency], spd could, under the right circumstances, provide a driving force for collapse. In order for DNA collapse to occur in the bacteriophage head due to neutralization of a critical DNA charge fraction by counterion condensation there would have to be a concentration increase for spd and reduction for all competing monovalent and divalent cations. This could be accomplished by an ion specific pump changing the cation composition in the ion impermeable phage head immediately preceeding DNA packaging. Following packaging, known ion permeability changes in the bacteriophage coat(2,38,39) could allow reequilibration of the phage head interior with the intracellular ion pool.

Any model that attempts to explain lambda DNA packaging(114) would have to avoid DNA entanglements during ejection, account for the ATP driven DNA pump located at the vertex of the phage head-tail junction(115) and involve the lambda terminase activity that recognizes and cleaves one genome length of DNA for packaging(116).

Prebiotic and chromosomal evolution. Appealing analogies can be constructed between some of the important DNA chromosomal structures under active investigation today and some of the condensed DNA tertiary structures we have discussed in this review. It may be that nature has designed chromosomal systems incorporating available thermodynamically stable DNA states intrinsic to the isolated molecule.

Ideas on the origin of life have centered on the adsorptive, concentrating and potentially catalytic properties of inorganic clay surfaces, likely to have been in abundance on the shores of the primitive oceans and subjected to fluctuating environments. There exists some evidence that oligonucleotides could have bound to positively charged sites on inorganic surfaces such as the crystal edges of kaolinite or the hydroxy interlayer of interstratified clays of the chloride type(117). Dehydration results in stronger molecular adsorption as well as changed surface properties like acidity that may contribute to alterations in conformation states of bound species(118). We have discussed in this report how various dehydration protocols, specific ions and sequence specific secondary structures influence DNA tertiary structure. It is clearly possible then that various types of DNA self-assembly, intramolecular(toruses) and intermolecular, could have occured under primitive earth conditions under protective enclosed local environments to provide selective advantages to systems undergoing furthur self-assembly into primitive protocells and protoviruses.

Eickbush and Moudrianakis(10) showed, by EM of alcohol dehyrated bacterial DNA, characteristic beaded structures that resembled in size, shape and linear contraction ratio the structures seen when E.coli. chromosomes are lysed directly onto EM grids(119). This beaded DNA tertiary structure resembles repeating nucleosomal arrays and it is tempting to consider that nature has utilized an available, thermodynamically stable DNA conformation during evolution to design present day chromosomal structures in bacteria and all higher organisms.

In the primitive dinoflagellates, eukaryotic precursor organisms, chromosomes are devoid of the nucleosomal organization, possess few chromosomal proteins and have a macroscopic organization resembling toroidally stacked chromonema(120). Lerman(62) has discussed some structural similarities between the EM appearance of dinoflagellate chromosomes and large psi aggregates of DNA that appear to be organized as twisted stacks of DNA. Again, a thermodynamically preferred DNA state, perhaps stabilized and organized into regions of constrained negative superhelix density, may have been utilized during evolution to produce a characteristic chromosomal structure intermediate between that of prokaryotes and eukaryotes.

Much work remains to be done to understand the range of structures thermodynamically accessible to DNA in vitro and their implications for our understanding of the involvement of DNA in all biological processes.

ACKNOWLEDGEMENTS

The author acknowledges support from CA 23108 NCI; GM 25886 and AI 17586, NIH; Research Corp. 8859; BRSG RR-05392, NIH and GeoM Co. It has been his privilege to collaborate with Dr. G.C. Ruben and T.C. Reynolds on much of the published experimentation presented in this review.

REFERENCES

1. Finch,J.T., Lutter, L.C., Rhodes,D., Brown,R.S., Rushton,B. and Klug,A. (1977) Nature 269, 29-36.
2. Earnshaw,W.C. and Casjens,S.R. (1980) Cell 21, 319-331.
3. Wang, A.H.-J., Quigley,G.J., Kolpak,F.J., Crawford,J.L., van Boom,J.H., van der Marel,G. and A. Rich (1979) Nature 282, 680-685.
4. Chattoraj, O.K., Gosule,L.C. and Schellman, J.A. (1978) J. Mol. Biol. 121, 327-337.
5. Gosule, L.C. and Schellman, J.A. (1978) J. Mol. Biol. 121, 311-326.

6. Post, C.B. and Zimm, B.H. (1979) Biopolymers 18, 1487-1501.
7. Wilson, R.C. and Bloomfield, V.A. (1979) Biochemistry 18, 2192-2196.
8. Widom, J. and Baldwin, R.L. (1980) J. Mol. Biol. 144, 431-453.
9. Skuridin, S.G., Kadykov, V.A., Shashkov, V.S., Evdokimov, Yu. M. and Varshavski, Ya. M. (1978) Mol. Biol.(Moscow) 12, 413-420.
10. Eickbush, T.H. and Moudrianakis, E.N. (1976) Cell 13, 295-306.
11. Allison, S.A., Herr, J.C. and Schurr, J.M. (1981) Biopolymers 20, 469-488.
12. Subirana, J.A. and Vives, J.L. (1981) Biopolymers 20, 2281-2283.
13. Ruben, G.C., Marx, K.A. and Reynolds, T.C. (1981) Proc. Ann. Electron Microscopy Soc. Amer. 39, 438-439 Claitor's Pub., Baton Rouge, LA.
14. Post, C.B. and Zimm, B.H. (1982) Biopolymers 21, 2123-2137.
15. Post, C.B. and Zimm, B.H. (1982) Biopolymers 21, 2139-2160.
16. Marx, K.A. and Reynolds, T.C. (1982) Proc. Natl. Acad. Sci. USA 79, 6484-6488.
17. Marx, K.A. and Ruben, G.C. (1983) Nuc. Acids Res. 11, 1839-1854.
18. Marx, K.A. and Reynolds, T.C. (1983) Biochim. Biophys. Acta. 741, 279-287.
19. Widom, J. and Baldwin, R.L. (1983) Biopolymers 22, 1595-1620.
20. Rhee, K.W. and Ware, B.R. (1983) J. Chem. Phys. 78, 3349-3353.
21. Thomas, T.J. and Bloomfield, V.A. (1983) Biopolymers 22, 1097-1107.
22. Benbasat, J.A. (1984) Biochemistry 23, 3609-3619.
23. Marx, K.A. and Ruben, G.C. (1984) J. Biomol. Struc. & Dyn. 1, 1109-1132.
24. Castleman, H., Hanau, L.H. and Erlanger, B.F. (1983) Nuc. Acids Res. 11, 8421-8429.
25. Castleman, H., Specthrie, L., Makowski, L. and Erlanger, B.F.(1984) J. Biomol. Struc. & Dyn. 2, 271-283.
26. Thomas, T.J. and Bloomfield, V.A. (1985) Biochemistry 24, 713-719.
27. Ruben, G.C. and Marx, K.A. (1983) Proc. Ann. Elec. Micros. Soc. Amer. 41, 714-715, San Francisco Press, S.F., CA.
28. Ruben, G.C. and Marx, K.A. (1984) Proc. Ann. Elec. Micros. Soc. Amer. 42, 210-211, San Francisco Press, S.F., CA.
29. Ruben, G.C. and Marx, K.A. (1984) Proc. Ann. Elec. Micros. Soc. Amer. 42, 212-213, San Francisco Press, S.F., CA.
30. Marx, K.A. and Ruben, G.C. (1985) in "The Molecular Basis of Cancer" 172A, 131-140, R. Rein, ed., Alan R. Liss, N.Y.
31. Ruben, G.C. and Marx, K.A. (1985) Proc. Ann. Elec. Micros. Soc. Amer. 43, 522-523, San Francisco Press, S.F., CA.
32. Marx, K.A. and Ruben, G.C. (1986) Biophys. J. 49, 53-55.
33. Ruben, G.C. and Marx, K.A. (1984) J. Elec. Micros. Tech. 1, 373-385.
34. Manning, G.S. (1978) Q. Rev. Biophys. 11, 179-246.

35. Marx, K.A., unpublished results.
36. Marx, K.A. and Ruben, G.C. (1986) J. Biomol. Struc. & Dyn. 4, 23-39.
37. Campbell, R.A. et. al.,eds. (1979) "Adv. in Polyamine Research" 1, Raven Press, New York.
38. Cohen, S.S. (1978) "Adv. in Polyamine Res." 1, 1-10.
39. Tabor, C.W. and Tabor, H. (1984) Ann. Rev. Biochem. 53, 749-790.
40. Cohen, S.S. and Greenberg, M.L. (1981) Proc. Natl. Acad. Sci. USA 78, 5470-5474.
41. Porschke, D. (1984) Biochemistry 23, 4821-4827.
42. Baase, W.A., Staskus, P.W. and Allison, S.A. (19 4) Biopolymers 23, 2835-2851.
43. LeBret, M. and Zimm, B.H. (1984) Biopolymers 23, 271-286.
44. LeBret, M. and Zimm, B.H. (1984) Biopolymers 23, 287-312.
45. Thomas, T.J. and Bloomfield, V.A. (1984) Biopolymers 23, 1295-1306.
46. Liquori, A.M., Costantino, L., Crescenzi, V., Elia, V., Giglio, E., Puliti, R., De Santis Savino, M. and Vitagliano, V. (1967) J. Mol. Biol. 24, 113-122.
47. Schellman, J.A. and Parthasarathy, N. (1984) J. Mol. Biol. 175, 313-329.
48. Aikens, D., Bunce, S., Onasch, F., Parker, R., Hurwitz, C. and Clemans, S. (1983) Biophysical Chemistry 17, 67-74.
49. Wilson, W.D. and Jones, R.L. (1981) Biopolymers 20, 141-154.
50. Pullman, B. (1985) in "The Molecular Basis of Cancer" 172A 55-69, R. Rein ed., Alan R. Liss, N.Y.
51. Lin, S.C., Thomas, J.C., Allison, S.A. and Schurr, J.M. (1981) Biopolymers 20, 209-230.
52. Thomas, T.J. and Bloomfield, V.A. (1985) Biopolymers 24, 725-729.
53. Krasnow, M.A. and Cozzarelli, N.R. (1984) Biochemistry 23, 3609-3619.
54. Morris, D.R. and Lockshon, D. (1981) Advs. in Polyamine Research 3, 299-307, Raven Press, N.Y.
55. Jorstad, C.M., Harada, J.J. and Morris, D.R. (1980) J. Bacteriology 141, 456-463.
56. Hingerty, B.E., Brown, R.S. and Klug, A. (1982) Biochim. Biophys. Acta. 697, 78-82.
57. Dante, R. and Niveleau, A. (1980) Eur. J. Biochem. 110, 539-544.
58. Fukuma, I. and Cohen, S.S. (1975) J. Virol. 16, 222-227.
59. Camerini-Otero, R.D., Sollner-Webb, B. and Felsenfeld, G. (1976) Cell 8, 333-347.
60. Ruben, G.C. and Marx, K.A. (1983) Proc. Ann. Elec. Micros. Soc. Amer. 41, 704-705, San Francisco Press, S.F., CA.
61. Manning, G.S. (1980) Biopolymers 19, 37-59.
62. Lerman, L.S. (1973) Cold Spring Harbor Symp. Quant. Biol. 38, 59-73, Cold Sring Harbor Lab., N.Y.
63. Maniatis, T., Venable, J.H. and Lerman, L.S. (1974) J. Mol Biol. 84, 37-64.
64. Evdokimov, Yu.M., Pyatigorskaya, T.L., Polyvtsev, O.F., Akimenko, N.M., Kadykov, V.A., Tsvankin, D.Ya. and

Varshavsky, Ya.M. (1976) Nuc. Acids Res. 3, 2353-2366.
65. Laemmli, U.K. (1975) Proc. Natl. Acad. Sci. USA 72, 4288-4292.
66. Shin, Y.A. and Eichhorn, G.L. (1984) Biopolymers 23, 325-335.
67. Shapiro, J.T., Leng, M. and Felsenfeld, G. (1969) Biochemistry 8, 3219-3232.
68. Haynes, M., Garnett, R.A. and Gratzer, W.B. (1970) Biochemistry 9, 4410-4416.
69. Carroll, D. (1972) Biochemistry 11, 421-426.
70. Fasman, G.D., Valenzuela, M.S. and Adler, A.J. (1971) Biochemistry 10, 3795-3801.
71. Sponar, J. and Fric, I. (1972) Biopolymers 11, 2317-2330.
72. Gottesfeld, J.M., Calvin, M., Cole, R.D., Igdaloff, D.M., Moses, V. and Vaughan, W. (1972) Biochemistry 11, 1422-1430.
73. Keller, D. and Bustamante, C. (1985) J. Chem. Phys. 84, 2961-2971.
74. Keller, D. and Bustamante, C. (1985) J. Chem. Phys. 84, 2972-2980.
75. Kim, M.H., Ulibarri, L., Keller, D., Maestre, M.F. and Bustamante, C.F. (1985) J. Chem. Phys. 84, 2981-2989.
76. Zehfus, M.J. and Johnson, W.C. Jr. (1981) Biopolymers 20, 1589-1603.
77. Zehfus, M.J. and Johnson, W.C. Jr. (1984) Biopolymers 23, 1269-1281.
78. Vengerov, Y.Y., Semenov, T.E., Streltsov, S.A., Makarov, V.L.,Khorlin, A.A. and Gursky, G.V. (1985) J. Mol. Biol. 184, 251-255.
79. Yengerov, Y.Y., Semenov, T.E., Streltsov, S.A., Makarov, V.L., Khorlin, A.A. and Gursky, G.V. (1985) FEBS LETTERS 180, 81-84.
80. Widom, J. and Baldwin, R.L. (1983) Biopolymers 22, 1621-1632.
81. Kapuscinski, J. and Darzynkiewicz, Z. (1984) Proc. Natl. Acad. Sci. USA 81, 7368-7372.
82. Kapuscinski, J., Darzynliewicz, Z., Traganos, F. and Melamed, M.R. (1981) Biochem. Pharmacol. 30, 321-240.
83. Klevan, L. and Wang, J.C. (1980) Biochemistry 19, 5229-5234.
84. Ohlendorf, D.H., Anderson, W.F., Fisher, R.G., Takeda, Y. and Matthews, B.W. (1982) Nature 298, 718-723.
85. Trifonov, E.N. and Sussman, J.L. (1980) Proc. Natl. Acad. Sci. USA 77, 3816-3820.
86. Stellwagen, N.C. (1983) Biochemistry 22, 6186-6193.
87. Wu, H.M. and Crothers, D.M (1984) Nature 308, 509-513.
88. Hagerman, P.J. (1985) Biochemistry 24, 7033-7037.
89. Singleton, C.K., Klysik, J. and Wells, R.D. (1983) Proc. Natl. Acad. Sci. USA 80, 2447-2451.
90. Thomas, T.J. and Bloomfield, V.A. (1985) Biopolymers 24, 2185-2194.
91. Osterberg, R., Persson, D. and Bjursell, G.(1984) J. Biomol. Struc. & Dyn. 2, 285-290.
92. Clarke, M.J., Jansen, B., Marx, K.A. and Kruger, R. (1986)

Inorg. Chim. Acta 124, 13-28.
93. Porschke, D. (1985) Biopolymers 24, 1981-1993.
94. Porschke, D. (1985) J. Biomolec. Struc. & Dyn., in press.
95. Manning, G.S. (1983) Biopolymers 22, 689-729.
96. Diekmann, S., Hillen, W., Morgeneyer, B., Wells, R.D. and Porschke, D. (1982) Biophys. Chem. 15, 263-270.
97. Porschke, D., these proceedings.
98. Griese, G.A., Serwer, P. and Horowitz, P.M. (1985) Biopolymers 24, 1635-1646.
99. Bloomfield, V.A., Wilson, R.W. and Rau, D.C. (1980) Biophys. Chem. 11, 339-343.
100. Oosawa, F. (1971) Polyelectrolytes, Marcel Dekker, N.Y.
101. Rau, D.C. and Bloomfield, V.A. (1979) Biopolymers 18, 2783-2805.
102. Rau, D.C., Lee, B. and Parsegian, V.A. (1984) Proc. Natl. Acad. Sci. USA 81, 2621-2625.
103. Earnshaw, W. and Harrison, S. (1977) Nature 268,598-602.
104. Hohn, T. and Katsura, I. (1977) Current Topics Microbiol. Immunol. 78, 69-110.
105. Klimenko, S.M., Tikchonenko, T.I. and Andreev, V.M. (1967) J. Mol. Biol. 23, 523-533.
106. Richards, K.E., Williams, R.C. and Calendar, R.J. (1973) J. Mol. Biol. 78, 255-259.
107. Furlong, D., Swift, H. and Roizman, B.J. (1972) J. Virol. 10, 1071-1074.
108. Haynes, M., Garrett, R. and Gratzer, W. (1970) Biochemistry 9, 4410-4416.
109. Haas, R., Murphy, R.F. and Cantor, C.R. (1982) J. Mol. Biol. 159, 71-92.
110. Widom, J. and Baldwin, R.L. (1983) J. Mol. Biol. 171, 419-437.
111. Hall, S.B. and Schellman, J.A. (1982) Biopolymers 21, 1991-2010.
112. Hall, S.B. and Schellman, J.A. (1982) Biopolymers 21, 2011-2031.
113. Virrankoski-Castrodeza, V. and Parish, J.H. (1980) Arch. Microbiol. 126, 277-283.
114. Harrison, S.C. (1983) J. Mol. Biol. 171, 577-580.
115. Hendrix, R.W. (1978) Proc. Natl. Acad. Sci. USA 75, 4779-4783.
116. Feiss, M. and Widner, W. (1982) Proc. Natl. Acad. Sci. USA 79, 3498-3502.
117. Lahav, N. and White, D.H. (1980) J. Mol. Evol. 16, 11-21.
118. Parthasarathy, R., Soriano-Garcia, M. and Cheda, G.B. (1976) Nature 260, 807-808.
119. Griffith, J.D. (1976) Proc. Natl. Acad. Sci. USA 73, 563-567.
120. Oakley, B.R. and Dodge, J.D. (1979) Chromosoma 70, 277-291.

NULL-DNA: A THEORY OF THE ELASTIC INSTABILITY AND SPONTANEOUS KINKING OF DNA WITH CHARGE-NEUTRALIZED PHOSPHATES

G.S. Manning
Department of Chemistry, Rutgers, The State University of New Jersey, New Brunswick, NJ 08903

INTRODUCTION

DNA aggregates when its phosphate charge is extensively neutralized by multivalent cations like polyamines or cobalt hexachloride.[1-3] The phenomenon is interesting as an example of <u>in vitro</u> DNA packaging. The form of the aggregated structure is striking. It is toroidal - a doughnut shape with a hole in the middle. The long axis of the double helix is circumferentially wound, like a stored garden hose, throughout the volume of the torus. Each torus may contain one or perhaps several high molecular-weight DNA molecules, but hundreds of 400 bp DNA fragments also pack into a curved torus structure. The curvature of the aggregated structure is all the more impressive when it is remembered that DNA stands out among polymers for its unusually high degree of stiffness. Under ordinary solution conditions the duplex axis tends, over distances long compared to other polymers, to be straight.

The variety of multivalent cations that may be used, ranging from histone H1 and polylysine to Mg^{2+} (in mixed solvents), suggests that both the aggregation and the toroidal shape of the aggregate are induced by efficient but nonspecific neutralization of the phosphate charge. A consistent explanation for the stability of the curved torus shape, therefore, is that the DNA duplex axis loses stability under charge-neutralization conditions. The stable form of the axis under charge-neutralization conditions possesses curvature. Curved DNA segments are then expected to pack into a torus.

It is difficult to understand torus stability without the hypothesis of intrinsic curvature. Of course, a self-aggregated long DNA duplex must contain regions in which it is bent back on itself. The side-by-side aggregation forces between segments can in principle compensate for the positive bending free energy. Thus the torus could be the compromise between favorable segmental parallelism and unfavorable curvature. A theory of this type[4] (the only serious attempt other than ours to explain the torus, as far as we are aware) predicts that DNA polymers over a few million molecular weight are <u>not</u> stable as toruses, in contrast to the routine observation of toruses for DNA of 20-30 million molecular weight.

But even qualitative reasoning appears to rule out a highly ordered circumferentially wound single DNA polymer as a plausible aggregate in the absence of intrinsic curvature. Many nucleated intermediate forms within the polymer would be hairpin structures, with a large (low-energy) loop and parallel aggregation of two duplex segments in the stem. These structures could not further aggregate into the form of a stored garden hose (torus) on the time scale of the experiment, even if the torus were the true equilbrium structure.

Still more difficult to explain is the packing of hundreds of 400 bp fragments into a torus, as observed,[5] if positive bending free energy intervenes. In this case a single long DNA polymer is not required to fold back on itself. Relatively short, intrinsically straight, fragments would pack into a "stack of logs" instead of a torus. But if the axis of each fragment were intrinsically curved, then the torus is immediately grasped as a stack of curved logs. No explanation other than ours for toroidal aggregation of relatively short fragments has yet been advanced.

We will also be able to see in subsequent discussion why very short 140 bp fragments do not pack into a torus.[5] Alternate theories must be able to account for the full range of observations as well. Whether through psychological bias or otherwise, we cannot discover a convincing reason for torus formation unless stable local curvature is invoked.

Therefore we coin a new name --null-DNA-- for DNA lacking most of its phosphate charge. Our expectation is that null-DNA, which shares a B-structure with ordinary DNA, is nevertheless a distinct form. Its most obvious characteristic is intrinsic axial curvature, but we will suggest as well that null-DNA is slightly contracted along its duplex axis. Moreover, localized defects in the regular duplex structure, perhaps open base pairs, which occur as transient high energy fluctuations in ordinary DNA with an average frequency of one in every 10 to 100 base pairs, are stablized in null-DNA.

Having accepted the existence of stable local curvature in null-DNA as strongly suggested by the observation of toruses, the challenge for the theoretician is to understand why the ordinarily straight duplex axis loses stability, and bends spontaneously, when the phosphate charge is efficiently neutralized. The classical theory of elastic instability of thin rods is ideally suited to serve as a framework for theoretical investigations of this question. We will see that the classical phenomenon of Eulerian buckling of the DNA rod is expected to occur when the phosphate charge is extensively abolished. On the level of classical theory, then, the packaging of null-DNA into a torus shape is probably a manifestation of Eulerian buckling of the axis (accompanied by aggregation).

As a compliment to the classical theory of elasticity, we have also fashioned a theory based on the molecular structure of DNA --a quasi-Eulerian theory of buckling. As a consequence of the molecular theory, we can offer an explanation of why ordinary DNA is such a stiff molecule in the first place. The conventional statement that DNA is stiff because of the stacking forces between base pairs is, in our view, an incomplete description. But perhaps the most important result of the molecular theory is the prediction of the existence of cooperative bending units of determinable average length and average angle of bend. From this result follows another, that null-DNA, buckled by phosphate neutralization, has a stable radius of curvature in the neighborhood of 16nm, a numerical prediction that turns out to agree with the observed dimensions of DNA toruses.

Both theories have been well-developed in a series of papers, and we refer to them for the detailed derivations.[1,6] Our intent here is to direct attention to the main ideas, and especially to emphasize the wide extent to which the theories agree with experiment. Indeed, we are sometimes asked how the theories can be tested. The answer is the same as for any theory. The predictions of a theory must be compared to the relevant experimental data, a confrontation that is a classical aspect of the scientific method. To an extent, our two theories have already been tested.

EULERIAN BUCKLING

Everyone should be familiar with the concept of buckling of a thin rod.[1] If a rod is subjected to a crushing load (a longitudinally - directed compressional force), its initial elastic response is to contract slightly. But as the load is increased, a critical threshold is encountered for which outward buckling is as efficient a response to the load as contraction. Under loads exceeding the critical value, the only stable form of the rod is bent.

Euler's formula is

$$L = 2\pi\sqrt{B/F} \tag{1}$$

where B is the Hooke's law bending constant for the rod, and F is the compressional force on it (crushing load). The meaning of the critical length L is that rods shorter than L merely contract under the load F, but rods longer than L are bent by F. To implement Euler's formula for DNA requires knowledge of B. This quantity is experimentally accessible, and values characteristic of DNA are available. In its application to DNA, the force F, however, is not as straightforward a concept as it is in the more familiar engineering context. When a column buckles under its own weight, it is clear that F is an externally applied mechanical compressive force (gravity pushing down and reacting with the base). The force F equivalent to the effect of phosphate neutralization, on the other hand, is not a directly applied crushing load like gravity. Instead, it is the equivalent annihilation of a tensile (stretching) force. The idea is simple, but experience suggests the appropriateness of detailed discussion.

Consider first a hypothetical diatomic molecule with bond length equal, say, to x_o. This stable structure we call the null structure. Imagine now that ionic charge of the same sign is loaded onto each atom, until the interatomic bond is significantly stretched to a length $x_1 > x_o$, thus providing us with a stable charged structure. What happens when the ionic charge of the charged structure is neutralized? The charged structure, with bond length x_1, is no longer stable. The stable structure is again the null structure. The bond contracts in length from x_1 to x_o. The effect is the same as the contraction that would be caused by application of a crushing load F exactly equivalent, but acting in the opposite direction, to the stretching ionic force in the charged structure.

A hypothetically bent triatomic molecule affords a more interesting example. Let the bent triatomic molecule be called the null structure. Imagine loading an amount of ionic charge on each of the end atoms sufficient to straighten the molecule by electrostatic repulsion of the ends. Now we have a stable linear triatomic molecule, ionized at the ends, which we call its charged structure. When the ionic charge is neutralized, the charged structure becomes unstable. The null structure emerges as the stable form, and the molecule bends. If we had started with the charged structure, that is, if we were unaware at the outset of the ionically neutral bent form, then we would have been "surprised" to find that charge neutralization can cause bending.

We may develop the example of the triatomic molecule a bit further. To achieve exactly the same effect, we need not actually neutralize the ionic charge of the charged structure. Instead, we can impose on the end atoms a crushing force F along their line of centers. The magnitude of F is chosen to equal the ionic stretching force, and hence has the effect of exactly nullifying it. The linear triatomic molecule (the charged structure) cannot support the crushing load F, and the molecule bends (reverts to the null structure).

We return to DNA. The null structure is what we have called null-DNA, a bent form with most of the phosphate charge neutralized by multivalent cations. The charged structure is ordinary DNA with its full complement of ionic charge. Consider first the ordinary charged structure. Phosphate-phosphate repulsion acts, among other ways, as a stretching force along the duplex axis. The ordinary DNA molecule is held by ionic repulsion in a state of tension, like the charged diatomic molecule and the charged linear triatomic molecule. Neutralization of the phosphate charge has the effect of imposition of a crushing load F, equal in magnitude to the electrostatic stretching force but oppositely directed. The stretching force is a polyelectrolyte force; it may be quite strong. Therefore, the crushing load may also be strong. The DNA charged structure (ordinary DNA) may not be able to sustain such a strong crushing load. It bends under the load and takes the shape of intrinsically curved null-DNA, as observed in the torus.

If we did not have the torus, and our interpretation of it, to guide our understanding, then we would have been faced with a problem. Although annihilation of the DNA phosphate charge undoubtedly has the effect of compressing the DNA structure along the duplex axis, how could we tell whether or not the DNA structure is strong enough to sustain the crushing without buckling? The answer is by reference to Euler's formula, eq. (1). The Hooke's law bending constant B for the ordinary charged structure of DNA is 1.3 - 4.0 x 10^{-28} J m, depending on ionic strength.[1] The crushing load F can be calculated from polyelectrolyte theory[1] as equal in magnitude to the polyelectrolyte tensile force. Over the same ionic-strength range, it works out to be 4.0 - 6.0 x 10^{-11} N. Substitution into eq. (1), with the conversion factor 3.4 x 10^{-10} m/bp, gives us L in the range 33-48 bp. Longer DNA segments cannot support the compression F and bend under it.

In the ordinary charged DNA structure the persistence length λ is a measure of the length of rod-like DNA segments. Internal segments of length about equal to λ behave like rods and, in particular, their natural form features a straight axis. The persistence length is in the range 94-290 bp, well above the critical length L (33-48 bp). <u>We conclude that the structure of a rodlike persistence-length segment of DNA cannot sustain the crushing load attendent on phosphate</u>

neutralization. Its straight axis becomes unstable, and it bends. Our Euler buckling calculation is thus consistent with the observation of the packaging of null-DNA into a torus shape.

Several further remarks may be helpful before passing to the molecular-level theory. The Euler calculation assumes that DNA bends uniformly with no change of length along its axis, like a macroscopic elastic rod. It shows that the axis of null-DNA is in a state of unstable equilibrium when straight. According to Euler's theory, when the applied load slightly exceeds the critical crushing load, the rod has lowest energy in the form of an arc of a sine wave of small amplitude. (The arc is short, so the curvature is always of the same sign). The interpretation for DNA is that a piece of null-DNA has lower energy when uniformly bent to a sine arc than when straight. We cannot conclude that this shape gives the lowest energy form of null-DNA, for DNA, unlike a slightly bent macroscopic rod, may be able to bend more easily in a nonuniform mode (for example, it may kink at the site of an open base pair). The Euler calculation shows that null-DNA cannot be locally straight; it does not predict its stable shape.

We may comment on the value of the constant 2π in Eq. 1. It actually defines only one of a family of Euler's formulas. The boundary conditions at the ends of the rod determine the particular value of the constant. The boundary condition appropriate for an internal segment of smoothly bent DNA is that of "clamped ends"; the ends of the segment are clamped by adjoining segments.[7]

Indeed, the Euler theory, as we have used it, can only be applied to internal segments. The crushing load on the segment can only be transmitted by adjoining segments. We have here a possible explanation of why aggregates of 140 bp fragments, in contrast to the 400 bp fragments discussed above, are not shaped like a torus.[5] The full polyelectrolyte crushing load may not be transmitted to the interior of such a short fragment because the ends encroach on the adjoining transmitting segments (in other words, there are no adjoining segments). We wonder whether explanations of the torus shape other than ours will be able to handle the full range of observations: an aggregate of 400 bp fragments is torus-shaped; an aggregate of 140 bp fragments is not.

Our final observation concerns the apparent identity of the forces that drive bending within a rodlike segment and the forces that drive aggregation of different segments. It is not surprising that the forces tending to crush the material within a segment should also serve as attractive side-to-side forces between segments. The degrees of neutralization of repulsive charge-charge interactions required to unmask the longitudinal forces that crush within a segment and the lateral forces that crush different segments together may be the same.

This consideration provides us with a closing argument that perhaps makes intrinsic local curvature of null-DNA trivially obvious. If environmental conditions favor aggregation, then, as indicated in the upper path of Fig. 1, there would be a tendency for a stiff local segment to break into two pieces and lie side by side. As an absurdly high cost of energy forbids this event, the segment does the next best thing (lower path of Fig. 1). It bends.

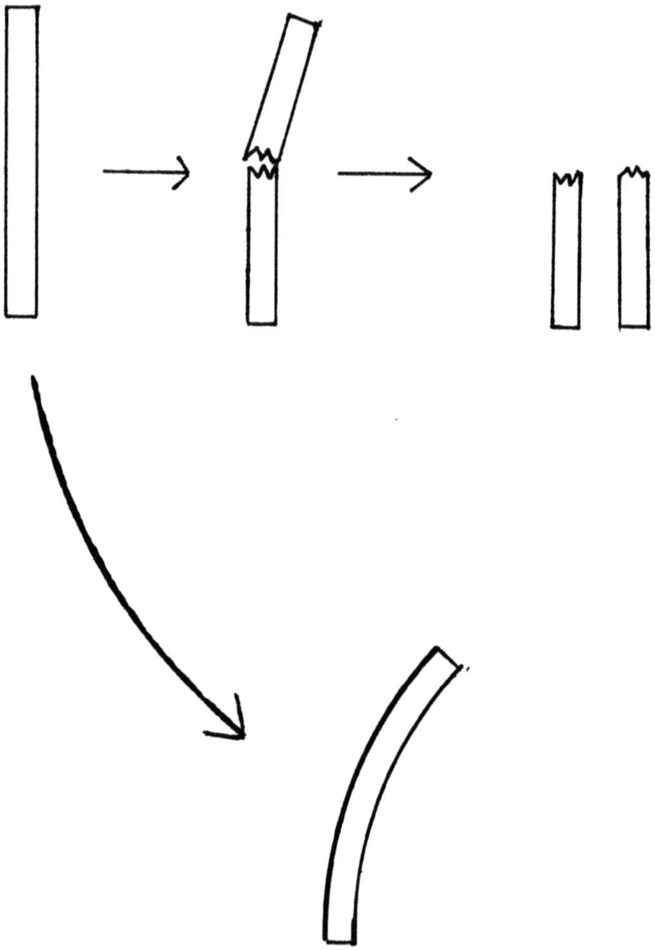

Fig. 1

THE KINK-CONTRACTION MODEL

In an attempt to explain the logarithmic salt dependence of the persistence length of DNA, we introduced in 1983 a model for DNA bending that postulates abrupt changes of direction at the interfaces between straight blocks of base pairs.[6] Two landmark papers in DNA crystallography,[8,9] published the following year, give additional motivation for consideration of such a kink model.

The double helix of nucleosomal core particle DNA does not follow a regular superhelical path as it wraps around the histone octamer, but "is bent fairly sharply or possibly kinked at several locations."[8] Indeed, to our eye, in looking at figs. 1 and 2 in ref. 8, the double helical axis is not badly represented by a nonplaner polygonal arc with average edge about 10bp and average kink angle about 30°-40°. "The departure from good stacking [in the kinks] is apparently spread over several base pairs."[8] The 7A resolution reported cannot distinguish between a series of concerted small-amplitude dislocations or more substantial disruption.

The structure of the DNA in a recognition complex with EcoRI endonuclease was determined to 3A and exhibits true kinks at the base pair level.[9] Four straight blocks of three base pairs each are separated by three kinks. The central kink is both torsional (25° unwinding) and bending (12° directional change of the double helical axes of the adjacent blocks). The two terminal kinks are primarily bending kinks, introducing angles of 23° between helical axes.

In both cases the DNA is interacting with protein. For our purpose, however, the point is that kinks are accessible structural features of DNA. It follows from first principles that kinks will occur as thermal fluctuations in free DNA, with probabilities that cannot be determined by strong opinion or incomplete energy maps.

We assume, then, that the ordered structure of DNA is interrupted with significant frequency by irregularities of transient thermal origin. The repair of these structural defects is further assumed to be slow on the time scale of bending motions. If we plausibly ascribe enhanced flexibility to these sites, we have the picture of a fluctuating kink angle at each of them. In summary, we postulate the dominant mode of DNA bending to be localized kinking at the sites of preexisting structural irregularities.

It is sometimes asserted that visualization of free DNA in the electron microscope has ruled out kinking as an important bending mechanism. EM representation of DNA, however, is insufficiently resolved to distinguish a kinked duplex axis from the smooth curve enveloping it (and the fate of a thermally-induced structural distortion in solution when subjected to the EM preparation procedures is unknown).

Our second postulate is that kinking motions are coupled to small contractions along the helical duplex axis, of the order of 1%. Motivation along two lines can be given. The persistence length of DNA is a measure of the Hooke's Law constant for local bending. It is observed to be a linear function of the logarithm of salt concentration in the range of low-to-moderate ionic strengths (see the discussion in the Appendix of ref. 6, which is not affected by subsequently published data that fail to decouple excluded-volume effects from local ones,[10] and which has in fact been further confirmed by light-scattering data from Harpst's laboratory[11]). We can account - and are otherwise unable to account - for this correlation, if we assume the accompaniment of kinking by lengthwise contraction.

The estimate of a 1% contraction is the result of a calculation to be discussed; it is not really part of the assumption. We mention it here only to forestall expressions of disbelieve; the contraction we need is very small, well within the range of step height variations observed by diffraction measurements. The contraction is indeed so small, that a reaction more sophisticated than the questioning of its existence is to wonder how it can affect an observable like the persistence length. Remember, however, that a 1% change in the very large polyelectrolyte field of DNA - determined by the number of phosphate groups per unit length along the axis - can be significant.

The second line of argument is purely deductive. It is reasonable to suppose that as the angle of bend is increased at the site of a base pair or between consecutive base pairs, space is cleared in the region of the angle's vertex, and water molecules enter the duplex to occupy the space. Consider two such bends and the intact DNA segment between them. The balance of stacking forces between base pairs has been perturbed at the ends of the segment; the attraction of internal base pairs for the base pairs at the ends of the segment is no longer exactly balanced by an equal attraction of base pairs external to the segment. The latter force has been diminished by substitution of water for base pairs at the ends. The segment contracts until the force balance is restored.

My opinion is that this argument is self-evident. I regret now having removed the corresponding discussion from a key reference[6] as the price required by an obstinate referee for his favorable recommendation. His perceived difficulties, despite a voluminous correspondence, remain a mystery to me. Perhaps the problem is in the smallness of the effect; in most contexts, but not the present one, I can agree that the effect need not be considered. Or perhaps some physical chemists do not think of the transmission of forces as I do. It is true that a stack of base pairs contains many structural degrees of freedom as potential absorbers of compressional forces applied at the ends of the stack. It is therefore true that forces applied at the ends of the stack are not necessarily transmitted with little attenuation throughout its length. Then the segment does not necessarily respond like a solid rod to applied forces, and our argument for a uniform contraction would break down. The answer to this argument is that its validity depends on the length of the stack. If applied to a segment of hundreds of base pairs, it would be correct, but it fails for the shorter stacks of interest to us. For we know that a DNA segment of tens of base pairs (ours will turn out to consist of about 40 bp) is much shorter than a persistence length and hence <u>is</u> rodlike. It may be that a stack of tens of base pairs contains many structural degrees of freedom but evidently not <u>sufficiently</u> many. The statement that DNA is locally stiff has a comfortably familiar ring. Its meaning may not have been as thoroughly assimilated.

The contraction postulate leads immediately to our <u>first prediction: the persistence length of DNA decreases as a linear function of log</u> (Na^+)

up to about 1M. This prediction follows from the proportionality between the persistence length and the work (free energy) required for local bending.[1,6] The stated dependence on salt is a well-known experimental and theoretical feature of polyelectrolyte work involving an increase in linear charge density, as postulated.[12] That the prediction agrees with persistence-length data regarded by most workers as the best currently available has already been mentioned.

To get numerical predictions, for example, to calculate the slope of the logarithmic salt dependence, requires another assumption. Stated roughly, it is that the lowest-energy path of the kinked and contracted axis should be as smooth as possible. This postulate will allow us to calculate the extent of contraction and how it is correlated with the kink angle. We try to make our meaning more precise.

The usual assumption in the field of polymers is that the wormlike chain is an adequate model of bending. Curvature is continuously distributed along the length of the polymer, and the length itself does not change. Justification of the continuous worm model lies in an intuition developed from geometry and the bending of macroscopic rods. We "feel" that the contour of a polymer should behave like a smooth curve in geometry, where bending means length-invariant directional change.

Our first two postulates suggest, however, that consensual intuition applied to the complex structure of DNA (which includes water and counterion structure in the grooves) may have to give way to more reliable guidelines. The path of the DNA axis may proceed as a series of variable kink angles. The length of the axis between two kinks may be contracted. The extent of contraction may depend on the angular amount at the adjacent kinks.

Our intuitive bias toward smooth length-invariant bending having been cast aside, we have stated another postulate that restores it. The low-energy shape of the duplex axis may not be smooth or of fixed length, but it should match these criteria as closely as possible. We proceed with our construction.

Consider a particular configuration of smoothly bent DNA with no structural irregularities. Its length does not depend on the curved shape of the duplex axis. Over sufficiently short lengths, the path of the axis is a circular arc. Our first two postulates imply that this DNA is a high-energy form. Its structure may be accessible, but DNA does not bend this way, at least not usually (Boltzmann factor argument). Smoothly bent, ideally structured DNA is not stable in solution.

What is the fate of this smooth but unstable configuration? Consider a sequence of sites along the chain. Any two nearest members of the sequence are separated (on the average) by N base pairs. Choose N to be the average number of regularly structured base pairs intervening between two nearest structural irregularities in DNA as it actually exists in solution. The sites, however, are regularly structured in our unstable smooth DNA configuration. They (or some other translationally equivalent set) pass to the "irregular state" as the DNA polymer slips into its stable form. The curvature all "drains" into these sites of enhanced flexibility. Kink angles form there, while the intervening stacks of N "regular" base pairs straighten and contract to compensate the force imbalance at their ends.

The constraint imposed on this spontaneous process is that the sites that are becoming irregular remain spatially fixed (think of them as fixed points on the smooth original curve of the duplex axis). We impose this constraint in order to preserve insofar as possible the directional variation of the smoothly curved but unstable form. From the mathematical viewpoint, the stable form would be described as merely a polygonal-arc approximation to a smooth curve. But the physical situation is important: there are structural irregularities at the vertices of the polygonal arc, and each polygonal edge is the axis of a stack of N regular base pairs that are compressed relative to their state along the smooth axis (because the polygonal edge is a chord of the smoothly curved axis, hence shorter).

The geometric relation between the kink angle α in the stable axis and the curvature R^{-1} of the unstable smooth curve is $\alpha = s_o R^{-1}$, where s_o is the invariant length 0.34N (nm) of N smoothly bent base pairs.[1,6] The free energy of the stable kinked form is proportional to α^2 (the contraction free energy is correlated to α^2 as well as the localized free energy stored in the neighborhood of the angle itself, since there is no contraction when there is no angle). Hence the free energy is proportional to R^{-2}, and Hooke's Law for smooth bending (see Eq. 5 below) is applicable to the kinked form. The Hooke's Law bending constant for the stable kinked form is not equal, however, to that of the smoothly bent high-energy form. It is less. We associate the smooth axis with the lower energy of the kinked one and say that the resulting abstract entity is <u>equivalent</u> to the actual kinked and contracted DNA.

The extent of contraction Δs is determined[1,6] by the requirement that the equivalent smooth curve be of fixed length, $\Delta s/s_o = (1/24)\alpha^2$. The maximum kink angle we will encounter in subsequent discussion is less than 1 rad, which implies a relative contraction less than 5%, or a change in step height between base pairs less than 0.02 nm. The corresponding change in linear <u>charge</u> density, however, has significant consequences (polyelectrolyte effect).

When DNA bends, it must contract along its length. The contraction, hence the bending, is resisted by phosphate repulsion. We can calculate the ionic contribution to the bending free energy from polyelectrolyte theory.[1,6] It is proportional to the contraction Δs, hence to α^2, hence to the square curvature of the equivalent smooth axis. Thus we obtain a polyelectrolyte contribution to the Hooke's Law bending constant for the equivalent smoothly bent polymer.

The persistence length λ of the equivalent smooth (wormlike) polymer is proportional to the Hooke's Law bending constant. From the foregoing we therefore have a polyelectrolyte contribution to λ. As a consequence of polyelectrolyte theory, it turns out that the polyelectrolyte contribution to λ consists of the sum of two terms, one independent of salt concentration, the other proportional to its logarithm. When the logarithm is common, and the units of λ are nm, the polyelectrolyte theory gives the expression $-0.015N^2$ for the slope of the linear correlation of λ with log [salt]. The quantity N is the same as before, the average number of regularly structured base pairs intervening between two kink sites.

Although we cannot predict the value of the slope, because N is not theoretically calculated, we can predict the density of kink sites by comparing $-0.015\ N^2$ with the measured slope 21.8 nm. We get N = 38 bp. Thus we have our second prediction: sites of structural irregularities (kink sites) occur with an average spacing along the duplex of about 38 base pairs.

In its recognition complex with EcoRI endonuclease, kink sites occur every three base pairs.[9] Our estimate of a 38 bp spacing in free DNA is therefore not unreasonable. In our visualization of the figures provided with the report of nucleosome core particle structure,[8] we see irregularities in the direction of the DNA about every ten base pairs. Richmond et al.[8] refer to four positions of "tightest DNA bending," which implies an average spacing of 29 bp, a figure similar to our estimate for free DNA.

Hydrogen exchange data for duplex polynucleotides free in solution suggest that one of every 20 bp in poly(rA)·poly(rU) and one of every 100 bp in poly(rI)·poly(rC) are in an "open" state.[13,14] Their hydrogen bonds are disrupted and perhaps one member of the pair is significantly unwound out toward the solvent. We will refer to such a site as an MKE open base, after the authors of the report.[13] The density of MKE sites is similar to our estimate of one per 38 bp for structural irregularities in free DNA. Moreover, the MKE open base pairs satisfy our requirement for a kink site. They are transiently formed as thermally induced fluctuations, but the time scale is slow, so that we may think of them as fixed sites of enhanced flexibility. We tentatively assign our "sites of structural irregularity" to MKE open bases but emphasize that much of the subsequent development does not depend on this assignment. We will make explicit statements when it does.

Porschke[3] has added free DNA to a solution of DNA previously stabilized as toruses. The added DNA depletes the reservoir of free multivalent cation. Multivalent cations dissociate from the toruses, which consequently unfold. The kinetics of unfolding can be followed. Proschke concludes that DNA unfolds from the torus in cooperative units of about 60 bp.

Our second prediction has suggested a bending unit of about 40 bp. Unfolding from the torus would proceed as these units straighten and disrupt the ordered torus structure (actualy the unit would have to be a little longer than 40 bp, since the kink angles at either end of the 40 bp segment would each have to have two "arms"). Our second prediction appears to be confirmed by the unfolding rates observed by Porschke.

We next pursue the problem of intrinsic, or natural, curvature. To avoid confusion, we explain the distinction between the natural path of the contour of a polymer and the myriad of shapes it can take in response to thermal energy fluctuations. A polymer that is naturally straight (like the duplex axis of ordinary DNA) is in fact bent (or kinked) in solution by thermal forces. The point is that work, of thermal origin or not, is required to bend it. Moreover, on a sufficiently short length scale, the polymer tends to preserve its initial direction even in the face of fluctuating thermal energy. But a naturally bent or kinked polymer requires energy to straighten. Directional change of the axis is part of its static structure and random thermal forces cannot completely eliminate the bias.

We state our third prediction: null-DNA is naturally kinked. This statement is the analog for the molecular model of our earlier conclusion stemming from application of Euler's buckling theory to DNA viewed as a smoothly bending macroscopic rod. We mentioned above that polyelectrolyte theory produces an expression with two terms for the polyelectrolyte effect on the persistence length λ. The term proportional to log [salt] has been discussed. The other term does not depend on salt concentration. Its calculated value is 140 nm. There is also a part of the persistence length, which we call λ_o, that originates from restoring forces other than the phosphate-phosphate polyelectrolyte repulsion. In all, the sum $\lambda_o + 140$ gives the salt-independent term in the theoretical expression for λ. The corresponding measured term equals 30 nm. Thus $\lambda_o = -110$ nm. The nonpolyelectrolyte part of the persistence length of ordinary DNA is negative. The positive overall (observable) value of λ for ordinary DNA reflects the dominance of polyelectrolyte restoring forces over the net resultant of all other effects.

Suppose now that the phosphate valence is reduced to zero. In practice, this situation can be approached by titration with multivalent cations (but not by increasing the Na^+ concentration).[12] For example, a trivalent cation can neutralize up to 92% of the phosphate charge (while Na^+ neutralizes only 76% regardless of concentration). In the ideal case of complete annihilation of the phosphate change, a state of DNA that we have referred to as null-DNA, the polyelectrolyte contribution to the persistence length vanishes, and $\lambda = \lambda_o$. But $\lambda_o = -110$ nm. Null-DNA has a negative persistence length, and the question of physical interpretation is immediately raised.

Its negative persistence length does not imply the absurdity that null-DNA has "turned inside out." The meaning is very simple and appealing. We have calculated the persistence length from the formula

$$\lambda = B/kT \qquad (2)$$

where B is the bending constant in Hooke's Law

$$\Delta G = (1/2)BLR^{-2} \qquad (3)$$

ΔG is the free energy of bending an (equivalent) segment of length L to a radius of curvature R. In other words, we have determined the bending energy ΔG by theoretical means, computed B from Eq. 3, and then calculated the persistence length λ from Eq. 2. A negative value of λ just means that B is negative, hence that ΔG for bending is negative. The bent segment has lower energy than when it is straight. The segment is stable when bent, unstable when straight. The segment is naturally, or intrinsically, curved.

The effect is strong. Ordinary DNA (with an intrinsically straight axis) has earned its reputation as exceptionally stiff on the basis of a persistence length in the range 30-80 nm, depending on salt concentration. The value -110 nm for null-DNA can only mean that its natural curvature (in the form of kink angles) effectively resists thermal disruption in solution.

There is a less direct, but fully equivalent and possibly more elegant, route to the same conclusions. It bypasses use of the abstract concept of negative persistence length. Instead, we derive the following formula for the free energy of bending the equivalent smoothly bent DNA segment of invariant length L_o to curvature R^{-1}:

$$\Delta G = (1/2)[B-(1/12)Fs_o^2]L_oR^{-2} \qquad (4)$$

The formula has the form of Hooke's Law but with an effective bending constant given by the expression in brackets, where B is the bending constant before application of a contraction force F directed along the duplex axis. The length s_o has been defined above as the length of the average segment of regularly structured base pairs intervening between two structural irregularities. When F exceeds the critical value $12\ B/s_o^2$, the energy of bending in Eq. 4 becomes negative. This expression for the critical value of F can be thrown into the form of Euler's law for buckling, Eq. 1; s_o replaces L and the numerical factors differ (but they are of the same order of magnitude). Therefore we have referred to the onset of negative ΔG in Eq. 4 as quasi - Eulerian buckling.

The force F can originate as the reaction to annihilation of the phosphate charge, just as in the Section entitled "Eulerian Buckling." Complete phosphate neutralization yields forces F sufficiently larger than critical to stabilize a kinked axis against thermal buffeting. A detailed free energy analysis in the general case of incomplete charge neutralization, with variable background Na^+ concentration and variable valence of the titrating cation, has been presented.[1]

If the energy of bending is negative, it would appear from Eq. 4 that the curvature R^{-1} spontaneously becomes greater and greater, finally corresponding to radius of curvature R = 0. But Eq. 4, and Hooke's Law generally, loses its validity long before this physically unreal limit is reached. The process that actually happens is a spontaneous increase of curvature from zero to a range of curvatures in which steric hindrance (or other short-range forces) increasingly resist further bending. The complete curve of energy vs curvature follows Eq. 4 for small R^{-1}, dropping from zero to negative values if $F > F_{critical}$, flattens in this range, then turns up, thus creating a potential energy well and a stable maximum value for the curvature. We have determined this value, our fourth predciction, in three independent ways, with similar results: the natural radius of curvature of null-DNA is about 12-19 nm, corresponding to kink angles of about 40°-60° between straight segments about 38 bp long.

Before indicating how we have arrived at this result, we point out its agreement with the observed dimensions of toruses. We should compare our value with the inner radius of the torus, because the predicted value is a maximum radius. The value of inner radii observed by Chattoraj et al.[15] are in the range 15-20 nm. Marx and Rubin[16] have observed a range of values with an abrupt lower cut-off at 16 nm.

The prediction is the common result of three independent lines of reasoning:

Method 1. There is a standard formula from which one may calculate the root-mean-square bending angle of the equivalent smoothly bent axis from its persistence length. The root-mean-square kink angle α_{rms} follows from a geometric identity. Thus we have computed α_{rms} as a function of $[Na^+]$ from measured values of the persistence length.[6] The values of α_{rms} increase from 21° at $[Na^+]$ = 0.001M to 37° at 1.0M, then level off at 37°-38° up to 4.0M. A consistent interpretation is that the rms kink angle relaxes to larger values with increased shielding of the phosphate repulsions, until contact forces prevent increase past 38°. We therefore take the natural kink angle in null-DNA as 38°. From the geometric relation $R = s_o/\alpha$ between the kinked-contracted axis and the equivalent smoothly bent axis, where R is the radius of curvature of the latter and s_o the length of a 38 bp segment (second prediction), we find R = 19 nm for the natural radius of curvature of null-DNA. (The common numerical value 38 for α in degrees and s_o in base pairs is a coincidence).

Method 2. We assume explicitly that the structural defect at the kink site is an MKE open base, and that the structural limitation on the magnitude of the kink angle involves the extent to which the base has unwound out into the solvent. The largest possible kink angle corresponds to complete unwinding of the base. From data on the temperature-dependence of the persistence length, combined with polyelectrolyte theory, we can estimate[6] the contribution to the entropy of kinking from nonpolyelectrolyte sources such as the increased conformational freedom at the kink site: $\Delta S° = 1.4 \times 10^4 R^{-2}$ (JK^{-1} mol^{-1}) where the radius of curvature R has nm units. An independent estimate of the maximum value of $\Delta S°$, assumed to correspond to complete unwinding of a base, is obtained from melting temperature data: ΔS^o_{max} = 46 (JK^{-1} mol^{-1}). Setting ΔS^o equal to ΔS^o_{max} when the curvature R^{-1} is maximal, hence equal to the natural curvature of null-DNA, we get the value 17nm for the natural radius of curvature R of null-DNA.

In Method 2, as described here, we have neglected a presumably small entropy generated by contraction of the segments between kinks. This assumption is not used in Ref. 6, where we combined Methods 1 and 2 to establish the result that the natural radius of curvature of null-DNA is greater than 17nm but less than 19-20 nm.

Method 3: In this method we assume that the rate mechanism of the helix-coil transition includes a preequilibrium step in which the MKE open base (or, more generally, some other structural irregularity) at a kink site serves as a nucleus. The kink angle must be sufficiently strained before large-scale melting can proceed. Melting is preceded by a preequilibrated maximization of the kink angle. The contact forces that prevent larger kink angles and preserve the large-scale structure in the temperature range of duplex stability occasionally give way under stress at the transition temperature, and the duplex secondary structure collapses.

The preequilibrium step is salt-dependent because segmental contraction accompanies kinking and generates a polyelectrolyte effect. We assume that a block of 2N base pairs melts cooperatively and independently (unwinding begins at an MKE open base and proceeds outward on both sides until interrupted by the flexibility of the two nearest MKE sites). Our second prediction tells us that N = 38. We then find, in agreement with experiment, that the logarithm of the transition rate constant is a linear function of the logarithm of salt concentration. With the maximum value α_{max} of the kink angle as the sole parameter, we can match the experimental slope of about 0.75 by setting α_{max} equal to 62°, corresponding to radius of curvature 12 nm.

Details of the calculation and references to the experimental literature may be found in Refs. 6 and 17. The result in Ref. 6 is roughly similar to the one given here; the origin of the difference lies in different assumptions about the length of the melting unit. We prefer our present interpretation. The method does not depend on the assumption of an MKE base at the kink but only on the idea that kinking involves some structural disruption.

When a base pair passes from the closed to the open state, the enthalpy increases by about 5 kcal mol^{-1}.[13] Therefore at higher temperatures (but still in the range of duplex stability) there are more MKE open bases. Since the opening of a base pair involves at least some unwinding, we are led to expect a decrease of the duplex winding angle (angular degree/bp) as the temperature increases. The latter effect is indeed well known,[6] but has always been interpreted as a uniform slight "loosening" of the duplex structure. That is, the measured value of about -0.01 (angular degrees/bp/degree Kelvin) for $d\Psi/dT$, where Ψ is the duplex winding angle, has been thought to reflect an unwinding of each base pair by about this amount (with the understanding of probable but relatively small base composition and sequence effects). Our point, however, is that $d\Psi/dT$ is more likely to represent the net effect of large-amplitude unwinding of relatively few base pairs (the open ones). We state and then outline the derivation[6] of our fifth prediction, which is in essentially exact agreement with measurement: the temperature coefficient of the duplex winding angle equals -0.01.

In our view the winding angle Ψ is an average of a uniform winding angle among N-1 intact base pairs and the winding angle characteristic of the single open pair among a total of N. The temperature derivative of this average angle therefore involves dN/dT, where N is the number of intact base pairs between two open pairs. Since 1/N is the equilibrium constant for the breathing reaction, intact bp ---> open bp, we can evaluate dN/dT from the measured enthalpy of this reaction. Values of other quantities appearing in the expression for $d\Psi/dT$ are 36° for the winding of the open base pair (see Ref. 6 for explanation), and 38 bp for N (second prediction). We then get $d\Psi/dT = -0.014$.

We stress again that the expectation of strongly nonuniform duplex winding follows from the hydrogen-exchange measurements[13] and is independent of theoretical considerations. The remarkably accurate fifth prediction is only frosting on the cake.

Bending energies for DNA are known to be small. A kink, on the other hand, may involve some unstacking of the base pairs and hence high energies. We are frequently asked, then, how a kink model of DNA bending can be a realistic description. The answer is that we have not proposed a kink model but a kink-contraction model. Bending is envisaged as a complex cooperative molecular event; a structurally regular stack of base pairs contracts slightly as kink angles form at both ends.

To illustrate the energetics of bending according to our model, we have constructed a table[6]:

i	ΔH_i	$T\Delta S_i$	ΔG_i
el	0.1	-7.0	7.1
k	9.2	6.3	2.9
c	-7.9	-0.4	-7.5
total	1.4	-1.1	2.5

All the values in the table refer to 25°C and 0.1 M salt. The units are kJ mol^{-1}. The changes in thermodynamic functions occur when the duplex axis of a 76 bp segment internal to an ordinary DNA polymer bends from zero curvature to its root-mean-square curvature $(27\ nm)^{-1}$ prevailing under the stated conditions.

The electrostatic (el) contributions are from polyelectrolyte repulsion in the contracted segments. The kinking contributions (k) are generated by molecular events localized to the immediate neighborhood of the site of kinking. It has been assumed explicitly that the kink site is at a preexisting MKE open base. The contribution from longitudinal contraction (c) is from the nonpolyelectrolyte energetics of contraction. Finally, the total, or overall, values are the sum of the three preceding entries in each column and represent the observable energetics. The value 2.5 kJ mol^{-1} for ΔG_{total} equals RT.

The overall values tell us little of the molecular nature of DNA bending. For qualitative purposes they may as well be replaced by a row of zeros. In each case - enthalpy, entropy, and free energy - they mask large compensating effects of different molecular events. Note, however, that the overall free energy (elastic work) of bending is positive, in agreement with the ordinary statement of Hooke's law. The phosphate groups are in their ordinary charged state in 0.1 M $[Na^+]$.

The kinking enthalpy ΔH_k equals 9.2 kJ mol^{-1}, almost an order of magnitude greater than the observed total. This contribution arises from unstacking of a base at the kink site and indeed does motivate the question posed at the beginning of this section. But improved stacking in the contracted segments is represented by -7.9 kJ mol^{-1}, compensating for unstacking. The effect of base pair stacking is dual: unstacking at the kink site makes DNA hard to bend; but better stacking in the segment between kinks (held in a stretched state by the phosphates, hence held away from optimal stacking) almost fully compensates.

An important contribution to the work of bending from increased entropic flexibility at the kink, -6.3 kJ mol^{-1}, helps make DNA easier to bend.

What happens if the phosphate charge is neutralized, and the polyelectrolyte effect disappears? The compensating enthalpic effects of base unstacking at the kinks and improved stacking in the contracted segments are unaffected. The large polyelectrolyte entropy contribution to the bending work, 7.0 kJ mol^{-1}, is wiped out. The only sizable effect left is the increased entropy from flexibility at the kink. The null-DNA segment spontaneously bends, driven by increased flexibility at the kink, despite base unstacking there.

The table also tells us why DNA under ordinary ionic conditions is stiff. The enthalpy of base unstacking at the kink sites (9.2 kJ) and the increase of polyelectrolyte repulsion in the contracted segments (7.1 kJ) act together as a strong resistance to bending. Compensating effects bring the work of bending down to observed levels.

Our molecular model of DNA bending[6] has stimulated Professor Schurr to a characteristic defense of the status quo.[17] His method begins with a misrepresentation of our analysis to create the false appearance of conflict with experimental data. He proceeds to "correct" the analysis but actually produces one identical to ours. He shows from his corrected treatment, which is the same as ours, that our model leads to the prediction of a root-mean-square kink angle decreasing with increasing temperature. Schurr finds this behavior incredible, even if no data are available to check it. He then administers the coup de grace, an exclamation point.

It is unfortunate that a reply to Schurr's ill-considered outburst has become necessary. He has not succeeded in formulating either a clear or correct criticism of our work, but he has misled many people. Word has gotten around of our "failure" to include the large observed enthalpy of breathing into our "total enthalpy" and, with a large leap, the consequent irreconcilability of our model with the real bending behavior of DNA.

We start with discussion of Schurr's only valid point (which, however, does not imply faultiness of either our model or its analysis). In our notation, the persistence length λ is proportional to N/α_{rms}^2, where N is the number of intact base pairs between two open ones, and α_{rms} is the root-mean-square kink angle at the site of an open pair. The temperature dependence of λ has been observed to be small (in a restricted range of ionic strengths). The temperature dependence of N, that is, the temperature dependence of the equilibrium constant for the closed bp ---> open bp reaction, has been observed to be large, with N decreasing as T increases. Therefore, the temperature dependence of α_{rms} must be large (but we will subsequently quantitate this adjective) and compensating; α_{rms} must also decrease as T increases. This conclusion follows both from our analysis (when unpolluted by Schurr's misrepresentation) and from Schurr's "correction" of the misrepresented analysis (which returns him to our original one).

No data are available to check whether or not α_{rms} actually does decrease with increasing T. Our model proposes a complex coupled event when DNA bends. Average kink angles at the sites of open base pairs at the ends of a stack of intact base pairs result in an average contraction of the stack. If the temperature is raised, so that the hydrogen bonds of another base pair are disrupted, and the new open pair partially unwinds, forcing its way outward from the interior of the contracted stack, it is entirely possible (in the absence of information to the contrary) for the resulting lowest-energy structure to feature less contraction, hence smaller average kink angles. To accommodate the new open pair, the old structure may have to relax.

Or here is another way to look at it. In our model the contraction of the stack of intact base pairs is less when the kink angle is less. A decreased average kink angle at higher temperatures means that the contraction is less at higher temperatures. Would Schurr's intuition be satisfied if we noted that an increased temperature results in less contraction?

We can easily estimate the magnitude of the effect from the data given in Ref. 6. With the assumption of an invariant mechanism of bending in the range 10°C to 40°C, α_{rms} decreases from 34° at 10°C to 21° at 40°C, both values referring to 0.1M salt. Schurr may find this behavior astonishing. We do not.

We have just mentioned an assumption inherent in this whole discussion, that the kink-contraction mechanism of bending is dominant at temperatures significantly different from 25°C. Perhaps at 40°C, or at 10°C, there is an exchange of stabilities between a kinked-contracted structure and a more smoothly bent one.

If the considerations were exclusively of a scientific nature, we could stop here. But the only aspect of Schurr's communication that people seem to have noticed is his insistence that we have incorrectly left the breathing enthalpy out of our analysis. To begin an answer, we first must make more precise an ambiguous definition used by Schurr. What Schurr means by "the enthalpy to form new kinks," is the equilibrium enthalpy of the closed bp ---> open bp reaction.

Our enthalpy change called ΔH is the same as the one measured by Gray and Hearst.[6] It is the enthalpy of the isothernmal bending process. The final state is at some temperature T and some nonzero value of the kink angle (and relative contraction). The initial state is at the same T but zero kink angle and relative contraction. No one has suggested that the closed bp ---> open bp equilibrium is significantly shifted as the result of the low-energy bend. The number of open base pairs in the final and initial states is therefore assumed, as part of the model, to differ only negligibly. A preexisting open base pair is structurally compatible with a range of kink angles, starting from zero, and bending is localized to the sites of each preexisting open pair. Our ΔH correctly does not include a contribution from the enthalpy of the closed ---> open equilibrium.

Schurr, in adding 5 kcal/mol to our ΔH (actually he doesn't, but he says he does), is adding the enthalpy of opening one mole of base pairs. He is saying that every time one of our segments bends at the site of a preexisting open base pair, another pair in the same segment opens. Assumption of this behavior cannot be motivated by any experimental or theoretical facts of which we are aware.

Schurr's assertion that our ΔH needs correction is thus refuted. His misrepresentation of our theory involves our use of ΔH and is a separate point. He makes the incomprehensible claim that ΔH governs the temperature dependence of α_{rms} (this is the meaning of his eq. 4a). He must know that ΔH governs instead the temperature dependence of the combined ratio N/α^2_{rms}, that is, of the persistence length λ. But his acknowledgment of this fact, would not have allowed him to announce a conflict of our analysis with the observed temperature dependence of λ. There is no conflict. Our numerical value of ΔH was obtained from the observed temperature dependence of λ.

Schurr does not know how to stop, and more misinformation may be gleaned from his paper. For example, he finds "extremely peculiar" our notion of a compressive force transmitted over the "colossal" distance of 38 bp. But 38 bp is much less than a persistence length, and segments up to a persistence length are rodlike, and rods do transmit forces throughout their lengths. We draw the line here.

References

1. G.S. Manning, Cell Biophys. 7, 57 (1985), and references therein.
2. J.A. Benbasat, Biochemistry 23, 3609 (1984).
3. D. Porschke, Biochemistry 23, 4821 (1984).
4. A. Yu. Grosberg, Biofizica 24, 32 (1979).
5. J. Widom and R.L. Baldwin, J. Mol. Biol. 144, 431 (1980).
6. G.S. Manning, Biopolymers 22, 689 (1983).
7. G.S. Manning, Biopolymers 20, 1261 (1981).
8. T.J. Richmond, J.T. Finch, B. Rushton, D. Rhodes, and A. Klug, Nature 311, 532 (1984).
9. C.A. Frederick, J. Grable, M. Melia, C. Samudzi, L. Jen-Jacobson, B. Wang, P. Greene, H.W. Boyer, and J.M. Rosenberg, Nature 309, 327 (1984).
10. G. Maret and G. Weill, Biopolymers 22, 2727 (1982).
11. J.A. Harpst and E.S. Sobel, Biophys. J. 41, 288a (1983).
12. G.S. Manning, Quart. Revs. Biophys. 11, 179 (1978).
13. C. Mandal, N.R. Kallenbach, and S.W. Englander, J. Mol. Biol. 135, 391 (1979).
14. R.S. Preisler, C. Mandal, S.W. Englander, N.R. Kallenbach, J. Frazier, H.T. Miles, and F.B. Howard, Biopolymers 23, 2099 (1984).
15. D.K. Chattoraj, L.C. Gosule, and J.A. Schellman, J. Mol. Biol. 121, 327 (1978).
16. K.A. Marx and G.C. Ruben, in The Molecular Basis of Cancer, R.R. Rein, ed., Liss, New York, 1985.
17. G.S. Manning, Biopolymers 19, 37 (1980).
18. J.M. Schurr, Biopolymers 23, 191 (1984).

THE DYNAMIC STRUCTURAL FLUCTUATIONS OF NUCLEIC ACID-DRUG COMPLEXES.

J. Ramstein[1], J. Markovits[2], B.P. Roques[3] and J.B. Le Pecq[2]

[1]Centre de Biophysique Moléculaire du CNRS, 45000 Orléans, [2]Laboratoire de Physicochimie Macromoléculaire (UA 147 CNRS, U 140 INSERM) Institut Gustave Roussy, 94805 Villejuif Cédex (France), [3]Département de Chimie Organique (UA 613 CNRS, U 266 INSERM) UER des Sciences Pharmaceutiques, Université Paris V, Paris (France).

INTRODUCTION

Different modes of structural fluctuations of the DNA double helix have been characterized. Several of them will be reviewed in this course. Measurements of hydrogen exchange kinetics have been extremely useful to demonstrate and study the structural fluctuations associated with the local melting of hydrogen bonds between base pairs, often referred as the breathing reaction. The studies of DNA breathing have recently been reviewed in detail by Englander and Kallenbach (1983). Most of the data have been obtained by tritium exchange. More recently hydrogen exchange in nucleic acids have been studied by a new approach taking advantage of the modification of UV absorption of the bases upon deuteration (Mandal et al., 1979). Schreir and Baldwin (1976) were the first to show the potential usefulness of hydrogen exchange measurements for the study of ligand macromolecule interactions. They were able to deduce in such measurements very small dissociation rate constants. Mandal et al. (1980) measured by absorption spectroscopy the hydrogen deuterium exchange of ethidium bromide in interaction with DNA.

Recently, we have prepared and studied a new class of DNA ligands, the poly functionnal intercalators (Le Pecq et al., 1975 ; Barbet et al., 1976 ; Gaugain et al., 1978a,b ; Capelle et al., 1979 ; Roques et al., 1979 ; Pelaprat et al., 1980 ; Reinhardt et al., 1982 ; Esnault et al., 1984 ; Laugaa et al., 1985).

Because several of these molecules elicit an extremely high affinity for DNA, associated with useful fluorescence properties, they represent a new class of fluorescence probes for the study of nucleic acid structure. In addition several of these molecules have strong antitumor properties. (Roques et al., 1979 ; Pelaprat et al., 1980 ; Esnault et al., 1984).

Ethidium dimer (EthDi) and acridine-ethidium dimer (AcEthDi) are molecules of special interest (Figure 1). Their fluorescence is strongly enhanced upon binding to DNA. The dissociation rates of the DNA complexes are small. Their fluorescence quantum yields are significantly increased upon deuteration. Therefore, the kinetics of the hydrogen-deuterium exchange of the amino group of these molecules can be very easily and accurately measured using spectrofluorometry. These observations prompted us to study the hydrogen-deuterium exchange of these molecules while they are in interaction with DNA (Markovits et al., 1983, 1985). These studies demonstrated that, in the DNA complex, the hydrogen exchange rate of the intercalating dimers was limited by dynamic structural fluctuations, the rate of which was of the same order of magnitude than the rate of the DNA

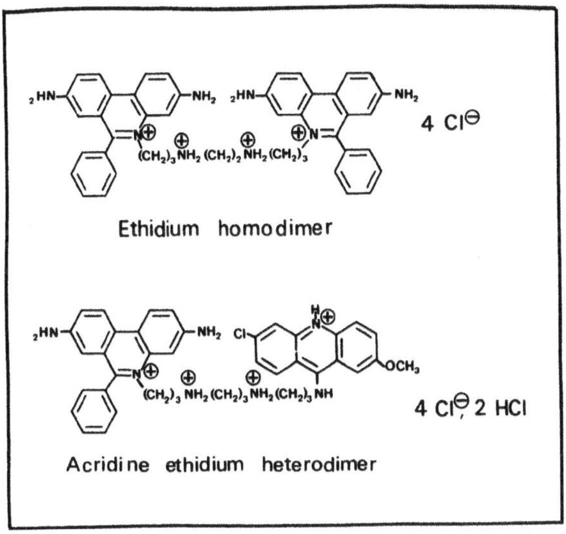

Figure 1 : Chemical Structures of ethidium dimer (EthDi) and ethidium-acridine dimer (AcEthDi) (Reprinted from Markovits et al. 1983).

breathing reaction. This raised the question whether the DNA structural fluctuations associated with the proton exchange of the bases are correlated with the structural fluctuations of the DNA-ligand complex associated with the amino group hydrogen exchange of EthDi and AcEthDi.

If this were really the case, molecules such as ethidium dimer could be used as fluorescent probes of the dynamic structure of various nucleic acids. In order to investigate this possibility, the characteristics of the Hydrogen-Deuterium (H-D) exchanges of ethidium dimer and ethidium acridine heterodimer, bound to nucleic acids of different structure, were investigated. Because the rate of proton exchange of Z-DNA has been found to be much slower than that of B-DNA (Ramstein and Leng, 1980), the study of the H-D exchange of EthDi in presence of Z-DNA was of special interest.

I - KINETIC ANALYSIS OF THE HYDROGEN-DEUTERIUM EXCHANGE OF A LIGAND INTERACTING WITH A MACROMOLECULE. THE SCHREIER AND BALDWIN (1976) MODEL.

The following scheme was considered.

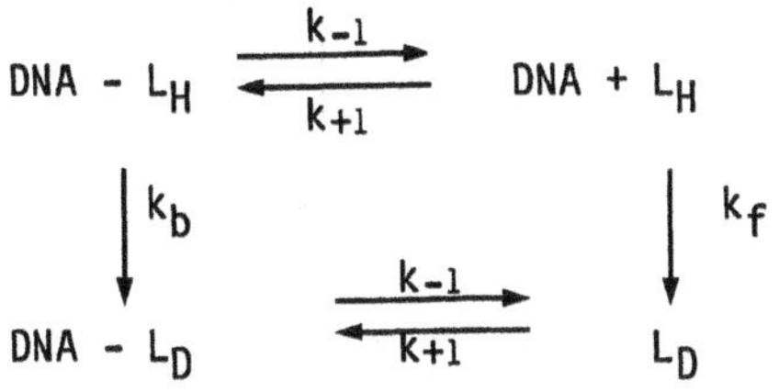

L_H and L_D refer to the ligand with hydrogen or deuterium respectively.

k_f and k_b are the rate of exchange in the free and bound state respectively.

k_{+1} is the on rate constant and k_{-1} the off rate constant associated with the kinetic of interaction of the ligand with DNA. It is assumed that deuteration of the ligand does not pertub the DNA binding.

The different equations applying to this system and their solutions are presented in the appendix. The exchange follows a two exponential kinetics with time constants λ_1, λ_2, (equations 8a, 8b, 8c, 9 of the appendix).

The λ_1 time constant is mainly associated with a preequilibration step reflecting the exchange of the ligand in the free state as the λ_2 time constant is mainly associated with exchange in the bound state. In the case of the EthDi-DNA system, the fluorescence of the bound ligand is much higher than that of the free ligand and the concentration of the free ligand is always very low because the binding constant for DNA is extremely high. Therefore in this system, the contribution of the free ligand to the variation of the fluorescence associated with the exchange is small.

The equation of λ_2 is close to that of k_{st}, obtained by doing the steady-state approximation (Schreier and Baldwin, 1976).

$$k_{st} = \frac{k_{-1}k_f + k_b(k_{+1}[DNA] + k_f)}{k_{-1} + k_{+1}[DNA] + k_f} \qquad (1)$$

$$\lambda_2 = \frac{k_{-1}k_f + k_b(k_{+1}[DNA] + k_f)}{k_b + k_{-1} + k_{+1}[DNA] + k_f} \qquad (2)$$

For reasons discussed in the appendix, the exchange can be described by a single exponential process of time constant $k_{ex} = \lambda_2$

if $k_b \ll k_{+1}[DNA]$ and $k_b \ll k_f$

$$k_{ex} = k_b + (k_{ex})_d \qquad (3)$$

with

$$(k_{ex})_d = \frac{k_{-1}\ k_f}{k_f + k_{+1}[DNA]} \qquad (4)$$

The rate of exchange is the sum of two terms :

k_b represents the direct exchange from the bound state.

$(k_{ex})_d$ represents the indirect exchange of the bound dye via the free state, during the dissociation-reassociation kinetic process at equilibrium. This term can be easily computed, once the binding parameters of the dye have been determined. Therefore, if the rate of exchange is measured for the free dye (k_f) and for the dye in presence of DNA, k_b can be easily deduced.

Two extreme cases can be envisioned

i) the dissociation rate of the complex (k_{-1}) is much larger than the hydrogen exchange rate from the bound state ($k_{-1} \gg k_b$).

In this situation :

$$k_{ex} = (k_{ex})_d$$

k_{-1} can then be deduced from hydrogen exchange measurement.

In the particular case of $k_f \gg k_{+1}[DNA]$, $k_{ex} = k_{-1}$.

ii) The exchange rate in the bound state is much larger than the dissociation rate ($k_b \gg k_{-1}$)

Then : $k_{ex} = k_b$

In the system studied by Schreier and Baldwin (1976), the former situation applied and the dissociation rate could be deduced from hydrogen exchange measurements.

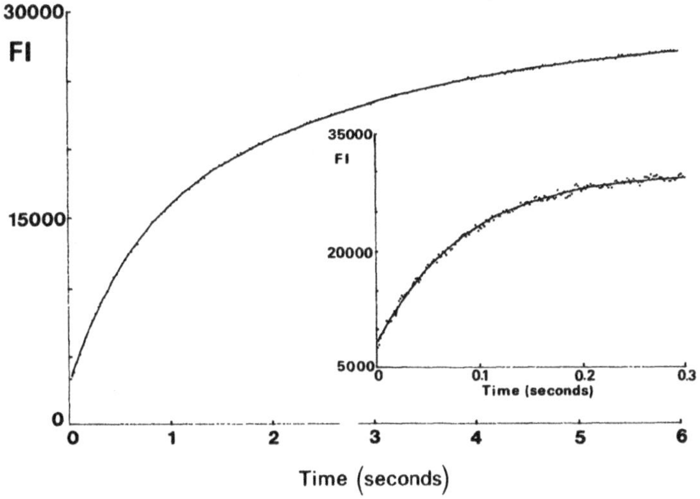

Figure 2 : An example of hydrogen-deuterium exchange of EthDi-DNA complex and free EthDi (inset) observed by fluorescence after stopped-flow mixing. The dots represent the experimental data and the continuous lines the best computed fit. Buffer was 0.05M tris(hydroxymethyl)aminomethane hydrochloride (Tris-HCl)-0.1M NaCl, pH 8.2. Temperature was 22°C. (Reprinted from Markovits et al., 1983).

Figure 2 shows an example of hydrogen-deuterium exchange of EthDi-DNA complex and free EthDi observed by fluorescence after stopped-flow mixing. The free-dimer kinetic of exchange can be fitted with a single exponential. In contrast, the fit of the DNA dimer H-D exchange data requires two exponentials. Statistical analysis of these data shows that both time constants and relative amplitudes can be estimated for a 95 % confidence limit, with an accuracy of ± 4 %.

The time constant for the first exponential process is much too large to correspond to the preequilibration step characterized by λ_1 (equation 8a appendix). Therefore the existence of at least two exponential processes in the exchange must be related to an heterogenity in the mechanisms of exchange in the bound state. This heterogenity does not probably reflect different rates of exchange for the two chromophores of EthDi since a two exponential process is also observed with AcrEthDi which has only one exchanging chromophore.

II - THE H-D EXCHANGE OF THE DYE-COMPLEXES TO DNA IS LIMITED BY STRUCTURAL DYNAMIC FLUCTUATIONS.

In figure 3, the values of the measured rates of exchanges at different ionic strengths are compared to the computed values of $(k_{ex})_d$ (equations 3 and 4). It is clear that, at the lower ionic strengths, where the rate of dissociation of the complex is very low, the exchange can proceed only from the bound state. In these conditions :

$$k_{ex} = k_b$$

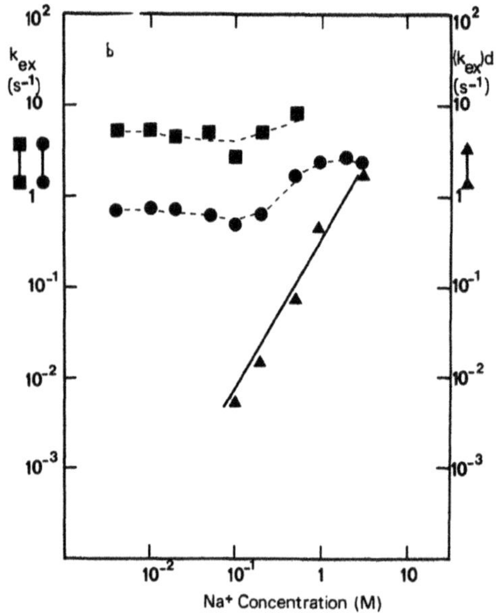

Figure 3 : Ethidium dimer-DNA complex. Comparison between observed ionic strength dependence of H-D exchange and computed contribution to exchange of dissociation-reassociation process [$(k_{Ex})_d$, eq 4]. (●,■) variation of the two observed rate constants, (▲) computed value of $(k_{Ex})_d$. (Reprinted from Markovits et al. 1983).

At high ionic strength, on contrary, the opposite situation is encountered. The exchange becomes limited by the dissociation. Interestingly, under these conditions, as expected, the exchange follows a single exponential kinetics, because the exchange is then limited by a single process, the dissociation of the dye from DNA.

At this stage, the point to be examined is to determine whether the exchange in the bound state is controlled by the limited accessibility of the amino exchangeable groups or by a dynamic structural fluctuation at the level of the DNA-ligand complex.

When the proton exchange of the DNA base pairs was first studied, the same basic question was raised (Mc Connel and Von Hippel, 1970). In that case, the question was answered when the effect of catalyst on the rate of exchange was studied.

If the rate of exchange is measured as a function of catalyst concentration, two extreme behaviors can be expected : i) the exchange increases linearly and continously with catalyst concentration. In that case, it would be concluded that the catalyst action, which results from a bimolecular process, is only hindered by limited accessibility of the catalyst to its target. ii) The exchange rate increases first linearly with catalyst concentration and later reaches a plateau at high catalyst concentrations. Such a behavior is expected when a dynamic structural fluctuations controls the exchange. The following scheme was proposed (Mc Connel and Von Hippel, 1970).

$$\text{closed} \underset{k_{cl}}{\overset{k_{op}}{\rightleftarrows}} \text{open} \xrightarrow{k_{CH}[C]} \text{exchanged}$$

The exchange cannot occur when DNA is in a closed conformation (hydrogen bonded). It only occurs when hydrogen bonds break and DNA base pairs open.

With k_{op} the rate of opening, k_{cl} the rate of closing and $k_{CH}[C]$ ([C] being the catalyst concentration) the rate of exchange in the open conformation, the following equation applies (Mc Connel and Von Hippel, 1970)

$$k_{ex} = \frac{k_{op}\, k_{CH}[C]}{k_{cl} + k_{CH}[C]} \qquad (5)$$

at high catalyst concentration $k_{ex} \rightarrow k_{op}$.
at low catalyst concentration :

$$k_{ex} \rightarrow k_{op}\, k_{CH}[C]/k_{cl}$$

The imidazole catalysis of the EthDi-DNA complex was then studied (Markovits et al., 1983). It was first observed that, as expected, the rate of exchange of EthDi in the free state was strictly proportional to imidazole concentration. At pH 8, k_{CH} was found equal to $5.6 \times 10^5 M^{-1} sec^{-1}$. As seen on figure 4, in the case of the EthDi DNA complex, it is clear that both time constants level off at high catalyst concentration.

Using the same arguments as for the study of DNA hydrogen exchange, we then concluded that the H-D exchange of the dimers in the bound state were also limited by dynamic structural fluctuations.

The effect of pH on the exchange was also studied (Markovits et al., 1983). The exchange of the free dye as function of pH followed the expected dependence for a process catalyzed either by H^+ or OH^-. The exchange of the dye bound to DNA was rather insensitive to pH suggesting that the local pH at the level of the polymer was very much different from the pH of the solvent.

III - IS THERE A CORRELATION BETWEEN THE DYNAMIC STRUCTURAL FLUCTUATIONS OBSERVED AT THE LEVEL OF DNA AND THOSE OBSERVED AT THE LEVEL OF A DNA BOUND LIGAND.

The H-D exchange kinetics of DNA bound ethidium dimer or ethidium acridine heterodimer resembles in several aspects that of DNA itself. The time range of the exchange processes, the effect of catalysts, the activation energies are comparable in the two cases. Moreover, as seen in figure 3, the H-D exchange of EthDi or AcrEthDi does not vary with the binding affinity of the dyes for DNA as this parameter is changed with

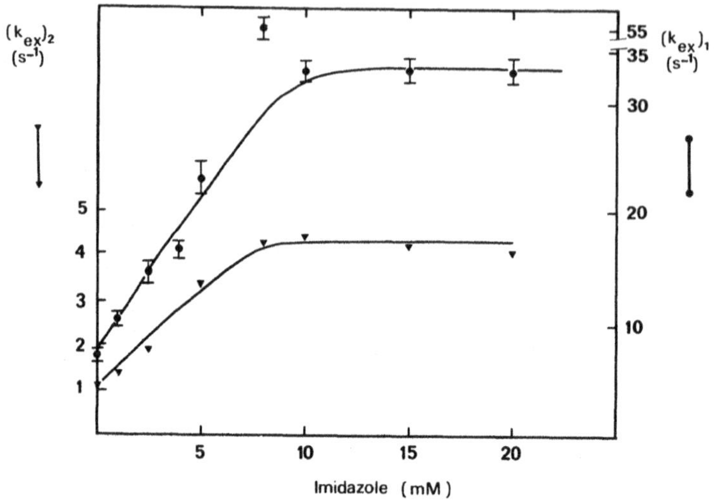

Figure 4 : Imidazole catalysis of the two H-D exchange rates of ethidium dimer-DNA complex. Rates were determined at pH 7, at 22°C. The total cation concentration was adjusted to 10^{-2}M by the addition of NaCl. (Reprinted from Markovits et al., 1983).

ionic strength by several orders of magnitude. These data suggest that a correlation between the two exchange processes could well exist. For instance, it could be envisioned that the displacement of a base pair, which must accompany imino proton exchange, could induce a concomitent displacement of an EthDi molecule bound adjacent to it.

To study whether such correlation could exist, the H-D exchange of EthDi bound to nucleic acids eliciting different structures and different hydrogen exchange characteristics was studied (Markovits et al., 1985).

The largest difference in hydrogen exchange kinetics between two nucleic acid structures is between B-DNA and Z-DNA (Ramstein and Leng, 1980). It was observed that EthDi can bind as well to Z-DNA and B-DNA. This is illustrated on Fig. 5.

An example of the fluorescence signal associated with the EthDi H-D exchange of poly(dGm^{5}dC) in Z and B forms is shown in figure 6. It is clear that the EthDi exchange is faster when bound to B-DNA than to Z-DNA.

The effect of catalyst on the exchange of EthDi on Z or B DNA was also compared. In both cases, the exchange appeared dependent on dynamic structural fluctuations.

The values of the exchange characteristics found for EthDi bound to nucleic acids of different structures are reported in Table I.

In Table II, the mean exchange rates of EthDi bound to different nucleic acids and the imino proton exchange rate of the corresponding nucleic acids as reported in the literature are shown for comparison.

Although the number of cases is small, it can be noted that the rates of exchange for the nucleic acid imino protons and bound EthDi amino

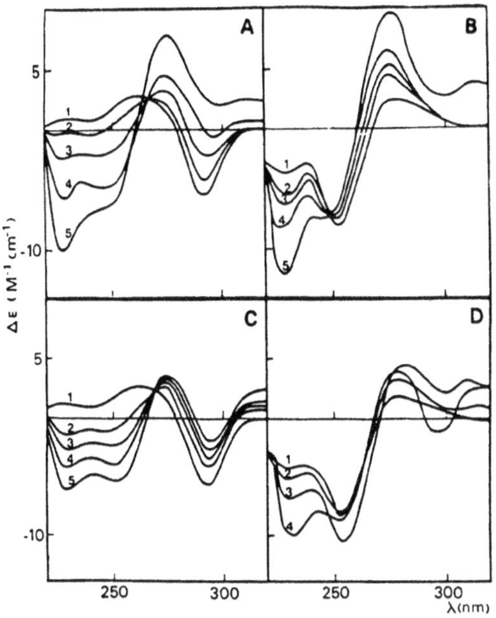

Figure 5 : Effect of EthDi and EthBr upon the circular dichroism spectra of poly[(dG-m^5dC)] in Z or B form. Temperature was 25°C. A) Ethbr + poly[(dG-m^5dC)] in Z buffer. The dye/base-pair ratios are 0, 0.10, 0.16, 0.26 and 0.56 for curbe n° 1, 2, 3, 4 and 5 respectively. B) EthBr + poly[(dG-m^5dC)] in B buffer. The dye/base-pair ratios are 0, 0.10, 0.16, 0.26 and 0.56 for curbe n° 1, 2, 3, 4 and 5 respectively. C) EthDi + poly [(dG-m^5dC)] in Z buffer. The dye/base-pair ratios are 0, 0.04, 0.06, 0.08 and 0.16 for curbes n° 1, 2, 3, 4 and 5 respectively. D) EthDi + poly [(dG-m^5dC)] in B buffer. The dye/base-pair ratios are 0, 0.5, 0.12 and 0.24 for curbe n° 1, 2, 3, and 4 respectively. In all experiments nucleotide concentration is 10^{-4}M.
(B buffer : 0.05M NaCl Tris-HCl 0.01M pH 7.4 - Z buffer : idem B buffer + 10mM $MgCl_2$).
(Reprinted from Markovits et al. 1985).

protons can be ranged in the same order. Nevertheless, it appears there is not a strong correlation between the two sets of value. Whereas Z-DNA and B-DNA exchange rates differ by a factor of 1000, exchange rates of EthDi bound to the corresponding DNAs differ only by a factor of 5 to 8.

Recently Pasternak et al. (1986) studied the interactions of water soluble porphyrins with Z poly(dG-dC). They observed that under some circumstances, the binding followed a first order kinetics. They accounted for their results according to the following scheme.

$$Z \underset{k_{-1}}{\overset{k_1}{\rightleftharpoons}} Z^* + P \rightarrow (PZ^*) \rightarrow P\text{-}B$$

Z-DNA is in equilibrium between two states Z, the more populated and Z* the less populated. Porphyrin binds only to Z* and the complex PZ* converts rapidly to B-DNA.

In some conditions, the kinetics is then limited by the isomerization step $Z \rightarrow Z^*$ whose constant $k_1 = 0.6$ sec^{-1}.

Pasternak et al. (1986) pointed out that this constant is very close to that we measured for the H-D exchange of EthDi intercalated in Z-DNA.

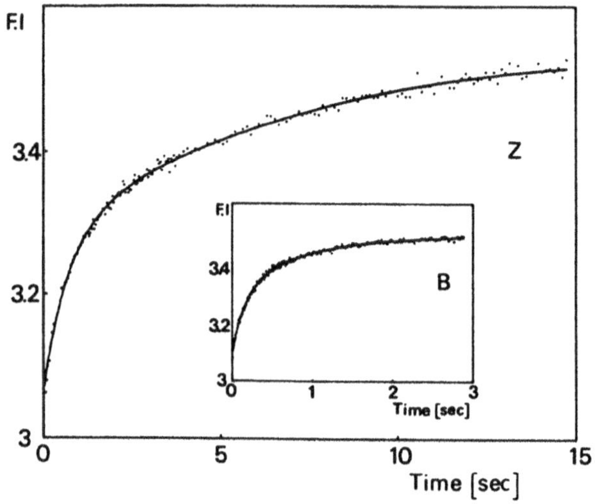

Figure 6 : Typical fluorescence signal for the H-D exchange reaction of EthDi bound to poly[(dG-m^{5}dC)] in Z buffer and B buffer (defined in figure 5) (inset). The dots represent the experimental determinations. The continuous line is the best fit for a two exponential process. (EthDi 1.8×10^{-6}M, poly[(dG-m^{5}dC)] 1.8×10^{-4}M in nucleotide). (Reprinted from Markovits et al., 1985).

They suggested that the structural fluctuation accompanying the Z → Z* transition could induce a motion of the intercalated dye and facilitate the exchange of its amino proton.

If such interpretation could be confirmed by further experiments, it would appear that indeed H-D exchange of DNA bound EthDi reflects a particular mode of DNA dynamic structural fluctuation. At this point, it must be emphasized that it is the H-D exchange of the EthDi amino protons which leads to fluorescence variation. Therefore the DNA dynamic structural fluctuations which are recorded by fluorescence must reflect fluctuations in the environment of the EthDi amino group in its binding site. These fluctuations reflect more likely the dynamic of the phosphate sugar backbone that that of the base pairs.

TABLE 1 : Relaxation times for the H-D exchange of EthDi bound to nucleic acids of different structures.

Polymer	$\tau_1$ex	$\tau_2$ex	$\bar{\tau}$ ex
Poly(dG-m^{5}dC) (Z buffer)	0.73	6.9	1.5
Poly(dG-m^{5}dC) (B buffer)	0.17	0.83	0.27
Poly(dG-dC) (Z buffer)	0.05	0.60	0.13
Poly(dG-dC) (B buffer)	0.05	0.55	0.14
C.T. DNA	0.13	1.08	0.28
Poly(dA-dT)	0.17	0.86	0.32
Poly(dA).poly(dT)	0.36	1.3	0.63
Poly(A).poly(U)	0.72	7.4	2.1
Yeast tRNA^{Phe}	0.16	1.65	0.50

τ1ex, τ2ex are the relaxation times corresponding to the fast and slow phase of the exchange respectively. $\bar{\tau}$ ex is the mean relaxation time computed according to Schwarz (1968). Calf thymus (C.T.) DNA, poly(dA-dT), poly(dA).poly(dT), poly(A).poly(U) were in B buffer (defined in figure 5). (From Markovits et al., 1985).

TABLE 2 : Comparison of mean exchange rate constants for EthDi bound to nucleic acid (measured at 25°C) and nucleic acid imino proton exchange rate constants.

	EthDi bound to nucleic acid $\bar{k}(s^{-1})$ 25°C	Nucleic acid imino protons $k(s^{-1})$ 25°C (a)	Ref.
B DNA	4-7	5-6	(b)
poly(dA-dT)	3.1	4	(b)
tRNA	2	1.5-5	(c)
poly(A).poly(U)	0.5	2	(c)
Z DNA	0.6	0.007	(d)

a) Values reported in the literature have been obtained at various temperatures. In order to compare them, they have been standardized to 25°C assuming an activation energy of 16 Kcal/mol.
b) Assa-Munt et al., 1984.
c) Johnston and Redfield (1981).
d) Ramstein, J. (unpublished result). From Markovits et al., 1985).

REFERENCES

1. Assa-Munt N, Granot J, Behling RW and Kearns DR : ^{1}H NMR relaxation studies of the hydrogen-bonded imino protons of poly(dA-dT). Biochemistry 23, 944-955, 1984.
2. Barbet J, Roques BP, Combrisson S and Le Pecq JB : DNA polyintercalating drugs. Proton magnetic resonance studies of a new acridine dimer. Conformations and interactions with mono-dinucleotides. Biochemistry 15, 2642-2650, 1976.
3. Capelle N, Barbet J, Dessen P, Blanquet S, Roques BP and Le Pecq JB : Deoxyribonucleic acid bifunctional intercalators : kinetic investigation of the binding of several acridine dimers to deoxyribonucleic acid. Biochemistry 15, 3354-3362, 1979.
4. Englander SW and Kallenbach NR : Hydrogen exchange and structural dynamics of proteins and nucleic acids. Quaterly Rev. Biophysics 16, 521-655, 1984.
5. Esnault C, Roques BP, Jacquemin-Sablon A and Le Pecq JB : Effects of new antitumor bifunctional intercalators derived from 7H-pyridocarbazole an sensitive and resistant L1210 cells. Cancer Res. 44, 4335-4360, 1984.
6. Gaugain B, Barbet J, Oberlin R, Roques BP and Le Pecq JB : DNA bifunctional intercalators 1. Synthesis and conformational properties of an ethidium homodimer and of an acridine ethidium heterodimer. Biochemistry 17, 5071-5078 (1978a).
7. Gaugain B, Barbet J, Capelle N, Roques BP and Le Pecq JB : DNA bifunctional intercalators 2. Fluorescence properties and DNA binding interaction of an ethidium homodimer and an acridine ethidium heterodimer. Biochemistry 17, 5078-5088 (1978b).
8. Johnston, PD and Redfield AG : Study of transfer ribonucleic acid unfolding by dynamic nuclear magnetic resonance. Biochemistry 20, 3996-4006 (1981).
9. Laugaa P, Markovits J, Delbarre A, Le Pecq JB and Roques BP : DNA tris-intercalation : First acridine trimer with DNA affinity in the range of DNA regulatory proteins. Kinetics studies. Biochemistry 24, 5567-5575 (1985).
10. Le Pecq JB, Le Bret M, Barbet J and Roques BP : DNA poly intercalating drugs : DNA binding of diacridine derivatives. Proc. Natl. Acad. Sci. USA 72, 2915-2919, 1975.
11. Mandal L, Englander SW and Kallenbach NR : Hydrogen-deuterium exchange analysis of ligand-macromolecule interactions : ethidium-deoxyribonucleic acid system. Biochemistry 19, 5819-5925, 1980.
12. Mandal C, Kallenbach NR and Englander SW : Base-pair opening and closing reactions in the double helix. A stopped-flow hydrogen exchange study in poly(rA).poly(rU). J. Mol. Biol. 135, 391-411.
13. Markovits J, Ramstein J, Roques BP and Le Pecq JB : Dynamic structure of DNA complexes. Fluorometric measurement of hydrogen-deuterium exchange kinetics of DNA-bound ethidium dimer and acridine-ethidium dimer. Biochemistry 22, 3231-3237, 1983.
14. Markovits J, Ramstein J, Roques BP and Le Pecq JB : Effect of B-Z transition and nucleic acid structure on the conformational dynamics of bound ethidium dimer by hydrogen deuterium exchange kinetics. Nucleic Acids Res. 13, 3773-3788, 1985.
15. Mc Connel B and Von Hippel PH : Hydrogen exchange as a probe of the dynamic structure of DNA. General acid-base catalysis. J. Mol. Biol. 50, 297-316, 1970.

16. Pasternach RF, Sidney D, Hunt PA, Snowden EA and Gibbs EJ : Interactions of water soluble porphyrins with Z-poly(dG-dC). Nucleic Acids 14, 3927-3943, 1986.
17. Pelaprat D, Delbarre A, Le Guen I, Roques BP and Le Pecq JB : DNA intercalating compounds as potential antitumor agents 2. Preparation and properties of 7H-pyridocarbazole dimers. J. Med. Chem. 23, 1336-1343, 1980.
18. Ramstein J and Leng M : Salt dependent dynamic structure of poly(dG-dC).poly(dG-dC). Nature 288, 413-414, 1980.
19. Reinhardt C, Roques BP and Le Pecq JB : Bindings of bifunctional ethidium intercalators to transfer RNA. Biochem. Biophys. Res. Commun. 104, 1376-1385, 1982.
20. Roques BP, Pelaprat D, Le Guen I, Procher G, Gosse CH and Le Pecq JB : DNA bifunctional intercalators. Antileukemic activities of new pyridocarbazole dimers. Biochem. Pharmacol. 28, 1811-1815, 1979.
21. Schreier AA, Baldwin RL : Concentration-dependent hydrogen exchange kinetics of ^{3}H-labeled S-peptide in ribonuclease S. J. Mol. Biol. 105, 409-426, 1976.
22. Schwarz G : Kinetic analysis by chemical relaxation methods. Rev. Modern Phys. 40, 206-218, 1968.

APPENDIX

HYDROGEN EXCHANGE KINETICS OF A LIGAND INTERACTING WITH A MACROMOLECULE

Let consider the general following scheme :

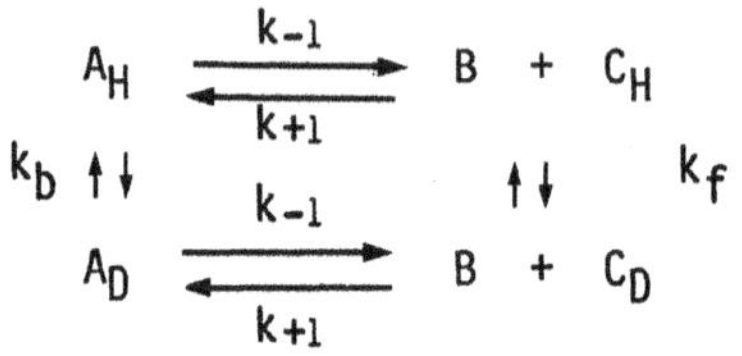

With : B The concentration of free sites for the ligand on the macromolecule
C_H, C_D the hydrogenated and deuterated concentration of free ligand respectively
A_H, A_D the hydrogenated and deuterated concentration of bound ligand respectively
At time 0, the solvent is 100 % H_2O, at infinite time the fraction of D_2O in the solvent is E.
Using the variables :

$$A_H = \bar{A}_H + \Delta A_H \qquad (1) \qquad C_H = \bar{C}_H + \Delta C_H \qquad (2)$$
$$A_D = \bar{A}_D + \Delta A_D \qquad\qquad C_D = \bar{C}_D + \Delta C_D$$

where $\bar{A}_H$, $\bar{C}_H$, $\bar{A}_D$ and $\bar{C}_D$ are the corresponding values of A_H, C_H, A_D and C_D at equilibrium. If, and only if, the H-D exchange reaction doesn't perturb the equilibrium, the total bound and free dye concentrations remain constant and :

$$\Delta A_H = - \Delta A_D \text{ and } \Delta C_H = - \Delta C_D$$

The differential equations which then applies to the system are (see for instance Cantor & Schimmel, 1980) :

$$-\frac{d\Delta A_H}{dt} = \Delta AH(k_{-1} + k_b E) + \Delta C_H(-k_{+1}B) + \Delta A_D(-k_b(1-E)) \qquad (3a)$$

$$-\frac{d\Delta C_H}{dt} = \Delta AH(-k_{-1}) + \Delta C_H(k_1 B + k_f E) + \Delta C_D(-k_f(1-E)) \qquad (4a)$$

Since $\Delta A_H = \Delta A_D$ and $\Delta C_H = - \Delta C_D$

This reduces to :

$$-\frac{d\Delta A_H}{dt} = \Delta A_H(k_{-1} + k_b) + AC_H(-k_{+1}B) \qquad (3b)$$

$$- \frac{d\Delta C_H}{dt} = \Delta A_H(-k_{-1}) + \Delta C_H(k_{+1}B + k_f) \qquad (4b)$$

with :

$$a_{11} = k_{-1} + k_b \qquad a_{12} = -k_{+1}B$$
$$a_{21} = -k_{-1} \qquad a_{22} = k_{+1}B + k_f \qquad (5)$$

The solution has the form (Cantor and Schimmel, 1980)

$$\Delta A_H = \Delta A_{H_1}\, e^{-\lambda_1 t} + \Delta A_{H_2}\, e^{-\lambda_2 t} \qquad (6)$$

$$\Delta C_H = \Delta C_{H_1}\, e^{-\lambda_1 t} + \Delta C_{H_2}\, e^{-\lambda_2 t} \qquad (7)$$

ΔA_{H_1}, ΔA_{H_2}, ΔC_{H_1}, ΔC_{H_2} being the first and second amplitudes for the variation of ΔA_H and ΔC_H respectively.

λ_1 and λ_2 are the roots of equation :

$$\lambda^2 - (a_{22} + a_{11})\lambda + (a_{11}a_{12} - a_{22}a_{12}) = 0 \qquad (8)$$

With :

$$\lambda_1 = a_{11} + a_{22} - \lambda_2 \qquad (8a)$$

or :

$$\lambda_1 = k_b + k_{-1} + k_{+1}B + k_f - \lambda_2 \qquad (8b)$$

Using the classical approximation which neglects the high order terms in the development of the square root of the determinant :

$$\lambda_2 \simeq \frac{a_{11}a_{22} - a_{12}a_{21}}{a_{11} + a_{22}} \qquad (8c)$$

or :

$$\lambda_2 \simeq \frac{k_{-1}k_f + k_b(k_{+1}B + k_f)}{k_b + k_{-1} + k_{+1}B + k_f} \qquad (9)$$

Let define D_t, the total concentration of the ligand.

$$\Delta A_{H_0} = D_t E(1 - f) \qquad (10)$$

$$\Delta C_{H_0} = D_t E\, f \qquad (11)$$

ΔA_{H_0}, ΔC_{H_0} being the values of ΔA_H and ΔC_H at time 0, and f the fraction of the free ligand.

Using the derivatives of (6) and (7) and equation (3) and (4) at 0 time together with expressions (10) and (11), it comes :

$$\frac{\Delta A_{H_1}}{\Delta A_{H_0}} = \frac{1}{\lambda_1 - \lambda_2}\left(a_{11} - \lambda_2 + a_{12}\frac{f}{1-f}\right) \quad (12)$$

$$\frac{\Delta A_{H_2}}{\Delta A_{H_0}} = \frac{1}{\lambda_1 - \lambda_2}\left(\lambda_1 - a_{11} - a_{12}\frac{f}{1-f}\right) \quad (13)$$

$$\frac{\Delta C_{H_1}}{\Delta C_{H_0}} = \frac{1}{\lambda_1 - \lambda_2}\left(a_{22} - \lambda_2 + a_{21}\frac{1-f}{f}\right) \quad (14)$$

$$\frac{\Delta C_{H_2}}{\Delta C_{H_0}} = \frac{1}{\lambda_1 - \lambda_2}\left(\lambda_1 - a_{22} - a_{21}\frac{1-f}{f}\right) \quad (15)$$

If k an instrument factor characteristic of the fluorometer.
ΔI the measured fluorescence variation,
V the ratio of the fluorescence of the bound ligand to the fluorescence of the free ligand when measured in identical conditions,
v_a and v_c the numbers by which the fluorescence quantum yield of bound and free ligand are multiplied, when going from H_2O to D_2O respectively :

$$\Delta I = k[V(v_a-1)\Delta A_{H_1} + (v_c-1)\Delta C_{H_1}]\ e^{-\lambda_1 t} + k[V(v_a-1)\Delta A_{H_2} + (v_c-1)\Delta C_{H_2}]\ e^{-\lambda_2 t} \quad (16)$$

Let consider two different cases :

1°) Ligand-macromolecules complexes characterized by high binding constant and which do not dissociate in the conditions of the exchange measurements (that is the condition met with EthDi-DNA complexes). From the equation (8a,b,c) and (9) it comes :

$$\lambda_1 = k_{+1}B + k_f \text{ and } \lambda_2 = k_b = k_{-1} \text{ at low concentration of B} \quad (17)$$

$$\lambda_2 = k_{+1}B \text{ and } \lambda_2 = k_b \quad \text{at high concentration of B} \quad (18)$$

$$\frac{\Delta A_{H_1}}{\Delta A_{H_0}} \longrightarrow \frac{-f}{1-f} \quad (19) \qquad \frac{\Delta A_{H_2}}{\Delta A_{H_0}} \longrightarrow 1 + \frac{f}{1-f} \quad (20)$$

The contribution of the free ligand can be neglected.
V is large. The terms in equation (16) which containing ΔC_{H_1} and ΔC_{H_2} can be neglected.
Furthemore, ΔA_{H_1} is small compared to ΔA_{H_2}, and equation (16) becomes :

$$\Delta I = k'[\Delta A_{H_2}\ e^{-\lambda_2 t}] \quad (21)$$

k' being a constant.

The solution is then very close to the steady state approximation of

Schreier and Baldwin (1976) and a single time constant is observed with :

$$k_b \leqslant \lambda_2 \leqslant k_b + k_{-1} \tag{22}$$

2°) Ligand-macromolecule complexes characterized by a low binding constant which are partly dissociated in the conditions of the H-D exchange kinetics measurement (that is the case of ethidium in most of the situations).

If the measurements are made at high concentration of macromolecules (concentrations of ligand sites being at least ten times greater than the dissociation constant) to minimize as far as possible the dissociation, it can be computed from equation (12) to (15) that the three amplitudes ΔA_{H_1}, ΔA_{H_2}, ΔC_{H_1} can be neglected and steady state approximation is still valid. On the contrary, when ligand in excess, conditions can be found where λ_1 and λ_2 can be very close to each other and ΔA_{H_1}, ΔA_{H_2}, ΔC_{H_1}, ΔC_{H_2} can have values which are of the same order.

If fluorescence is used, the free dye contribution can be neglected in some conditions. In contrast, when absorption spectroscopy is used to follow the H-D exchange of DNA-ethidium complex, such as in Mandal et al. (1980), the free and bound dye contribute equally to the signal. Under these conditions, the steady state approximation might not always apply and two relaxations can be observed. But it must be underlined that ΔA_{H_1} is negative and ΔA_{H_2} positive. Therefore, if λ_1 and λ_2 values are sufficiently far apart, this case can be experimentally detected.

REFERENCES

Cantor, C.R. & Schimmel, P.R. (1980) Biophysical Chemistry, W.H. Freeman and Compagny Eds, San Francisco, Part 3, p.887-937.

Mandal, C., Englander, S.W. & Kallenbach, N.R. (1980) Biochemistry 19, 5819-5825.

Schreier, A.A. & Baldwin, R.L. (1976) J. Mol. Biol. 105, 409-426.

THE HIGHER ORDER STRUCTURE AND DYNAMICS OF CHROMATIN -DNA

C.A. NICOLINI , A. DIASPRO and P. GERMANO,
Chair of Biophysics , School of Medicine ,University of Genova

Early in this book distinguished scientists have summarized the experimental evidences on the orderly organization of the genetic material within mammalian cells from the primary structure (base sequences) to the secondary (the various forms of DNA double helix) structure. While the tertiary structure - wrapping of DNA around octamer histones to form the nucleosome (1) - is rather clear in details , large uncertainty still exists on the exact three-dimensional geometry of the higher order DNA foldings (from the supernucleosomes on) . In this chapter we will attempt to summarize the present understanding of chromatin -DNA (2 ,3) higher order structure and dynamics ,emphasizing recently introduced biophysical probes - the only ones capable to cope with the inherent difficulties of an "in situ" characterization.

THE HIGHER ORDER STRUCTURE

What has been recently emerging is the critical role of total DNA length (2,3), in maintaining the native quaternary chromatin structure. Nuclease - isolated chromatin having DNA of about 2.5×10^7 Daltons yield at low ionic strength fibers of 100 Å , which, at 20 mM NaCl or 0.3 $MgCl_2$, takes - but only in presence of H1 - a more condensed form of 300-400 Å diameter and 0.2 nucleosomes/10 Å (4). This and the ionic strength dependence is exactly what is predicted by polyelectrolyte theory for the folding and bending angle of a single nucleofilament (5). Shearing of this chromatin preparation, even at high ionic strength, causes the disappearance of the quaternary "solenoidal" 300 Å fiber into the 100 Å nucleofilament (2), as confirmed by X-ray diffraction data where the 110 and 55 Å peaks, reflecting the organization of nucleosomes, are replaced by 38 and 27 Å peaks, reflecting the structure of the nucleosome itself. Isolated chromatin of larger molecular weights were also reported by gentle cold water swelling of isolated nuclei without any shearing ; this preparation was shown to display even at low ionic strength differential scattering of circular polarized light . Similar conclusions were achieved by studying the higher reactivity of certain chromatin regions with respect to EB intercalation when unsheared chromatin at low ionic strength was utilized. These DNA regions appear then maintained under physical constraints favorable to intercalation namely in presence of a negative number of superhelices (2,3 and other chapters in this book). In Drosophila and mammalian genome isolated as compact mass 10^9 Daltons long (6,7), only DNAse I nicks, whenever more frequent than every 6 x 10^9 Daltons (or 400 nucleosomes), appear capable of fully relaxing the genome to a form sedimenting more slowly without any negative superhelical turns. Since nucleosomes are preserved, even in the presence of DNAse I, it would then appear that 6 x 10^7 Daltons is the minimum length of DNA that can be considered as a closed structure with high negative superhelical turns (see the "fibrosome " model discussed in reference 3).

It appears certain also the presence of a skeleton, i.e. a scaffold of dense network of fibers, surrounded by "halo" of loops of "fibrosome" DNA anchored at both ends at the scaffold: these results seem to be reproducible both in metaphase chromosome and interphase chromosome leading also to the rather controversial nuclear matrix concept (8,9).

Keeping all the above evidence in the background, how DNA goes from the tertiary nucleosomes structure, constituting the basic repeating unit of chromatin, to the higher order structure? How are

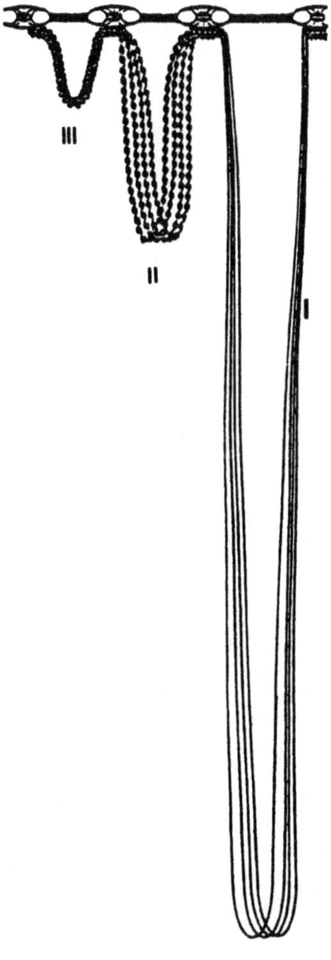

Figure 1

Rope-like and drapery-like model for quaternary and quinternary chromatin-DNA organization in the intact nuclei during interphase (3). Examples of the states I, II and III of chromatin superpacking defined in reference 3 are also given.

nucleosomes arranged in native conditions to yield the frequently observed 300 Å quaternary fiber? What is the exact arrangement of the two-order superhelical chromatin structure suggested earlier ? An answer came from recent electron-microscopic observations obtained on a phospholipid monolayer , which suggested that the 300 Å fiber from mammalian nuclei might be formed not by a coiled single nucleofilament but by up to four nucleofilaments (Fig. 1) properly interlinked at periodic fixed sites unattackable by DNAse or by very high ionic strength (10,11,12). Incidentally , the multifibrilar nature of chromatin does not necessarily imply a departure from the canonic concept of a single giant DNA molecule making the genome , since it could well be the same fibre going back and forth few times. The folding and unfolding occuring in the giant chromatin-DNA molecule (we have measured up to several micron long fibers) never propagate along the entire length, but remain confined to narrow regions of about 4600 Å or larger ("fibrosomes") , limited by fixed points identifiable as irregular pieces of "nuclear membrane" periodically attached along the chromatin fiber (11). All the above and the anular rings apparent with Feulgen- staining points to the suggestion of a "drapery-like" quinternary structure for the regular folding of the 300 Å fiber within the nucleus,confirmed by freeze-etching (13) . Furthermore the suggested fibrosome=gene equation - where each of the 75,000 fibrosomes has its own degree of supercoil locally maintained - has offered a coherent and not circumstantial hypothesis, both in structural and functional terms, which however call for further probatory experimentations.

Continuing along this line of search , we have recently implemented new theoretical and experimental undestructive biophysical probes (hereby summarized) and carried out new measurements on intact nuclei and cells in various functional states . This has permitted to attack the still numerous unanswered questions on the native high order structure of this complex biopolymer and on its dynamics with different and complementary methodics.

Microcalorimetry

A detailed analysis, carried out by continuous measurement of viscosity performed at 30 minute intervals over a period of more than 48 hours yielded for the first time a direct experimental evaluation of the three discrete unfoldings, corresponding likely to the three step-like changes in higher order superstructures previously described (13). Similar higher order DNA foldings and topological constraints are apparent in histone-depleted interphase nuclei and "nucleoids" cells from which much protein has been released by identical lysis in 2M NaCl, detergent and chelating agents.

An earlier calorimetric study confirmed these findings (14). For isolated unsheared chromatin, in the range from 300° to 400°K, three transitions (I, II, III) are indeed clearly and reproducibly identifiable at all ionic strengths, whose temperature values change only slightly with increasing salt concentration. As summarized in (14), the enthalpy changes for isolated chromatin are instead dramatic with changes in salt concentration: namely the enthalpy changes are decreasing between 1 and 10 mM to then remain constant for the transition I, are steadily increasing for the transition III, while sharply increasing up to 50 mM to then sharply decrease at 600 mM for

the transition II.
Namely, the effect of a 600 mM NaCl wash, known to remove the H1 histone from chromatin and to disrupt the quaternary structure, combined with the critical dependence on ionic strength, as expected also from polyelectrolyte theory , seemed to point to a possible assignment for the transition II to the quaternary rope-like structure which is then present even at 1 mM Tris-HCl. For the third (III) transition its melting temperature, combined with the salt dependence of the enthalpy, and with the effect of carcinogen - which alters chromatin structure toward a more disperse uncoiled state (15) apparently causing a rearrangement of part of the genome from the quaternary structure (II transition) to a lower order structure (III transition), - points to a possible assignment of the third (III) transition to the nucleosomal DNA, i.e. helix-coil stabilized in the nucleofilament. This assignment is compatible with the C.D. and O.D. thermal profiles which show that above 365°K helix-coil dominates the denaturation process.
The first (I) transition in isolated chromatin cannot be related

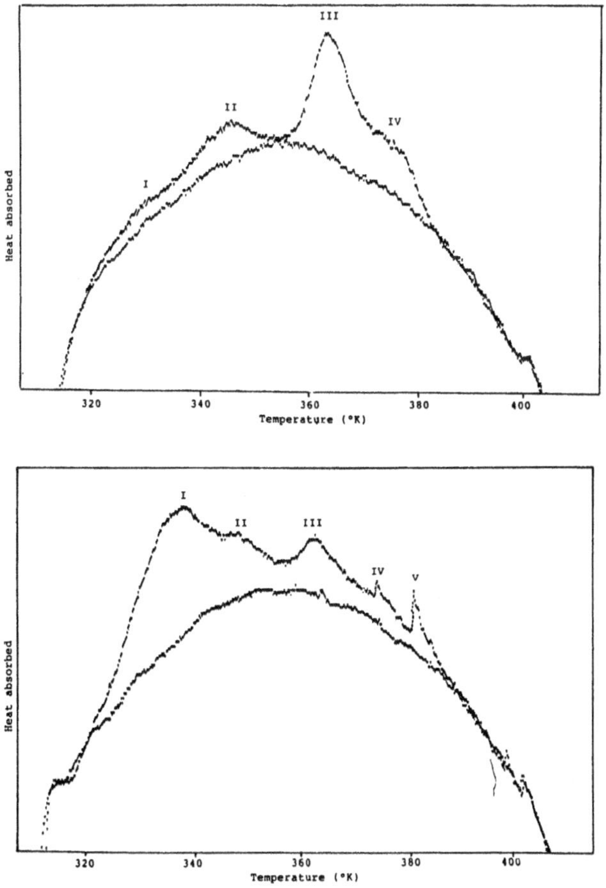

Figure 2

Heat capacity versus temperature of isolated nuclei (above) and cells (below) from rat liver. In both figures the inferior line represents the thermal profiles of the same samples after heat denaturation. Rough-data are acquired on line with a minicomputer as described in Fig. 4.

to the linker DNA as previous suggested, not only because of its independence from salt concentration (which instead should stabilize this type of structure), but namely because its pronounced increase of the enthalpy of transition in isolated GO resting nuclei where the DNA present in the linker conformation, if any, should decrease. Tentatively this transition and the other (0) consistently observed at lower temperature (332°K) only in native nuclei, could then be related either to the quinternary drapery-like chromatin structure and/or to the melting of macromolecules (i.e. membrane lipo-proteins,nuclear proteins and/or RNA) strongly bound to native chromatin (I).
As showed in Fig. 2 thermal profiles of GO resting cells reveal the presence of five endothermic transitions, where the first centered at about 338° K is indeed by far the most prominent , quite more than in the corresponding isolated nuclei; such a large broad background enthalpy in the low-melting region is then probably due to the melting of numerous cytoplasmic and nuclear components (16). For a more comprehensive study and assignment of transitions, calorimetry is applied to calf thymus DNA in various ionic strength media (Fig.3).

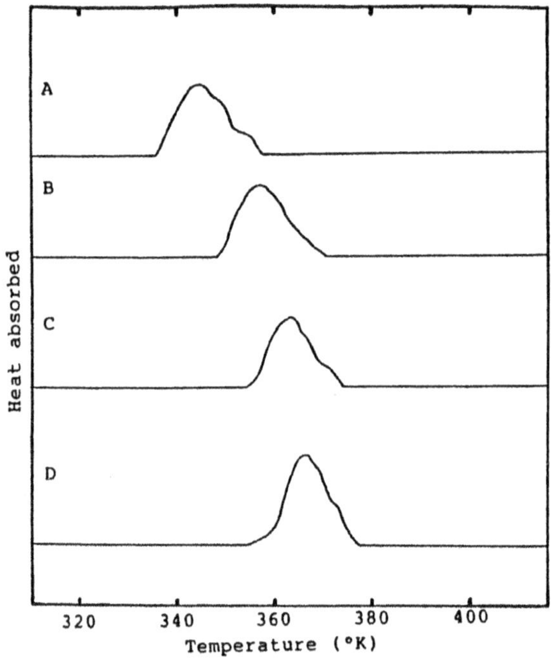

Figure 3

Differential scanning calorimetry of calf thymus DNA in various ionic strength solutions: A) dH_2O; B) 50 mM NaCl, 0.7 mM Na_2HPO_4; C) 150 mM NaCl, 2 mM Na_2HPO_4; D) 150 mM NaCl, 300 mM KCl, 2 mM Na HPO , 5 mM $CaCl_2$, 5 mM $MgCl_2$.
Heat absorption profiles as a function of temperature are obtained by subtracting the base line.
All solutions contain 2% DNA (w/w).

Melting temperature shifts from 346° to 366°K and the enthalpy of transition (summarized in Tab. 1) increases with the salt concentration. Furthermore experiments of DNAse digestion , being carried on both cells and nuclei at different times and enzyme concentrations , display a striking disappearence of the two higher temperature transitions , leaving present in nuclei only two broad transitions centered at 345°K and 365°K. The thermal profiles of nuclei and cells in the G1 cycling state , even pointing to the complexity of any interpretation of the observed enthalphy changes , confirm the dynamics of the higher order chromatin structure and further clarify the involvement of other macromolecular structures , such as ribosomal RNA and membranes (17).

DNA	T (°K)	ΔH(cal/g)
A	346	6.26
B	356	6.71
C	362	7.33
D	366	7.45

Tab. 1 Temperatures of the thermal transitions and melting enthalpies of calf thymus DNA in solutions of increasing ionic strength (see Fig. 3 for the meaning of the symbols). The melting enthalpies were determined from the area under each peak in thermal profiles in Fig. 3.

In order to increase the reproducibility and the sensitivity of the measurements with respect to the weak and weaker signals that can be detected when phase transitions occur in chromatin, cells and nuclei, it is mandatory to use instead of a standard data recovery signal, such as a chart-recorder, a computerized system, electronically interfaced (18), that offers great possibility for the acquisition, treatment and display of calorimetric data (Figure 4). The capabilities of digital acquisition and control card and its real time software, developed by the authors, coupled with the architecture of the IBM microprocessor allows us to have a powerful workstation that can support data acquisition, control and analysis in order to analyze a great number of data and to perform, with high speed, statistical operations on them.
The great number of data and their automated analysis allows us to evaluate a final experimental result, filtered both hardware (active filters) and software (statistical noise reduction) with low noise interference . Schematically, the interface system can be divided into three main hardware sections:

- output circuitry from DSC-2C;
- voltage amplifier with gain from G=1, voltage follower, to G=10 , differential amplifier;
- A/D channels card.

The amplifier output is sent to 4 analog input channels with 12-bit resolution with a sampling rate specified by the user in samples for second; the channels conversion time is 35 microseconds maximum and the throughput to memory is 15,000 conversions/second minimum.

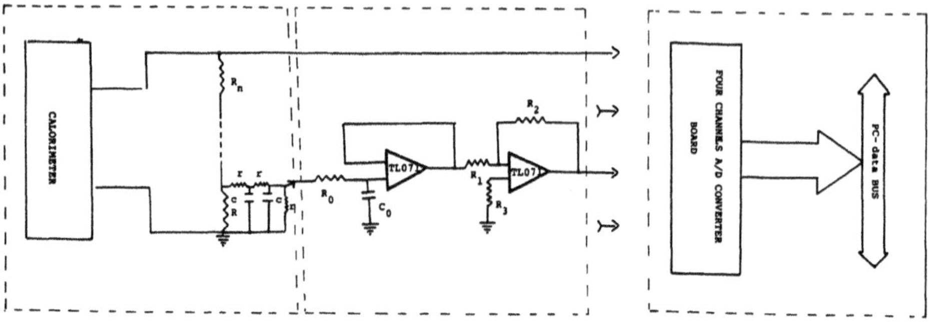

Fig. 4

Schematic of the "in house" interface between an AT-PC IBM Computer and the Perkin-Elmer calorimeter. The interface is divided in three main sections: output circuitry from DSC-2C, voltage amplifier, A/D channels card.

Circular Intensity Differential Scattering (CIDS)

Since the differential scattering of left versus right circularly polarized light has been observed (2,19-21) in biological specimens, different theoretical approaches have been presented to explain its origin (22-26).

For this phenomenon, has been adopted here and in the literature the term CIDS, Circular Intensity Differential Scattering. In the study of chromatin (2,19) the CD spurious signal, outside the absorption band of either proteins or DNA, appeared to depend from the wavelength of the incoming light as the fourth negative power; this signal furthermore disappeared after chromatin shearing (2,3,19).

While the former experimental evidence was compatible with the presence of a dipole component in the scattered electromagnetic field, the latter pointed to the existence of a high order structure in native chromatin (2,3).

In the past two distinct and opposite theoretical approaches have been proposed to explain the origin of CIDS in chromatin, namely:

- a multiple dipole approximation which treates nucleosomes like spheres, thereby ignoring the effects due to their geometrical anisotropy (24);
- the Born approximation, which assumes the real ellipsoidal shape for the nucleosomes, but does not take into account the interactions among them (23).

The results obtained (3,23,24), indicate that most of the signal is not due to dipole-dipole interaction of the scattering centers but rather to their ellipsoidal shape; namely the Born approximation more closely approaches the experimental data (3,23).

In a realistic model of the superhelix (1,2,3), the spacing between nucleosomes appears to be of about 5.5 nm (with about 6

nucleosomes per turn, a radius of 10 nm and pitch of 11 nm); such a close packing arrangement suggests that we cannot ignore the effects due to mutual interactions.
The approach recently developed (22) thereby combines the two previous ones, also in order to evaluate possible synergistic effects, considering both the geometrical anisotropy and the interactions among the dipolar scattering centers, i.e. nucleosomes; for a more complete and critical comparison with the other methods see (22). Theoretical predictions of the expected differential light scattering of circularly polarized light, have been made for chromatin within a coupled dipole approximation, treating nucleosomes as interacting ellipsoids. Within

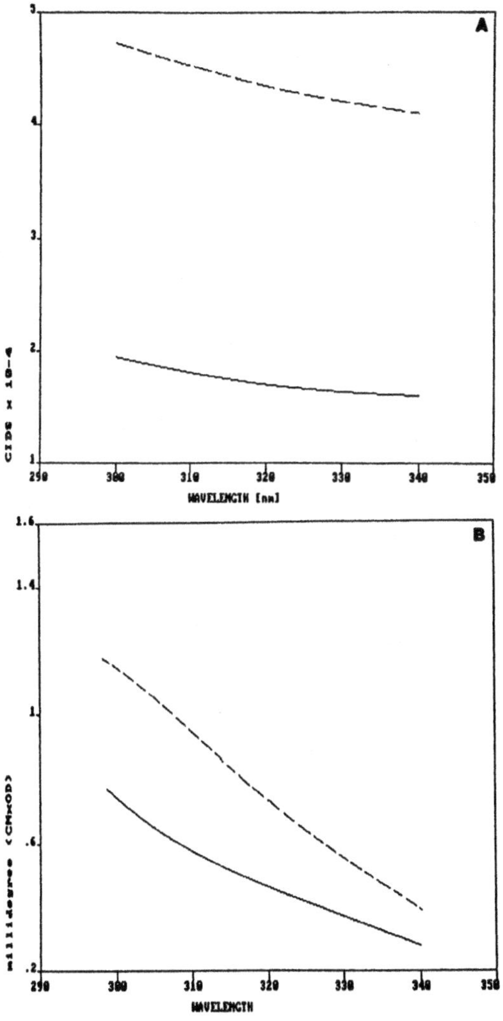

Figure 5

Wavelength dependence: (A) Theoretical CIDS of 2 different geometries P=15 nm, R = 10 nm (----) and P = 5 nm, R = 10 nm (——), evaluated using coupled dipole approximation of interacting ellipsoidal nucleosomes; (B) experimental data: Circular Dichroism spectra of unsheared chromatin from WI-38 cells quiescent (----) and stimulated to proliferate (——) by tryposonization and replating (ref. 2).

the framework of classical electrodynamics (27,28) chromatin is modeled as an helical array of ellipsoidal nucleosomes, each one corresponding to a dipole oscillator; each nucleosome, considered a scattering center, is impinged both by the source light and by the radiation scattered from the neighbours. We calculate the coefficients of the corresponding equations and then solve the system assuming either left or right circularly polarized incident light as known terms.
The amount of light scattered at a given angle ϑ is determined by computing the electric field at a certain point resulting from all radiating dipoles using the far field approximation; in terms of intensity, CIDS is evaluated as

$$\text{CIDS}(\vartheta) = \frac{\langle I(\vartheta)\rangle - \langle I(\vartheta)\rangle}{\langle I(\vartheta)\rangle + \langle I(\vartheta)\rangle}$$

after averaging, until convergence of the method, over more than 8000 possible orientations of the helix with respect to the laboratory frame.

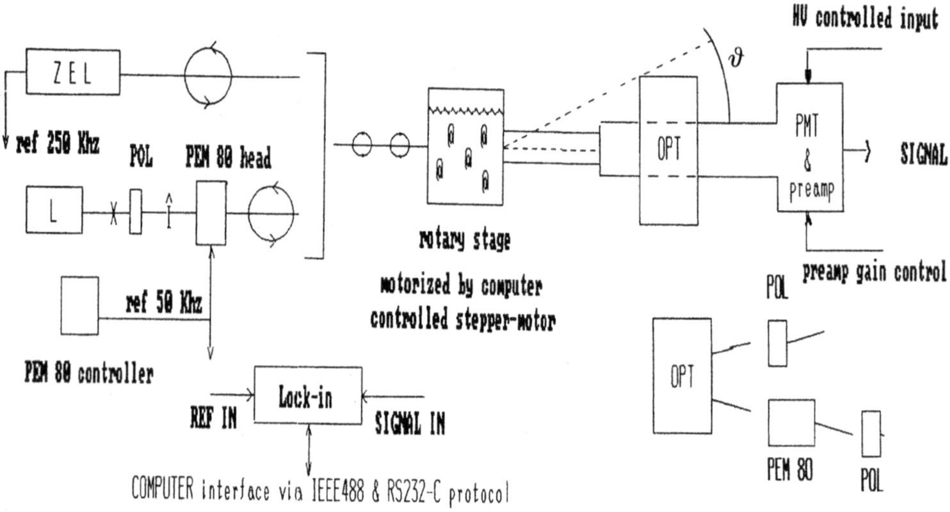

Figure 6

Simplified diagram of the experimental set-up used for CIDS measurements

The results obtained (22) have confirmed the CIDS sensitivity to variations in the geometric properties of the helix.
This theoretical approach (22) provides a coherent theoretical framework to explain and predict experimental observations on native chromatin.
The disappearance of CIDS for non-chiral structures and the predicted order of magnitude, very close to the experimental data, is consistent with the fact that the combined model is the best choice to predict patterns experimentally obtainable.
The high sensitivity shown in term of fixed orientations, not lost after averaging, let us think that CIDS is able to give important informations related to the higher order structures of complex chiral macromolecules, isolated or in situ, with more sensitivity than that obtained by classic spectroscopic or imaging techniques.
CIDS can then be used in monitoring minor changes in conformation of the chiral structures, like chromatin DNA, involved during cell cycle, aging and neoplastic transformation. A result worthy of notice (Figure 5) in this context is the predicted wavelength dependence of the CIDS for two distinct geometries of the polynucleosomal superhelix that closely parallels experimental data obtained at a fixed wide angle (with a standard spectropolarimeter) for two distinct chromatin structures isolated respectively from G0 and G1 mammalian cells . We are now developing the proper experimental apparatus in order to have a more critical and meaningfull comparison with experimental values obtained at all angles for both total scattered light and CIDS, using chromatin either isolated or in situ . The experimental set-up is schematically shown in figure 6 , see also chapter by Johnston et al. in this book.

Phase contrast optical microscopy

Phase contrast technique, based on the Zernike's filter principle (30), is a powerful method to observe, in a non invasive manner, living systems, such as mammalian cells, which are transparent in the visible range of the e.m. spectrum used in conventional optical microscopy. This made possible to observe, in their natural state, and even dinamically, many important cellular events, such as cellular proliferation, differentiation and so on.

Recent works have been in the direction of making this qualitative method a quantitative one (31) by modelling the instrument, i.e. a classical Zernike phase-contrast microscope, and by coupling the conventional optical data processing with a digital image system engineered to recostruct, using appropriate software, the phase information from the intensity acquired through a TV-based analysis system (32).

Rat hepatocytes were fixed in different glutaraldheide and AAC (alcohol - acetic acid) solutions and images were acquired by using the ACTA 500 image-processing system, through a Leitz microscope (objective magnification 25 X), and stored on magnetic tape for further processing on a VAX11/750 computer running under UNIX 4.2 BSD operating system (31); a suitable algorithm for the extraction of "phase" informatin from the phase-contrast images, implemented in FORTRAN, was used (32). The phase of the light passing through the optical system and the specimen is then correlated with physical and chemical characteristics of the sample itself. In order to have a calibration for the

reconstruction software and a test for the system modelling we have performed contrast matching measurements, using known specimens under all points of view: shape, dimension, refraction index, and so on. The test has been performed on spherical homogeneous particles against homogeneous media on the system described earlier (31), and presently we are reducing the data (33).
Actually we are performing new consistent recalibration using a PC-based image analyzer system described in this chapter. After eliminating artifacts arising from the specific imaging system employed, we have attempted to establish the difference in the refractive index of the various cell biopolymers (i.e. RNA, DNA and proteins) by matching living cells and/or native nuclei with different suspending media of known index of refraction (Figure 7).

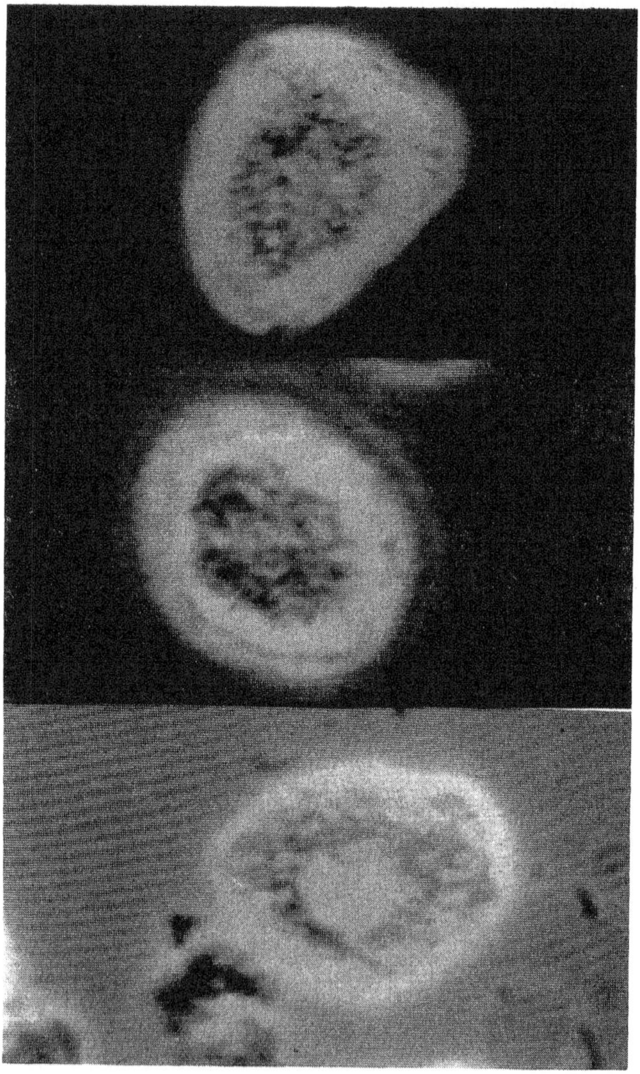

Figure 7

Rat hepatocytes processed by phase contrast optical microscopy and suspended at different saccarose concentration, namely: 1% (up), 30% (middle), 50% (down)

DYNAMICS OF CHROMATIN-DNA

Multiparameter analysis of sinchronized Feulgen -stained He-La nuclei revealed clear clusters around homogenous labeling patterns (following 3H-thymidine pulse) and morphometric data, unique for each time interval, as cells progress through two cycles of nuclear DNA condensation-decondensation (29). It is interesting that the high level of nuclear condensation is achieved both in very early G1 (120-71 IOD per unit area) and late G1 (63 IOD per unit area), just prior to a dramatic increase in nuclear DNA dispersion (24-25 IOD per unit area) which in turns precedes initiation of RNA synthesis (early G1) and DNA synthesis (early S). All of the above findings point to a role of "higher order" chromatin-DNA organization in the control of DNA replication and transcription.

On reviewing the autoradiography slides (defining as labeled only those nuclei with grain numbers well above the background), labeling patterns can be grouped independently of grain density into only two categories: those in which the grains were distributed circularly and sparing a central region (pattern 1), and those in which the grains were variously distributed throughout the nucleus (patterns II-IV). By analysing the area distributions at each time point of labeling Patterns I and II-IV, it appears that only nuclei of a characteristic size (or intranuclear distribution of the replicating DNA) will undergo the pattern of DNA synthesis such as to produce Pattern I, and that this would determine S-phase initiation. Furthermore the sites of DNA initiation, contrary to previous suggestions must be mostly near the nuclear envelope to induce a circular ring of 3H-thymidine incorporation around the nuclear border: as it can be shown from rigorous modeling a homogenous distribution within the nucleus would yield patterns similar to types II-IV, which are found only at later times (34).

We must emphasize that the abrupt changes in nuclear-DNA morphometry at the beginning of G1 and at the end, prior to S phase, so pronounced at the single cell level (3-6 fold increase in area and dye uptake) reduce to only 40-70% alterations in average value of "synchronized" populations: similarly the structural alterations between late G1 and middle S-phase cause significant but finite differences in differential light scattering in isolated bulk chromatin which then becomes a yes versus no phenomenum in individual nuclei. The sign and magnitude of the differential scattering of left versus right circularly polarized light (CIDS), outside the DNA absorptive band, is indeed explainable with a "left-handed" helical folding of one or more nucleofilaments (with the multifibrilar "rope" giving a higher CIDS signal than the unifibrilar "solenoid"); transition to a linear "unfolded" nucleofilament would be accompanied by a null differential light scattering , as indeed found for S phase nuclei.

Recent theoretical simulations (see earlier and reference 22) based on a multiple dipole interaction of ellipsoidal scattering centers (arranged as a superhelical folding of polynucleosomes) agree quite well (assuming given geometric parameters for the superhelix) with experimental CIDS measurements on isolated G1 and S chromatin, carried out between 300 and 340 nm in the forward angle (2).

The modulation of the two-order superhelical foldings of DNA around the nucleosomal axis and around a quaternary "rope-like" axis can be detected in the bimodal derivative plot of positive molar ellipticity changes with temperature and more clearly in the multimodal microcalorimetric enthalpy changes in the same chromatin isolated from cells in different functional states (15).

Finally a higher number of negative superhelical turns per 10 base pairs is observed in middle G1 with respect to middle-S phase unsheared chromatin,or in G0 versus early G1, or in transformed versus normal chromatin , as monitored by the intercalation kinetics of low concentrations of Ethidium Bromide . Their functional significance, in terms of additional free energy, has been addressed elsewhere (2,3; see also this book) .

Following initiation, replication of eukaryotic DNA proceeds by means of a continuous sequence of disassembly, reproduction and reassembly, which passes through each DNA sequence and which originates simultaneously in many sites near the envelope, propagating in both directions as a replication fork (35).

The "fibrosome" with its drapery-like structure (see Figure 1) may provide an excellent topological constraint for this process, warranting that prior to replication chromatin locally relaxes to condense again after replication, but in such a way to yield daughter chromosomes quite disantagled to allow their segregation into the daughter cells of mitosis .

The basic intranuclear DNA distribution in three orders of superpacking and the existence of discrete transitions among them appear similar in all of the most different conditions of ionic strength , pH and for many alternative procedures of chromatin characterization (either isolated or in situ) , being preserved even after complete removal of both H1 and HMG (for a review see the book in reference 3) .

The first level of control could then be exercised by other factors, namely:

a) by the strongly bound non-histones, still present at 2 M NaCl, and maintaining the DNA supercoil;

b) by nicking-closing enzymes such as topoisomerase which is known to vary its activity during the cell cycle and is critically determining negative DNA superhelix and therefore transcription;

c) by local nuclear changes in ions (those interacting with DNA phosphates and thereby affecting DNA bending and cell function);

d) by the local level of DNA methylation or hydration (critically determining secondary structure and gene activation).

It would then seem that H1 histone phosphorylation, H4 acetylation and HMG non-histones represent only the second level of control of gene expression, being a sufficient but not necessary co-factor for the lower order transitions, possibly between the nucleofilament and the free-DNA leading to local gene activation.

Recent data on the effect of nuclease digestion on nuclear DNA organization (thereby reported later) are indeed compatible with the above hypothesis (42).

Higher order chromatin structure and gene activation

In all physico-chemical and functional characterization of isolated chromatin, there is of course a large degree of artifacts introduced during the isolation process. In this respect abundant evidence exists in the literature that the structural and functional properties of native chromatin are modified by physical alterations introduced during chromatin preparation (2,19). In order to bypass such limitations, two different approaches has been used: the first is to analyze the geometry and the densitometric textures of the nuclear images produced by differential staining of chromatin DNA in situ (either by a Feulgen reaction or by Acridine Orange staining); the second is the use of nucleases to carry out mild digestion of chromatin either before or after cell fixation.

In recent years much attention has centered about the selective nuclease activities of such enzymes as micrococcal nuclease and DNAse I. Many investigators have reported the preferential digestion by DNAse I of genes that had been activated at some time point in the epigenetic history of the cell system under study (36). Due partially to release

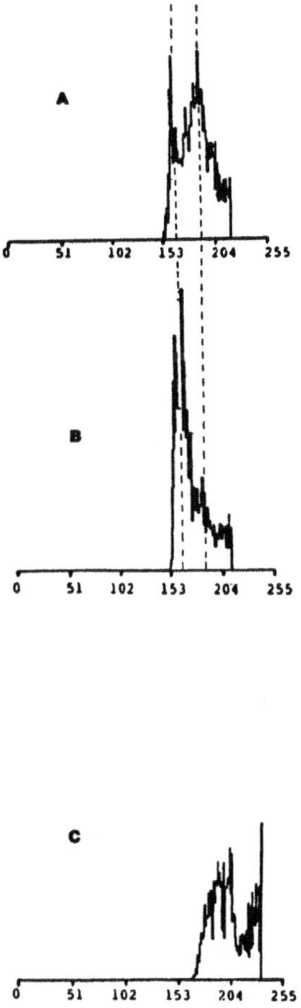

Figure 8

Transmittance histograms (number of picture points versus gray levels ranging from 0 to 255) of G1 nuclei control (A), G1 nuclei digested and then fixed (B), G1 nuclei fixed and then digested (C).

after limited digestion by micrococcal nuclease of relative amounts of histone 1, HMG proteins and other nonhistone proteins differing from that in bulk chromatin, a preferential digestion of transcriptionally activé chromatin has also been suggested for micrococcal nuclease (37). Also reported has been the organization of large 25-50 kb DNA domains with increased DNAse I sensitivity adjacent to potentially active genes (38).

Recently we have studied the digestion (following fixation) by micrococcal nuclease of HeLa S3 nuclei at the GI-S boundary through image analysis of Feulgen stained nuclei (39,40). Within the possible experimental error for the experimental design employed, no preferential digestion of the subfraction of dispersed chromatin ("euchromatin") within the fixed nucleus was demonstrated.

Indeed, both a comparison of percent digestion at given times versus nuclear area (for 2C DNA nuclei) and a comparison between experimental optical density histograms (of picture points vs. given optical density level) from individual "fixed" nuclei versus theoretical histograms generated using a hypergeometric probability distribution model for enzyme digestion (41) showed, if anything, a preferential digestion of denser nuclei or denser regions ("heterochromatin") of chromatin.

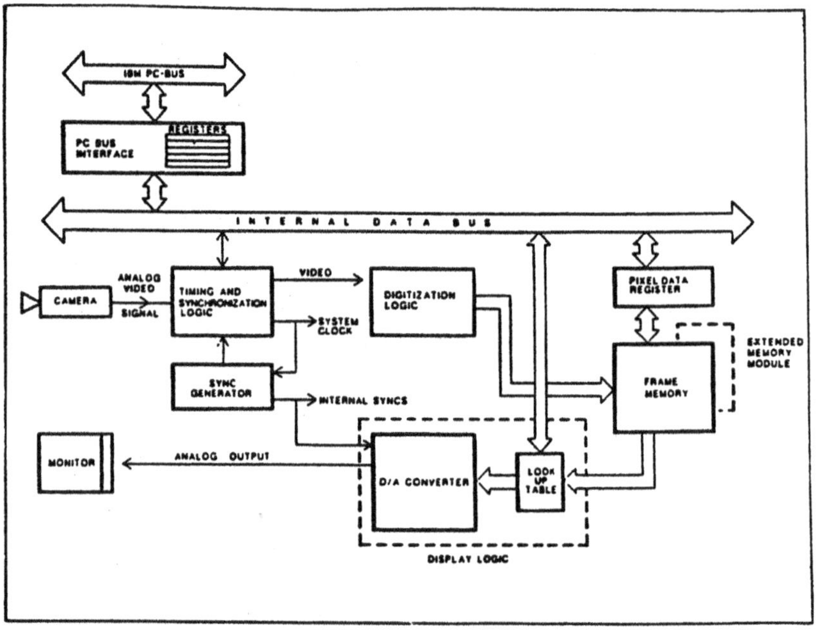

Figure 9

Architecture of the automated image analyses system, IBM microprocessor based and equipped with PC vision frame grabber board.

Questions, however, have been frequently raised concerning the actual specificity of micrococcal nuclease for active genes, and reasonable doubts remained as well concerning the existence of a very small fraction of "active" chromatin preferentially digested at early times and at low enzyme concentration yet undetected by image analysis in our early experiments (where digestion is being carried out also following rather disruptive cell fixation). Therefore, lately we extended the above experiments to the study of very limited digestion by DNAse I and micrococcal nuclease before and after cell (or isolated nuclei) fixation. Mostly by DNAse and before fixation we demonstrate a significant change in the intranuclear DNA organization and an occasional preferential digestion of a dispersed, low optical density chromatin subfraction, whose native intranuclear organization is such as to allow detection at optical levels of resolution (42).
The effect is readily apparent only when mild DNAse digestion is carried out on unfixed metabolically active G1 nuclei (Figure 8).
The configuration of the whole system consists of an AT-IBM 80286 microprocessor with mathematic 80287 coprocessor, color graphic card, board-level video digitizer and display frame grabber (Imaging Technology Inc.), an intensified vidicon tube TV-camera. All software is completely implemented in Assembly, Pascal, C-lattice and Fortran 77, under DOS. 3.10 operating system. The images acquired are stored in a fast on-board 512 x 512 x 8 bit frame memory and simultaneously displayed on the external monitor. The digitalization of analog video is made with 8 bit of precision using a 10 MHz flash A/D converter, each pixel represents one of 256 possible gray or over 16 million pseudocolor levels (Figure 9).

Acknowledgments

This work has been supported by Grant from National Research Council - Finalized Project Oncology and from the Ministry of Public Education - Project "Biophysical Control of Cell Growth"

BIBLIOGRAPHY

1. Finch,J. and Klug,À., P.N.A.S. USA , 73 ,1897-1901 (1976).
2. Nicolini,C. and Kendall,F. ,Physiol. Chem. Phys., 9 ,265-282 (1977).
3. Nicolini,C.,"Biophysics and Cancer", Plenum Press, New York (1986)
4. Bradbury,E.M. and Matthews,H., in "Cell Growth",editor C. Nicolini, Plenum Press, New York, NATO-Life Sciences Series, 411-454, (1982)
5. Belmont, A. and Nicolini, C., J.Theor.Biol., 90-169, (1981)
6. Benyajati, C. and Worcel, A., Cell , 9 , 393 (1976)
7. Cook, P.R. and Brazell, I.A., J.Cell Sci., 22 , 287 (1976)
8. Pardoll,D.,Volgenstein,B. and Coffey,D., Cell, 19 ,527-536 (1980).
9. Lebrowski,A. and Laemmli,U., J.Mol.Biol., 156 ,309-314 (1982).
10. Cavazza,B., Trefiletti,V., Pioli, F., Ricci,E., and Patrone,E., J.Cell Sci., 62 ,89-102 (1983).
11. Nicolini,C., Cavazza,B., Trefiletti,V., Pioli, F., Beltrame, F., Brambilla, G., Maraldi, N. and Patrone,E., J.Cell.Sci., 62 ,103-115 (1983).
12. Agard,A. and Sedat,J., Nature, 302 ,676-681 (1983).
13. Nicolini,C., Vernazza, G., Chiabrera, A., Maraldi, I.N. and Capitani, S., J.Cell.Sci., 72 , 75-87 (1984).
14. Nicolini,C., Trefiletti,V., Cavazza,B., Patrone,E. and Carlo,P., Science, 219 ,176-178 (1983).

15. Nicolini,C., Carlo,P., Martelli,A., Finollo,R. and Brambilla,G., J.Mol.Biol., 161 , 155-178 (1982).
16. Touchette, N.A. and Cole, R.D., P.N.A.S. USA, 82 , 2642-2664 (1985)
17. Germano P., Diaspro A. and Nicolini, C.A., unpublished data, in preparation
18. Millman, S., Halkias, C.C., Integrated circuits: analog and digital circuits and systems, Mc Graw-Hill, New York (1978)
19. Nicolini C., Baserga R. and Kendall F., Science, 192 , 796-798 (1976)
20. Dorman, B.P. and Maestre, M.F. (1973), P.N.A.S. USA, 70 , 251
21. Maestre, M.F., Bustamante, C., Hayes, T.L., Subirana, J.C. and Tinoco, I. Nature, 298 , 773, (1982)
22. Diaspro, A. and Nicolini, C., Cell Biophysics, in press (1986)
23. Belmont A., Zietz, S. and Nicolini, C., Biopolymers, 24 , 1301 (1985)
24. Zietz, S., Belmont, A. and Nicolini, C., Cell Biophysics, 5 , 163 (1983)
25. Patterson, C.W., Singham, S.B., Saltzman, G.C. and Bustamante, C., J. Chem.Phys. 84 ,1916 (1986)
26. Keller, D., Bustamante, C. and Tinoco, I., J.Chem.Phys. 80 , 4917 (1984)
27. Jackson, J.D., Classical Electrodynamics, Wiley (1975)
28. Shurcliff, W.A., Polarized Light, Harvard, Cambridge, Mass. (1962)
29. Belmont, A. and Nicolini, C., Cell Biophysics, 5 , 79 (1983)
30. Zernike, F., Z.Tech. Physik., 16 , 454, (1935)
31. Beltrame, F., Bianco, B., Castellaro, G., Diaspro, A. and Nicolini, C., Medical & Biological Engineering & Computing, 23, Supplement Part 1, (1985)
32. Beltrame, F., Bianco, B. and Chiabrera, A., Cell Biophysics, 6 , 103-116 (1984)
33. Diaspro, A., Germano, P. and Nicolini, C.A., unpublished data, in preparation
34. Belmont,A., Kendall,F. and Nicolini,C.,J.Cell.Sci., 65 ,123-138 (1984).
35. Huberman, J.A. and Tsai, A., J.Mol.Biol. 75 , 5 (1973)
36. Weintraub, H. and Groudine, M., Science 193 , 848-856 (1976)
37. Levy-Wilson, B., Watson, D. and Dickson, G.H., Nucleic Acids Res., 6 , 259-274, (1979)
38. Stalder, J., Seeheck, T. and Brown, R., Eur.J.Biochem., 90 , 391-396 (1978)
39. Nicolini, C., J.Submicrosc.Cytol. 12 , 475 (1980)
40. Kendall, F., Beltrame, F. and Nicolini, C., in "Chromatin Structure and Function", editor C. Nicolini, Plenum Press, New York, NATO-Life Sciences Series, 265-292 (1979)
41. Kendall, F., Beltrame, F., Belmont, A. and Nicolini, C., Cell Biophysics, 4 , 19-38 (1980)
42. Nicolini, C., Diaspro, A. and Germano, P., J.Cell.Sci. (submitted paper)

ON THE BEHAVIOUR OF IONIC POLYSACCHARIDES IN DILUTE AQUEOUS SOLUTIONS

M. RINAUDO
Centre de Recherches sur les Macromolécules Végétales
B.P. 68, 38402 St-Martin d'Hères Cedex (France)

INTRODUCTION

The polysaccharides form a very important class of biopolymers as much for their biological activities than the industrial applications.

This paper concerns mainly the thermodynamic properties of some polysaccharides from plants (natural or modified) and from microorganisms. Previously, much of our work was devoted :

- first to ionic cellulose derivatives, the carboxymethylcelluloses (CMC), to point out the role of the charge density on the thermodynamic (1,2),
- secondly to polyglutamic acid oligomers to demonstrate the role of the number of neighbour ionic sites (3,4). Here, we would like to demonstrate how thermodynamic and transport parameters in the field of polysaccharides allow to get information on the local conformation of these stereoregular polymers ; some examples will be described.

Two types of systems will be investigated :

- first, ionic polysaccharides dispersed in dilute or semi-dilute solution ;
- secondly, cooperatively connected chains forming thermoreversible or ionic gels. In these last systems, the specific role of divalent counterions will be discussed.

Some information on the hydrodynamic of the different polymers will be also given and discussed.

EXPERIMENTAL

The polysaccharides are usually commercial samples ; they are purified and prepared as monovalent salt form (generally Na form) before experiments to avoid interchain bonds and aggregation process due to possible multivalent ions interactions.

Pectins are extracted from apples (5,6), alginates and κ-carrageenans from algae (7,8) ; xanthan and succinoglycan are exopolysaccharides produced by bacteria (9,10). The chemical structures are given in Figures 1 and 2.

The activity coefficient γ_x given in this work is equal to γ/γ_o which reflects directly the polyelectrolyte effect ; γ is the effective activity coefficient of the counterions (x) measured in the polyelectrolyte solution and γ_o the activity coefficient of the same counterions in a simple salt solution at the same ionic concentration.

The activity coefficients are determined from potentiometry using specific electrodes Tacussel for Na and K and ORION (model 92.32) for Ca ; the potentiometer is a Tacussel Minisis 6000.

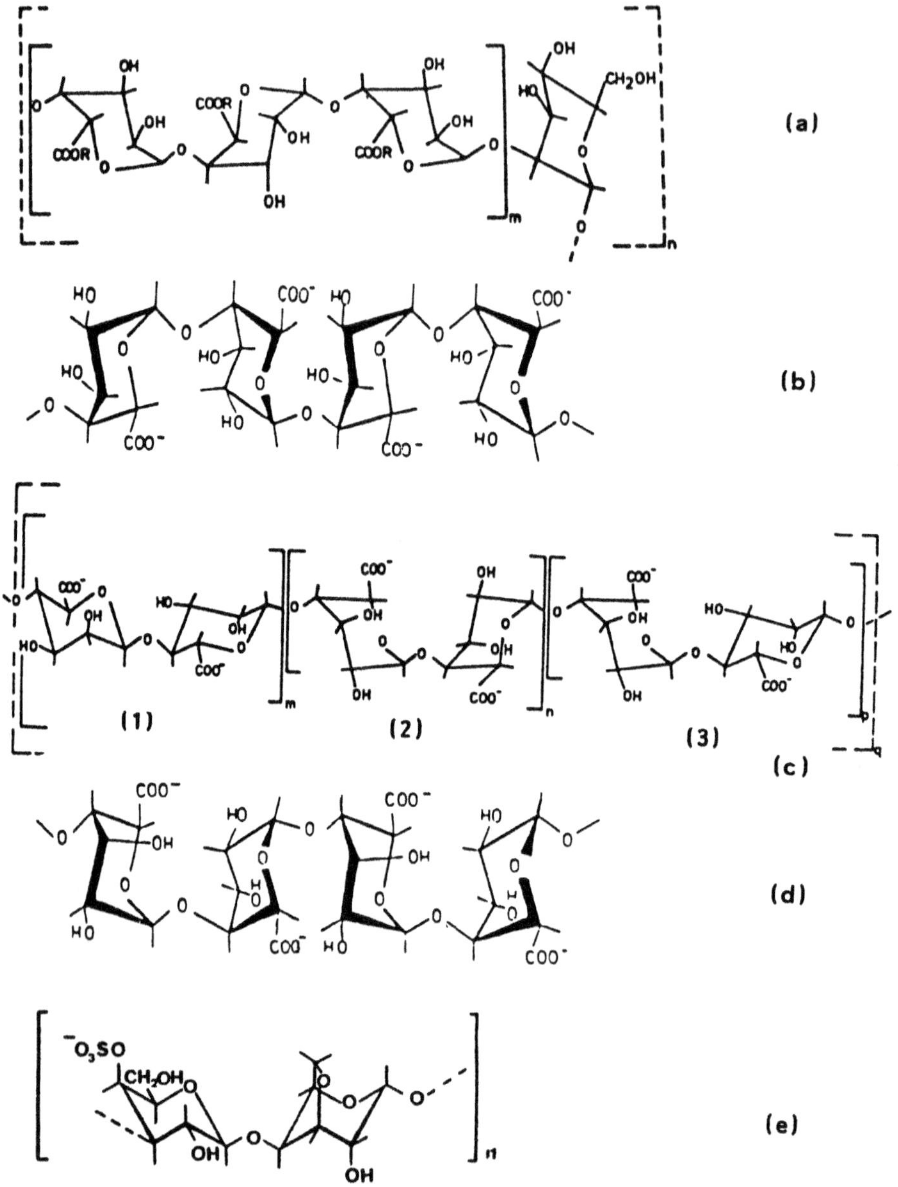

FIGURE 1 : Chemical structure of some ionic polysaccharides.
(a) Pectin = poly α-D-galacturonate partially esterified R = -H or $-CH_3$; kinks are L-rhamnose. (b) Poly α-D-galacturonate block.
(c) Alginate = poly β-D-mannuronate (1) and poly α-L-guluronate (2) blocks ; alternating sequence (3). (d) Poly α-L-guluronate block.
(e) κ-carrageenan = (1 → 3) β-D-galactose-4-sulfate (1 → 4) 3,6-anhydro α-D-galactose.

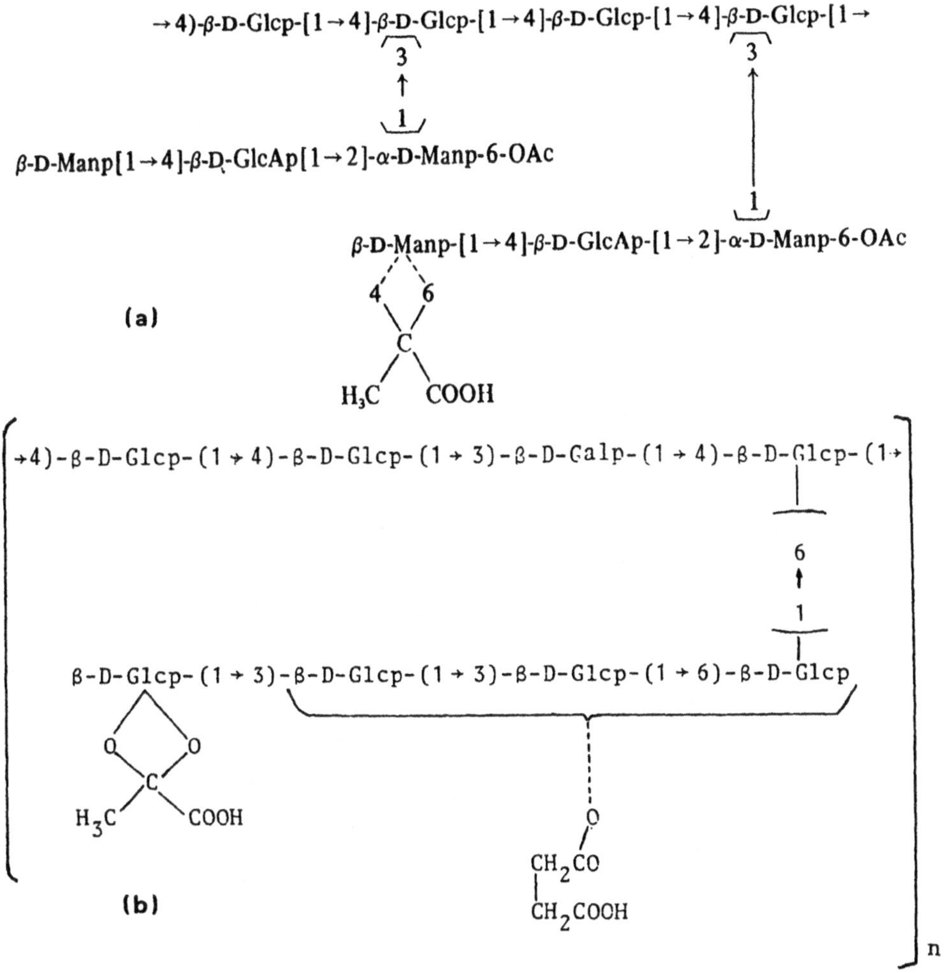

FIGURE 2 : Chemical structure of two bacterial polysaccharides.
(a) Xanthan
(b) Succinoglycan.

The transport coefficient f is determined from specific conductivity (χ) using a Tacussel CD 78 bridge equipped with platinum electrodes. The equivalent conductivity of the polyelectrolyte solution is related to the ionic equivalent conductivity of the counterions (λ_c) and that of the equivalent of ionic sites of the polyelectrolyte (λ_p) by :

$$\Lambda = f\ (\lambda_p + \lambda_c) \qquad [1]$$

When the change in f as a function of the temperature is considered, usually a differential method is used : the ratio χ_1/χ_2 (T) is measured using two cells having same characteristics introduced respectively in a simple electrolyte solution (χ_2) and in a polyelectrolyte solution (χ_1) which have exactly the same initial specific conductivity at 25°C.

The enthalpy of conformational melting ΔH is determined from microcalorimetry in a differential method using a microcalorimeter CALVET from SETARAM. ΔH is obtained from isothermal or isoionic calorimetry. In isoionic calorimetry the conformational change is induced by a variation of temperature but in isothermal calorimetry the transition is induced by a salt jump.

The optical rotation (α) is measured to follow the conformational changes using a Spectropol 1b from FICA.

The average molecular weights and the molecular weight distributions are obtained using a low angle laser light scattering Chromatix KMX6 or CMX 100 in static experiments or as detector on line in steric exclusion chromatography. The viscosity (η) at different shear rate ($\dot{\gamma}$) is obtained using a Low Shear viscometer and a Rheomat 30 from Contraves ; $[\eta]_o$ is the intrinsic viscosity determined in the low shear rate limit.

A. RELATION BETWEEN THERMODYNAMIC PARAMETERS AND LOCAL STRUCTURE

General properties of polyelectrolytes

First of all, it seems to be important to recall some of the parameters introduced in the theories of dilute solutions of polyelectrolytes. The thermodynamic parameters involved are the activity coefficient γ and the osmotic coefficient ϕ (11) ; they are controlled by the charge parameter λ of the polymer introduced in the different theories by Katchalsky (11,12), Manning (13) or Oosawa (14). The parameter λ is proportional to the charge density with

$$\lambda = \frac{\nu e^2}{D\ h\ kT} \qquad [2]$$

in which ν is the number of ionic sites along a chain of length h, kT the Boltzmann term, e the electronic charge and D the dielectric constant of the solvent.

The Manning theory is valid only for an infinite thin chain and infinite dilution ; it introduces the fundamental idea of counterions condensation when λ is larger than z^{-1} for counterions of valence z (15). The prediction of the properties in solution depends on λ value and are respectively in absence of external salt for counterions of valence z :

$$\ln \gamma = \begin{cases} -1/2\ |z|\lambda & \lambda < |z|^{-1} \quad [3] \\ -1/2 - \ln\ |z|\lambda & \lambda > |z|^{-1} \quad [4] \end{cases}$$

$$\phi = \begin{cases} 1 - 1/2\ |z|\ \lambda & \lambda < |z|^{-1} \quad [5] \\ 1/2\ |z|^{-1}\ \lambda^{-1} & \lambda > |z|^{-1} \quad [6] \end{cases}$$

For univalent counterions, the "fraction of free counterions" obtained from transport measurements are :

$$f = 1 - [0.55\ \lambda^2/(\lambda + 3.14)] \qquad \text{and} \qquad f = 0.87\ \lambda^{-1} \qquad [7]$$

In the model proposed by Katchalsky in absence of external salt $\gamma \equiv \phi$; a dependence with the polymer concentration is predicted but the extrapolation to infinite dilution leads to ϕ values identic to that given by Manning treatment.

On carboxymethylcelluloses, the experimental values are larger than the theoretical ones especially for high charge densities ; with sulfonic polymers the agreement is always better. This difference is usually found in the litterature. An other property of these polyelectrolytes is the ionic selectivity which appears in the range of $\lambda = 1$ with monovalent counterions as previously demonstrated. Due to large electrostatic potential, ion pairs are formed inducing the ionic selectivity mainly in the series of monovalent counterions ; it affects at least one part of the condensed counterions introduced by Manning.

Role fo the conformation

In stereoregular polymers, one may assume that their charge parameters (λ) are directly related to their local structure and degree of extension of the backbone.

As example, the activity coefficients γ_{Na} and γ_{Ca} were compared for a CMC ($\overline{DS}$ 0.7 ; $\lambda = 0.97$) and the bacterial polysaccharide xanthan with about the same contour length and charge density (9).

	γ_{Na}	γ_{Ca}	Local conformation
CMC	0.640	0.305	Coil
Xanthan	0.650	0.295	Na form coil $\lambda_c = 1.03$ {Ca form single helix $\lambda_h = 1.13$
Manning prediction	0.589	0.294 0.268 0.134	for λ_c for λ_h single helix for λ_h double helix

The apparently too low γ_{Ca} for xanthan compared to that of CMC was attributed to the increase of λ when single helix is formed ; this ordered conformation is confirmed by optical rotation measurements. From these data, one excludes the double helix formation for the xanthan even if this point is still in discussion in the litterature.

On succinoglycan, from conductivity measurements on the Li and K salt forms of the polymer using relation 1, one deduces

$$\Delta\Lambda_{K,Li} = f\ (\lambda_K - \lambda_{Li}) \qquad [8]$$

From $\Delta\Lambda$, the values of f are calculated for the both conformations depending on the temperature ; one obtains :

$$f_h = 0.93 \qquad \text{and} \qquad f_c = 0.97$$

The charge parameter of this polymer in the extended form is λ = 0.68 for which the theoretical prediction (relation 7) is 0.933 ; then the experimental value agrees with a single helical extended structure when the ordered conformation is stabilized at low temperature.

These results were confirmed by conductivity dependence during conformational change ; χ_1/χ_2 increases only of few per cent in the thermally induced helix-coil transition for xanthan and succinoglycan (Figure 3).

The increase in χ_1/χ_2 is directly related to $(f_h - f_c)$ corresponding to the helix coil transition for the both polysaccharides. Then the conformational transitions observed are single chain process.

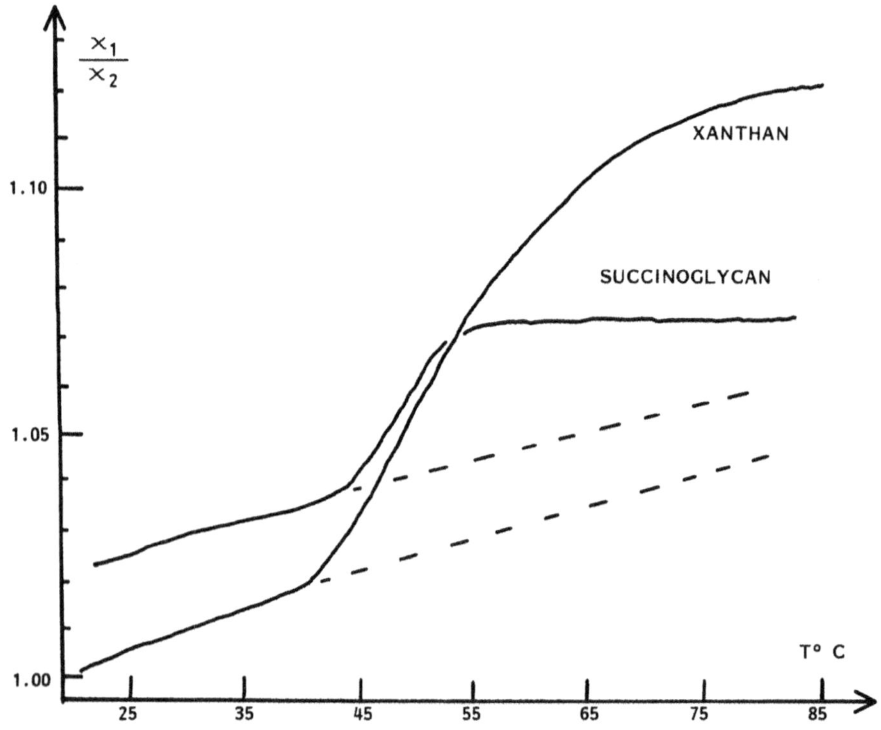

FIGURE 3 : Ratio of specific conductivities of the polyelectrolyte χ_1 and sodium chloride χ_2 as a function of temperature in the range of conformational transition.

On κ-carrageenan, activity coefficients and optical rotation measurements allow us to demonstrate :

1. that a large ionic selectivity exists betwen K^+ and Na^+ counterions even if $\lambda < 1$;

2. that K^+ induces helical conformation and dimer formation corresponding to a charge parameter twice that of the Na-form (16) :

Infinite dilution	γ_{Na}	0.72	Coil
	γ_K	0.32 - 0.39	Helical structure
Manning prediction	γ_c	0.72	$\lambda_c = 0.68$
	γ_h	0.37	λ_h dimer = 1.65

Rochas also determined the enthalpy of conformational transition ; the ΔH are in agreement with a transition helical dimer → coil in the sol condition and confirmed the activity coefficients measurements (17,18).

At end, an original result was obtained with pectins with low methoxyl content (5) ; f_{Na} and f_{Ca} are much more different than usually (theory predicts $f_{Na}/f_{Ca} = 2$). The results obtained when $\lambda = 1.58$ are the following :

	Experimental	Theoretical (relation 7)
f_{Na}	0.620	0.551
f_{Ca}	0.110	0.276 (Ca and Mg)
f_{Mg}	0.280	

Our conclusion was that under Ca form a dimer is formed which allows to justify the very low value f_{Ca} extrapolated to infinite dilution. In addition, ϕ_{Ca} varies largely with the polyelectrolyte concentration and aggregation occurs. On the opposite, under Mg form, the behaviour is normal and single chain conformation is suggested without any aggregation (19) (Figure 4).

<u>Conclusion</u>

These experimental data concerning potentiometric and conductimetric measurements show clearly how they can help to conclude about multiple helical structure formed in given conditions.

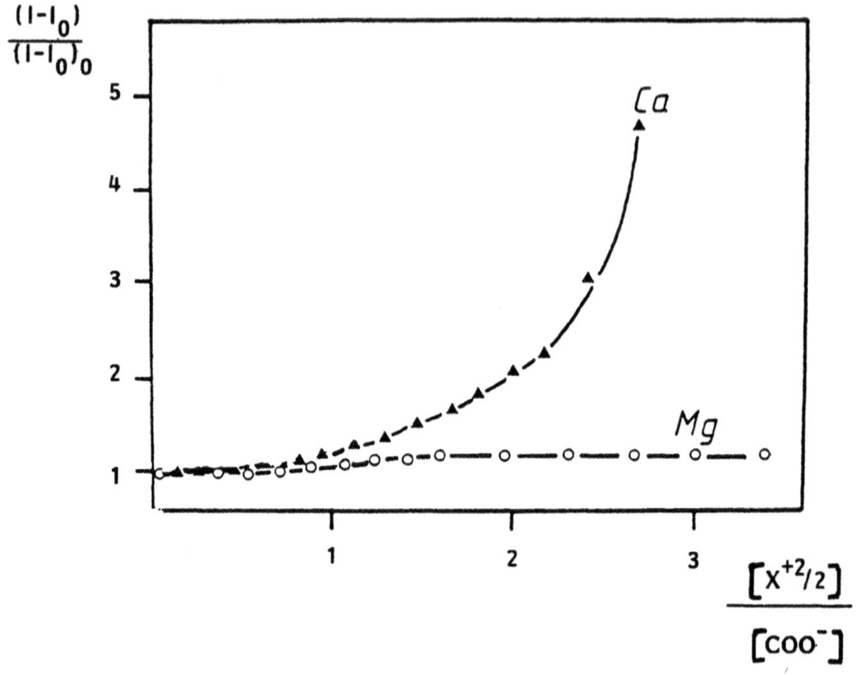

FIGURE 4 : Ratio of the intensity scattered by the solution $(I-I_o)$ on the initial intensity scattered $(I-I_o)_o$ during addition of divalent electrolyte on Na-polygalacturonate (polymer concentration 0.15 g/l).

B. CONFORMATIONAL TRANSITION

As monovalent counterion forms, a thermal increase often induces a change in the conformation as shown from optical rotation variation. The electrostatic properties are involved directly and by example when the degree of neutralization increased in salt free solution of xanthan, the ordered conformation is destabilized (Figure 5) (20).

When divalent counterions are added they decrease the net charge and following the ordered conformation is more stable (Figure 6) (21) ; as Ca form, the thermal stability is large and the melting temperature must be much larger than 100°C if it exists.

For the ordered conformation of the biopolymers in the monovalent form, it exists a direct relation between the temperature of conformational transition T_m and the total ionic concentrations (C_T) taking into account the free counterions $(\overline{\gamma}Cp)$ at half transition and that from the external salt C_s :

$$T_m^{-1} = C^{st} + k \log C_T$$

In fact, the electrostatic properties of the polymer and the change in the free counterions on both sides of T_m allows to obtain the following relations from the Manning theory :

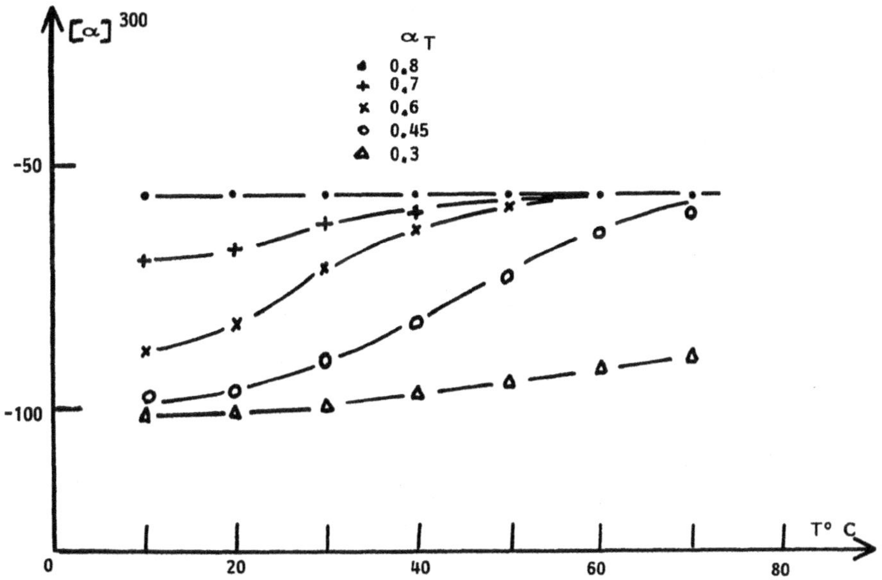

FIGURE 5 : Specific optical rotation at 300 nm as a function of temperature for different degrees of dissociation α_T.

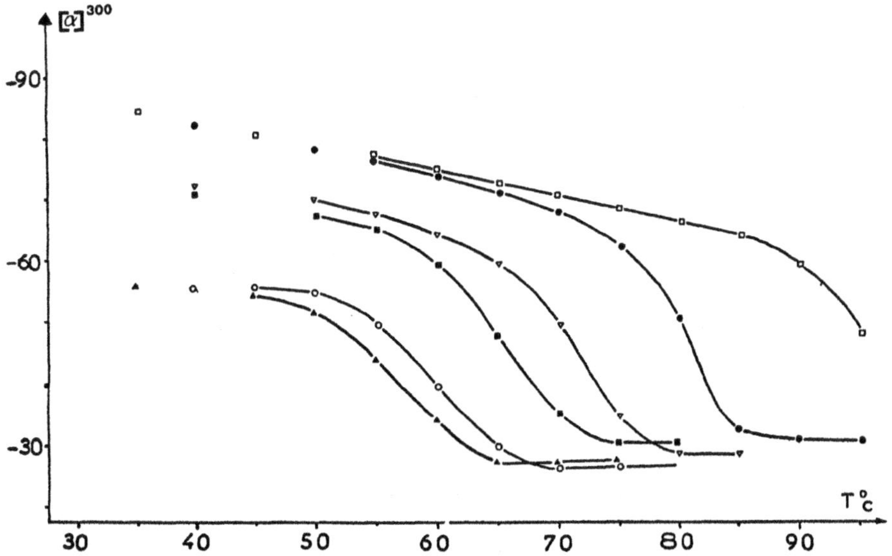

FIGURE 6 : Specific optical rotation at 300 nm as a function of temperature for Na-Xanthan (C_p = 1 g/l) in presence of different amounts of calcium (21).

▲ 4×10^{-4} N ■ 6×10^{-4} N ● 10^{-3} N

○ 5×10^{-4} N ▽ 8×10^{-4} N □ 1.5×10^{-3} N

$$\lambda_c, \lambda_h < 1 \qquad \Delta H = -R/2\ (\lambda_h - \lambda_c)\ d\ln C_T/d\ T_m^{-1} \qquad [9]$$

$$\lambda_c < 1 < \lambda_h \qquad \Delta H = R/2\ (2 - \lambda_c - \lambda_h^{-1})\ d\ln C_T/d\ T_m^{-1} \qquad [10]$$

$$1 < \lambda_c < \lambda_h \qquad \Delta H = R/2\ (\lambda_c^{-1} - \lambda_h^{-1})\ d\ln C_T/d\ T_m^{-1} \qquad [11]$$

From the slope of the experimental curve ln C_T (T_m^{-1}), the ΔH of conformational change can be deduced and compared with the value obtained by microcalorimetry (Figure 7).

	$d\ln C_T/d(T_m^{-1})$	ΔH calculated relations 9-11	ΔH measured by microcal.
κ-carrageenan (K^+)	8080 K^{-1}	24 KJ/disacch.	22 ± 2 KJ/disacch.
Succinoglycan (Na^+)	24534 K^{-1}	1.47 KJ/eq.	18 KJ/eq.
Xanthan (Na^+)	5 110 K^{-1}	single helix 1.83 KJ/eq. double helix 11.25 KJ/eq.	4.8 KJ/eq.

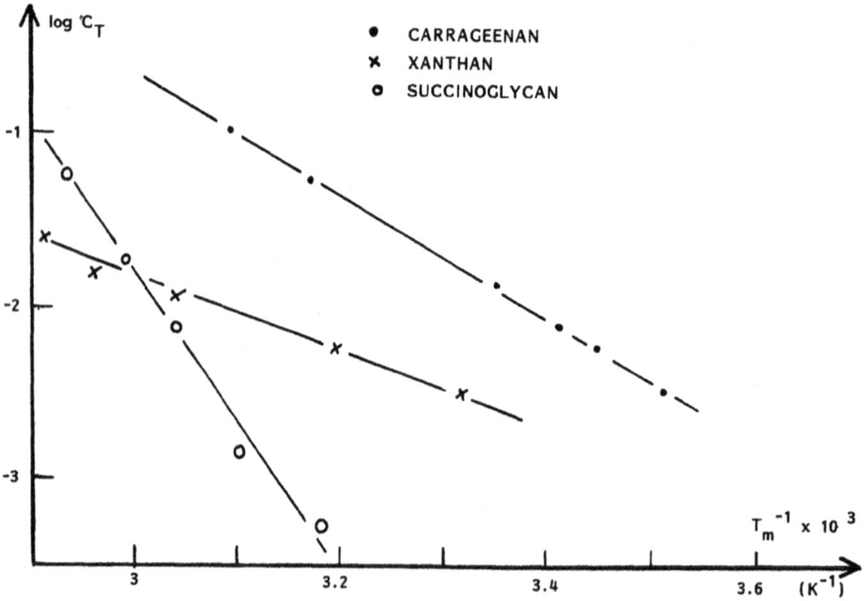

FIGURE 7 : Dependence of T_m^{-1} with the total ionic concentration

• Carrageenan   X Xanthan   o Succinoglycan

In dilute solution using isothermal calorimetry, the ΔH obtained for -carrageenan is very close the theoretical value of 24 kJ predicted for a helical dimer (18).

The values obtained for xanthan from $\ln C_T$ (T_m^{-1}) and ΔH determined by microcalorimetry excludes the double helix structure dissociation during denaturation.

In opposite, a much larger ΔH was measured for succinoglycan compared to the values deduced from electrostatic. It was attributed to a different mechanism based on stabilization of the side chain along the backbone by secondary forces.

C. GELATION OF SOME POLYSACCHARIDES

Polysaccharides form often physical gels which are thermoreversible gels as in gellan gum or κ-carrageenans or ionic gels as alginates or pectins in presence of calcium (22). More recently, it was found that new type of gels may be discovered in mixed systems as galactomannan-xanthan. The gel formation is based on the stereoregularity and on the stiffness of the molecules.

Thermoreversible gels of κ-carrageenans are only formed when the double helical structure is stable -this means at low temperature in presence of K^+ counterions by example. A phase diagram describing the sol-gel equilibrium conditions was established (8). The gel seems to be due to aggregation of the double helices forming sorts of cristalline domains which imposed the mechanical properties of the gel. We have clearly demonstrated that the thermodynamic in the sol phase is directly related to the gel formation : K^+ counterions form preferentially ion pairs under coil conformation ; that induces the dimer formation and their aggregation to gel.

On the bacterial polysaccharide called gellan, the presence of external salt induces gel formation related with a conformational change (23) ; till now, it was not proved if the polymer adopts a single or double helical conformation.

Ionic gels are formed in presence of divalent counterions with some specificity in plant polysaccharides. The ionic selectivity was directly related with the configuration of the monomeric units :

- in alginate, poly L-guluronate blocks bind Ca^{+2} in a cooperative complex called the egg box model (Figure 8) (24) ;
- in pectin, poly D-galacturate blocks which have much similitudes with poly L-guluronate are engaged in the crosslink points.

This specific interaction of calcium was demonstrated also on oligomers by activity coefficient measurements (25) and potentiometry (26).

Gel formation may be induced by addition of Ca excess on the dimeric form of the calcium salt form of polygalacturonate. At end, it can be demonstrated that the gel point is obtained for a divalent concentration which depends much on the type of counterions (6) ; as example, one gets the order

$$Ba^{+2} > Sr^{+2} > Ca^{2+}$$

for gel formation with polygalacturonate but Mg^{2+} does not form gel.

Conclusion

Cooperative networks are formed with stereoregular polysaccharides ; the mechanisms of crosslink depend directly on the chemical structure.

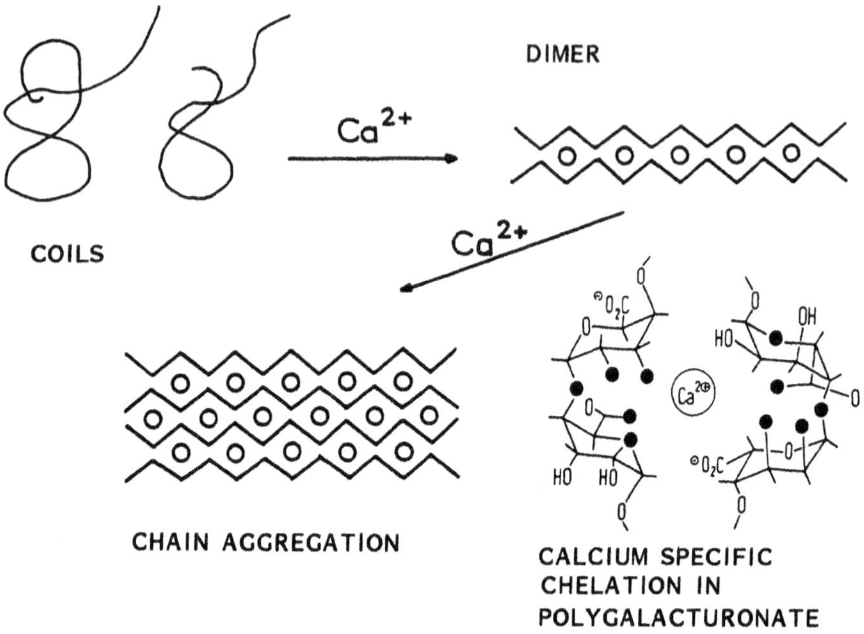

FIGURE 8 : Dimer and crosslink point formations in the egg box model proposed by Rees (24).

D. VISCOSITY BEHAVIOUR AND CHAIN DIMENSIONS

General behaviour of polysaccharides

M. Mandel recently discussed the use of the wormlike chain model to derive average dimensions of polyelectrolytes in solutions (27). At constant salt concentration the increase of the intrinsic viscosity $[\eta]$ with the net charge reflects expansion of the average dimension whereas the decrease of $[\eta]$ with increasing salt concentration C_s is related to the reduction of the dimensions due to the electrostatic screening of the charge interactions along the macromolecular chain. A new treatment was recently proposed by ODIJK (28, 29). He introduces the persistence length L_T, which is a measure for the local stiffness of the chain ; he expressed L_T the total persistence length as the sum of the intrinsic contribution L_p due to structure of the chain itself (limit of L_T when $C_s \rightarrow \infty$) and L_e the electrostatic contribution :

$$L_T = L_p + L_e \qquad [12]$$

The prediction is that L_T increases linearly with C_s and with the square of the charge density in controlled conditions. Taking into account the treatment suggested by Yamakawa and Fujii (30), Odijk related the intrinsic viscosity with L_T and predicts the dependence $[\eta]$ (C_s) (29). In presence of external salt, the model from Odijk predicts the dependence :

$$[\eta]_{C_s} = [\eta]_{\infty} + k\, C_s^{-1/2} \qquad [13]$$

usually found for experimental results (7). In relation 13, $[\eta]_{C_S}$ is the intrinsic viscosity at a given external salt concentration, $[\eta]_\infty$ the value extrapolated to infinite salt content (nearly the θ-conditions).

Relation with ordered conformation

When monovalent salt is added to some polysaccharide in dilute solution, the apparent viscosity first decreases rapidly as usually for polyelectrolyte. Then, for xanthan or succinoglycan, the viscosity levels off at nearly constant values. In the same conditions it is proved that the conformation changes and that helical conformation is stabilized (Figure 9).

More, as Ca form in presence of excess of calcium salt, the viscosity remained nearly unchanged. This behaviour is quite original.

The stiffness of the molecule depends largely upon the conformation ; this is very clear with the succinoglycan for which we have obtained the following values :

under T_m $[\eta]_o = 5\ 300$ ml/g

over T_m $[\eta]_o = 2\ 130$ ml/g

at the same ionic strength (NaCl 0.1 M) and molecular weight (31). This large difference may be due to cooperative interactions of the side chains along the backbone in the helical conformation as demonstrated by the large enthalpy of melting ; then the helical conformation is much more stiff that the disordered one.

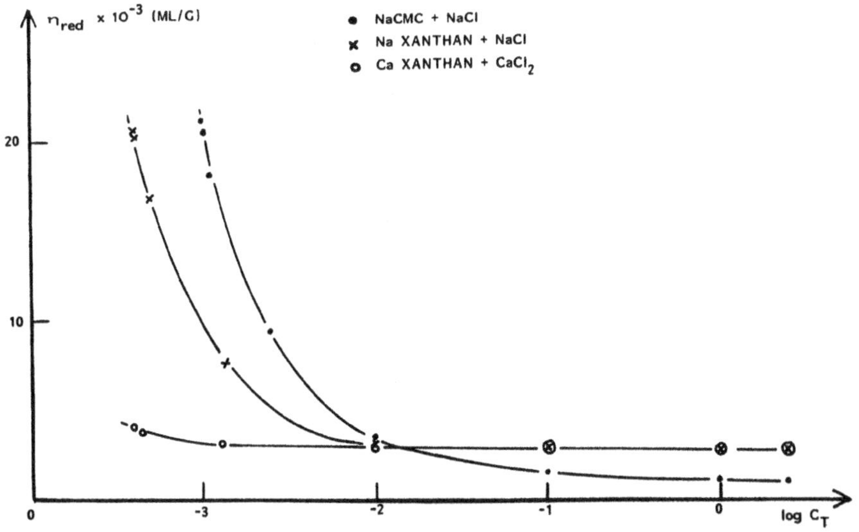

FIGURE 9 : Reduced viscosity as a function of total ionic concentration.
• NaCMC + NaCl X Na Xanthan + NaCl o Ca Xanthan + $CaCl_2$

Dependence of $[\eta]$ with the ionic strength and the molecular weight.

The influence of the external salt concentration on the intrinsic viscosity of xanthan is under discussion. The variation is less important than for synthetic polymers or even CMC due to the stiffness of the helical molecules whose expansion is nearly no more influenced by the electrostatic screening.

The behaviours of xanthan, alginate and succinoglycan in NaCl were compared from the dependence of the intrinsic viscosity $[\eta]_o$ (determined at shear rate $\dot{\gamma} \to o$) with the weight average molecular weights at constant ionic concentrations. The Mark-Houwink relation $[\eta] = KM^a$ is valid (Figure 10) and the exponent a are respectively :

- for alginate - NaCl 0.05 N (7) 1.04
- for alginate - $C_s \to \infty$ (7) 0.82
- for succinoglycan- NaCl 0.1M (31) 1.09
- for xanthan - NaCl 0.1M (32) 1.14
- for xanthan - $C_s \to \infty$ (33) 1.00

This means that these molecules are relatively stiff. Nevertheless, the application of the Yamakawa-Fujii treatment on xanthan and succinoglycan is not satisfactory and we have till now no good values for the persistence length ; this may be due to the role of the long side chains interacting with the main chain.

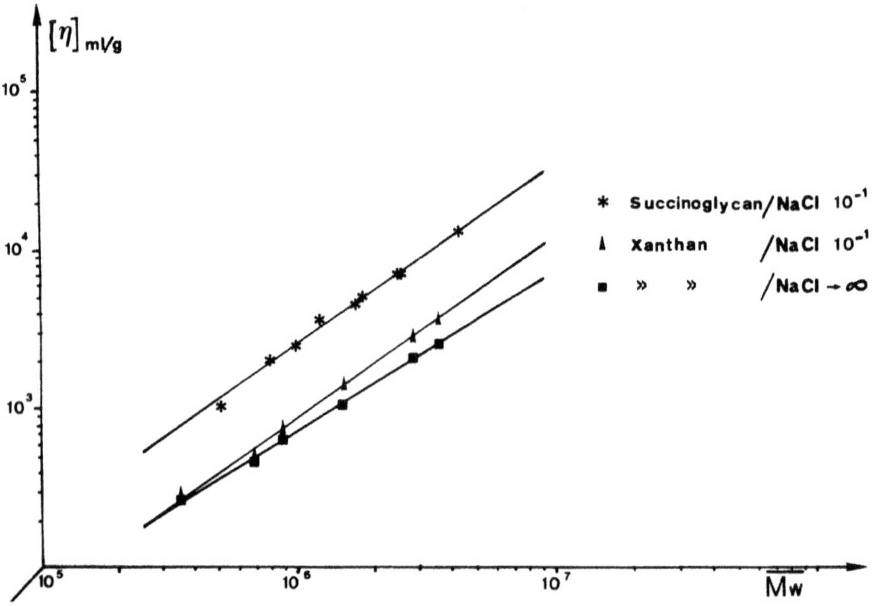

FIGURE 10 : Dependence of the intrinsic viscosity with the weight average molecular weight for xanthan and succinoglycan.

Conclusion

The dependence of the viscosity of stereoregular ionic polysaccharides with the external salt concentration is not that of general polyelectrolytes. It is controlled by the conformation of the polymers and the stiffness of the molecules in connection with the side chain interactions.

CONCLUSION

This paper concerns some original properties of stereoregular polysaccharides isolated from plant or produced by microorganisms. The properties are discussed in terms of the general behaviour of polyelectrolytes.

From the structural characteristics of the polysaccharides, the theory allows to predict the activity coefficient of counterions (γ) determined by potentiometry, transport parameter (f) obtained by conductivity or enthalpy of conformational change given by microcalorimetry. Comparison between experimental values and theoretical predictions allows to conclude on the local structure and multichain ordered conformation, confirmed by other technics.

The properties in dilute solutions are directly related to some specific behaviour as the network formation by cooperative aggregation. The very specific mechanism of crosslink by Ca^{2+} counterions is pointed out.

At end, the viscosity of dilute solutions is discussed ; its dependence with the external salt concentration is described to show the specificity of the stereoregular polymers whose local conformation controlls the viscosity. Stabilization of the helical conformation justifies the stability of viscosity in excess of monovalent or even divalent neutral salt.

The experimental results given in this work are important to explain also the biological behaviour of the polysaccharides in vivo ; the thermoreversible gels produced by κ-carrageenan present a swelling degree and mechanical properties depending directly on the molecular structure and on the nature and concentration of counterions. In the same order, the properties and stability of fruits or beetroots pulps depend on the degree of esterification of the pectins and Ca contents.

As this work progresses, some general view concerning the relation between the chemical structure of the polysaccharides and their physical properties seems to develop.

ACKNOWLEDGEMENTS

The author thanks M. Milas, C. Rochas and B. Tinland for fruitfull discussions on their experimental results.

BIBLIOGRAPHY

1. M. Rinaudo in "Polyelectrolytes", vol. I, 157-193, Edit. E. Selegny, M. Mandel, U.P. Strauss ; Reidel Publ. (1974).
2. M. Rinaudo, M. Milas in "Polyelectrolytes and their applications", vol. 2, 31-49, Ed. A. Rembaum, E. Selegny ; Reidel Publ. (1975).
3. M. Rinaudo, A. Domard, Biopolymers, 12, 2211-2224 (1973) and 18, 2341-2351 (1979).
4. A. Domard, M. Rinaudo, Macromolecules, 14, 620-625 (1981).
5. J.F. Thibault, M. Rinaudo, Biopolymers, 24, 2131-2143 (1985).
6. J.F. Thibault, M. Rinaudo, British Polym. J., 17, 181-184 (1985).
7. D. Graebling, M. Rinaudo, Polym. Bull., 15, 253-256 (1986).
8. C. Rochas, M. Rinaudo, Biopolymers, 23, 735-745 (1984).
9. M. Rinaudo, M. Milas, Biopolymers, 17, 2663-2678 (1978).
10. A. Heyraud, M. Rinaudo, B. Courtois, Int. J. Biol. Macromol., 8, 85-88 (1986).
11. S. Lifson, A. Katchalsky, J. Polym. Sci., 13, 43 (1953).

12. A. Katchabsky, Z. Alexandrowicz, O. Kedem in "Chemical Physics of Ionic Solutions", Ed. B.E. Conway, R.G. Barradas ; Wiley N.Y. 1966, p. 295.
13. G.S. Manning, J. Chim. Phys. 51, 924 (1969).
14. F. Oosawa, Polyelectrolytes, M. Dekker, N.Y. (1970).
15. G.S. Manning in "Polyelectrolytes", Vol. I, 9-37, Edit. E. Selegny, M. Mandel, U.P. Strauss ; Reidel Publ. (1974).
16. M. Rinaudo, C. Rochas, 367-378, in "Solution properties of polysaccharide" A.C.S. Symp. ser. n° 150 Edit. D.A. Brant (1981).
17. C. Rochas, M. Rinaudo, Carbohydr. Res, 105, 227-236 (1982).
18. C. Rochas, J. Thermal Analysis, 30, 1375-1380 (1985).
19. J.F. Thibault, M. Rinaudo, Biopolymers, 25, 455-468 (1986).
20. M. Milas, M. Rinaudo, 25-30 in "Solution properties of polysaccharides" A.C.S. Symp. ser. n° 150, Edit. D.A. Brant (1981).
21. F. Lambert, Thesis, Grenoble 1983.
22. M. Rinaudo in "Physics of finely divided matter" 16-20, Edit. N. Boccara, M. Daoud ; Springer-Verlag Publ. (1985).
23. V. Carroll, M.J. Miles, V.J. Morris, Int. J. Biol. Macromol. 4, 432-433 (1982) and P.T. Attwool, E.D.T. Atkins, C. Upstill, M.J. Miles, V.J. Morris, in "Gums and Stabilisers for the Food Industry", vol. 3, 135-145 ; Elsevier Publ. (1985).
24. D.A. Rees, Pure & Appl. Chem., 53, 1-14 (1981).
25. R. Kohn, O. Luknar, Czech. Chem. Commun. 42, 731-744 (1977).
26. G. Ravanat, M. Rinaudo, Biopolymers 19, 2209-2222 (1980).
27. M. Mandel, Angew. Makromol. Chem. 123/124, 63-83 (1984).
28. T. Odijk, Chem. Phys. Letters, 100, 145 (1969).
29. T. Odijk, Biopolymers, 18, 3111-3113 (1979).
30. H. Yamakawa, M. Fujii, Macromolecules 7, 128-135 (1974).
31. G. Gravanis, Thesis, Grenoble 1985.
32. M. Milas, M. Rinaudo, B. Tinland, Polym. Bull., 14, 157-164 (1985).
33. B. Tinland, Private communication.

GELS

Toyoichi Tanaka

Department of Physics and
Center for Materials Science and Engineering
Massachusetts Institute of Technology
Cambridge, MA 02139

ABSTRACT

Gel is a cross-linked polymer network swollen by absorbing a liquid. Many gels occur in the biological world. They are widely used in the vast area of chemical and food industries. The overall properties of a gel, determined by its equilibrium, dynamic, and kinetic properties, depend on the interaction of polymer network and liquid. In particular, the relative position of the gel state relative to the phase transition boundary crucially influences the gel characteristics.

The phase transition of a gel is represented by a discontinuous volume-phase transition in response to minute changes in temperature, solvent composition, pH, ionic composition, or applied electric field. The phenomena are universal to any gels. The physical principles underlying in the phase transition of gels enables a more universal understanding of the system. They will be the guiding principles with which to design and develop various functional devices using gels.

I. INTRODUCTION

Gel is a cross-linked polymer network swollen in a liquid medium. Its properties depend strongly on the interaction of these two components. The liquid prevents the polymer network from collapsing into a compact mass, and the network, in turn, retains the liquid in its interstial space.

There are great many natural and artificial gels. A familiar gel is the desert Jello, wherein the network is made up of polymers derived from animal protein gelatin. The polymeric network constitutes only about 3% of the volume of the gel; the remainder is colored, flavored, and sweetened water. The vitreous humor that fills the interior of the eye is a gel, as is the material of the cornea. Blood vessel walls, connective tissues, and the synovial fluid that lubricates the skeleton joints contain gels. In such biological gels, the liquid component allows the free diffusion of oxygen, nutrients, and other molecules, whereas the polymer network provides a structural framework that holds the liquid in place. The surface of internal tracts such as the stomach and intestine is covered by mucopolysaccharide gels. The epithelial cells in the stomach are protected by the gel from the extremely acidic gastric juice. The lung surface is also covered by the similar gel. The change in the viscoelastic properties of the gel may lead to a serious desease called cystic fibrosis. Some plants' roots are covered by gels produced by bacteria, which substantially affect the plants' metabolic functions. Microbial films, synthesized by bacteria on the surface of sewer pipes, thermal exchangers in reactors and other systems, may cause malfunctions.

Gels are important intermediates in the manufacture of polymers such as rubber, plastics, glues, films, and membranes. They are used as absorbent in disposable diapers, for water purification, and in chromatography and electrophoresis, where molecules are separated according to the speed with which they pass through or are expelled by the pores. Gels are widely used as implants in plastic surgery and as for artificial and soft contact lenses.

Understanding this form of material, therefore, has substantial scientific, medical, and technological importance. It requires experimental determinations and theoretical explanations of the various characteristics of the gels. Gels are characterized by their swelling, mechanical, thermal, optical, and many other properties. These characteristics of gels may be classified by the equilibrium, dynamic and kinetic properties of each characteristics. For example, the swelling of a gel is understood not only by knowing its equilibrium volume in a solvent, but also by the thermal fluctuations of the local swelling and shrinking, and by the time course of its volume change in response to various forms of external disturbances. These properties of gels depend on the gel state represented by its osmotic pressure, temperature, solvent composition, and degree of swelling. In particular, the position of the gel state in the phase diagram relative to the phase boundary crucially influences all the equilibrium, dynamic, and kinetic properties of the gel.

It is important to understand the processes by which the gel is formed or dissolved into a solution. The process of gel-sol transition has significant effects on the structure and properties the formed gel. The understanding of the gel-sol transition, therefore, is not only scientifically important and challenging but also has substantial technological significance.

II. GEL-SOL TRANSITION

The olymer network of a gel can be formed in various ways [1]. In a typical condensation polymerization, a polymer network is formed by polymerizing bifunctional units and polyfunctional units, as shown in Figure 1. The bifunctional units form linear chains, and the polyfunctional units serve as cross-links. A polymer network can also be formed by cross-linking polymers formed from bifunctional monomers.

Gels are also classified by the strength of the cross-linkages [1,2]. Some gels are cross-linked chemically by covalent bonds, whereas other gels are physically cross-linked by weak forces such as hydrogen bonds, van der

Waals forces, hydrophobic and ionic interactions. and ionic interaction. Examples of the chemically cross-linked gels are the styrene and acrylamide gels used for electrophoresis and chromatography.

An example of a physically cross-linked gels is a Jello, which can be reversibly dissolved by moderate heating. The reversible gelation process is called a gel-sol transition. It has been one of the major research area in the chemistry of gels since the pioneering works by Flory and Stockmayer [3,4]. A remarkable progress has been made recently in understanding of gel-sol transition with the help of the new theoretical concepts and techniques taken from the physics of phase transition and percolation phenomena [2,5,6], and the field theory [7-9].

III. EQUILIBRIUM PROPERTIES

Phase Properties

In order to characterize a physical system, its phases must be known. A system consisting of identical molecules, for example, can assume a gas, liquid, or solid phase, depending on its state represented by a set of state variables such as temperature, pressure, and volume. The system at each phase should be characterized in terms of its mechanical, optical, thermal and other properties. The phase diagram and the transitions among the phases give information concerning the interaction of molecules. For this reason, it is important to know the phases and phase transitions in polymer gels, as well as the gel properties at each phase.

Phase transition in polymer gels has been discovered only recently [10]. Drastic changes in the state of the gel can be brought about by small changes in the external conditions. For example, when the temperature is lowered the polymer network loses its elasticity and becomes increasingly compressible. When a certain critical temperature is reached, the compressibility becomes infinite. Gel can swell or shrink by a factor of as much as 1000, when the temperature is varied. Under some conditions the swelling or shrinking is discontinuous, therefore a minute change in temperature causes a large change in volume. Such transition can also be brought about by altering solvent composition, pH, or ionic composition, or by imposing a small electric field across the gel.

These phenomena can be understood in terms of the phase transition and critical phenomena. A gel can assume two phases, one corresponding to a gas phase, and the other to a liquid phase [10]. Phase transition fundamentally influences the dynamic and kinetic behaviors of polymer gels.

Volume-Phase Transition

The equilibrium swelling curve of the acrylamide gels in acetone-water mixtures is shown in Figure 2. At low acetone concentrations, the gel is quite swollen and the volume is large. As the acetone composition increases, the volume gradually decreases until, at a certain composition, the gel collapses discretely into a small compact mass. Further increase of acetone concentration results only in a slight and continuous decrease in the volume. The phenomenon is reversible, and the shape of the gel remains the same before and after the transition. The volume change at the

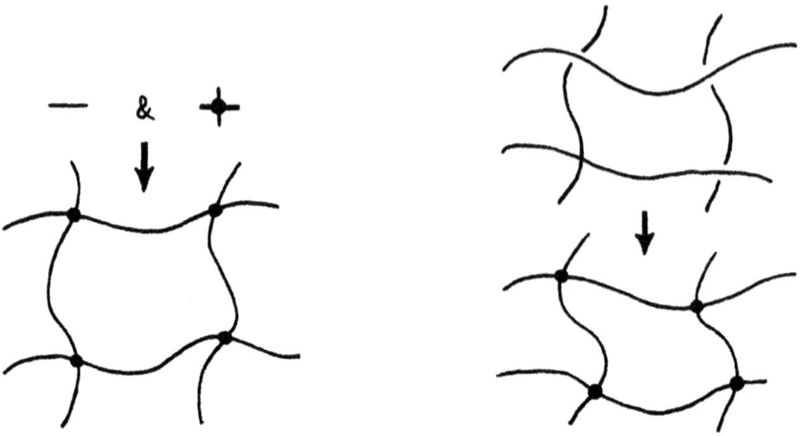

1 Polymer network of a gel can be formed in various ways. In the typical condensation polymerization, bifunctional units are connected to form long chain molecules with polyfunctional units which serve as cross-links. Polymer Network can also be formed by cross-linking polymer chains which are already formed from bifunctional units.

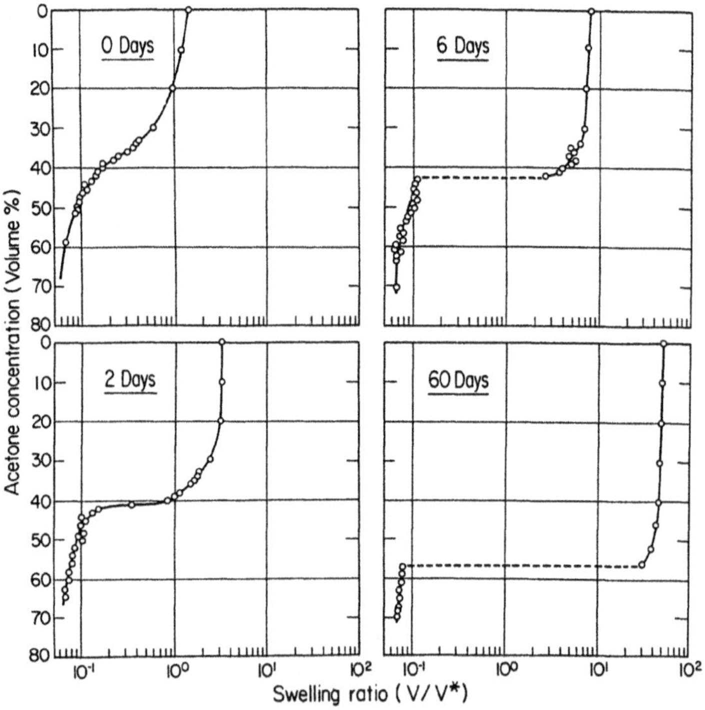

2 Swelling curves recording the change in the volume of an acrylamide gel placed in solvents with varying concentrations of acetone. The extent of the change in volume and the shape of the swelling curve depend strongly on the degree of hydrolysis. (a) No hydrolysis, (b) hydrolysis for 2 days in a slightly basic solution, (c)hydrolysis for 6 days, and (d) hydrolysis for 60 days.

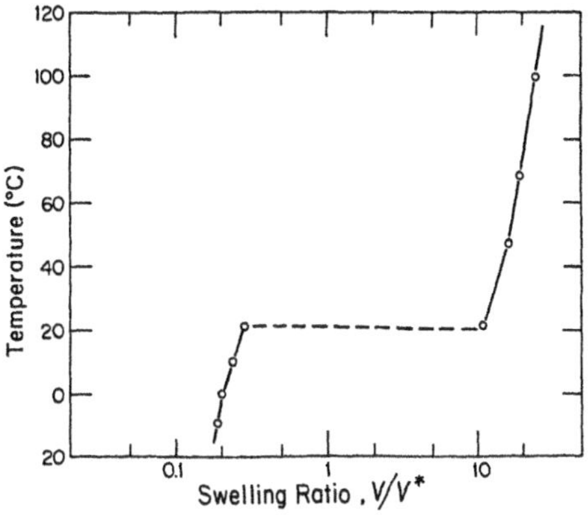

3 Temperature dependence of the ionized acrylamide gel in the 40% acetone-water mixture. The gel undergoes a discontinuous volume transition.

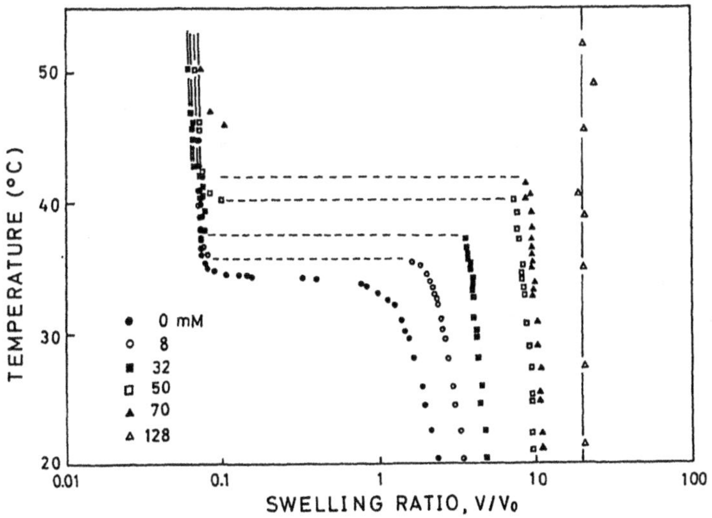

4 Equilibrium volume of N-isopropylacrylamide-sodium acrylate copolymer gels in pure water is shown as a function of temperature for various concentrations of ionizable sodium acrylate up to 120mM. The total monomer concentration is 700mM.

transition increases with the degree of hydrolysis of the acrylamide; the gel without hydrolysis (0 days) shows only a continuous volume change, whereas at 60 days of hydrolysis in an alkaline solution, the volume change is as large as 500-fold [11,12]. The acetone concentration at which the volume transition occurs increases with the degree of hydrolysis. Volume-phase transition also occurs when the temperature is varied. At higher temperatures, the gel is swollen; at lower temperatures it is collapsed (Figure 3) [10].

The volume transition of N-isopropylacrylamide-sodium acrylate copolymer gels in pure water are shown in Figure 4 as a function of temperature for various ratios of copolymerization [13]. With the increase of ionizable groups (sodium acrylate), the volume change at the transition increases and the transition temperature rises.

Equation of State of Ionic Gels

The equation of state of gels relates the osmotic pressure, polymer-network concentration, temperature, and solvent composition [1,11,14]. Most phase properties of gels can be theoretically deduced from it. The discrete volume transition is understood as a manifestation of the competition among three osmotic pressures acting on the polymer network: positive pressure of counterions, negative pressure due to the affinity among polymers, and the so-called rubber elasticity which keeps the network in a moderate expansion. The balance of these three pressures determines the equilibrium volume of the gel. The temperature, pH, and added salt ions affect only the former two pressures, whereas the solvent composition influences only the negative pressure of polymer-polymer affinity. Changing one of these factors creates an excess nonzero pressure, positive or negative, which results in a new equilibrium volume. For example, in ionized acrylamide gels, at higher temperatures the positive pressure of counterions dominates, and network is swollen. At lower temperatures, the negative pressure prevails, and the network shrinks. At a particular temperature, two different gel volumes gives zero total osmotic pressure. This corresponds to the temperature at which the discrete volume transition takes place.

These descriptions can be quantitatively understood with the help of the equation of state of ionized polymer gels [10,11], which relates the osmotic pressure (Π), polymer-network concentration (ϕ), temperature, and solvent composition. Most phase properties of gels can be theoretically obtained if the equation of state is known. Using the Flory-Huggins derivation [1,15], phase and phase transition can be analyzed and understood [11,14]:

$$\Pi = -\frac{NkT}{v}\left[\phi + \ln(1-\phi) + \frac{\Delta F}{2kT}\phi^2\right] + \nu kT\left[\frac{\phi}{2\phi_0} - \left(\frac{\phi}{\phi_0}\right)^{\frac{1}{3}}\right] + \nu fkT\left(\frac{\phi}{\phi_0}\right) \qquad (1)$$

where N is Avogadro's number; k the Boltzmann constant; T the absolute temperature; v the molar volume of the solvent; ϕ the volume fraction of the polymer network; ΔF the free-energy decrease associated with the formation of a contact between polymer segments; ϕ_0 the network volume fraction at the condition that the constituent polymer chains have random walk configurations; ν the number of constituent chains per unit volume at $\phi = \phi_0$; and f the number of dissociated counterions per effective polymer chain. The first two terms in Eq.(1) denote the osmotic pressure due to mixing entropy of the polymer network at $\phi = \phi_0$ with liquid. The third term is the attractive interaction among polymer segments, and the last term indicates the pressure due to concentration imbalance of counterions in and out of the gel.

The osmotic pressure of a gel must be zero for the gel to be in equilibrium with the surrounding solvent. Zero osmotic pressure corresponds to the minimum free energy of the gel with respect to its volume. From Eq.(1), this equilibrium condition is expressed as

$$\tau = 1 - \frac{\Delta F}{kT} = - \frac{\nu v}{N\phi^2}\left[(2f+1)\left(\frac{\phi}{\phi_0}\right) - 2\left(\frac{\phi}{\phi_0}\right)^{\frac{1}{3}}\right] + 1 + \frac{2}{\phi} + \frac{2\ln(1-\phi)}{\phi^2} \tag{2}$$

The left side of Eq.(2) corresponds to the reduced temperature, τ, which depends on the temperature and the solvent composition. Thus, the equilibrium network concentration ϕ is determined in Eq.(2) as a function of the reduced temperature. For certain values of the reduced temperature, however, Eq.(2) is satisfied by three values of polymer-network concentration, ϕ, corresponding to two minima and one maximum of the free energy. The value of ϕ corresponding to the lower minimum represents the equilibrium value. A discrete volume transition occurs when the two free-energy minima have the same value.

The calculated equilibrium swelling ratio, ϕ/ϕ_0, are plotted as a function of the reduced temperature for various degrees of ionization f. The theoretical curves are qualitatively in excellent agreement with experimental results shown in Figures 2 and 4. The influence of the ionization on the phase transition can be clarified by expanding the term $\ln(1-\phi)$ in Eq.(2) and retaining the terms to order ϕ^3. This expansion yields

$$t = S\left(\rho^{-\frac{5}{3}} - \frac{\rho^{-1}}{2}\right) - \frac{\rho}{3} \tag{3}$$

where

$$t = \frac{(1-\Delta F/kT)(2f+1)^{\frac{3}{2}}}{2\phi_0} \tag{4}$$

$$S = \frac{v\nu}{N\phi_0^{\frac{3}{2}}}(2f+1)^4 = S_0(2f+1)^4 \tag{5}$$

and

$$\rho = (\phi/\phi_0)(2f+1)^{\frac{3}{2}} \tag{6}$$

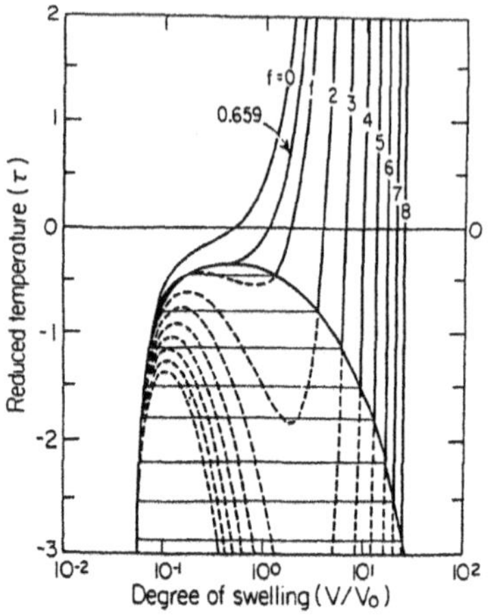

5 Theoretical curves are plotted for the reduced temperature versus equilibrium volume of a gel having various f-values, the number of counter-ions per effective polymer chain. The curves are constructed using Eq. (2) in the text with $S_0 = 10$.

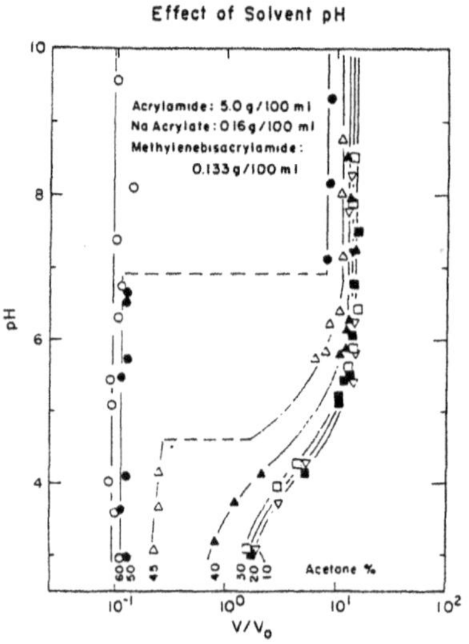

6 The pH dependence of the equilibrium degree of swelling of acrylamide-sodium acrylate gels in acetone-water mixtures. The gel having 700mM acrylamide, 34mM sodium acrylate and 8.6mM N,N'-methylenebisacrylamide in 100ml of water was prepared. The line are theoretically calculated using Eq. (2) and assuming a dissociation equilibrium.

In this approximation, the form of the equation is determined by a single parameter S. S solely determines whether the volume change is continuous or discontinuous [11]. The theoretical calculation shows a critical value of S below which the phase transition is continuous and above which it is discontinuous. The ratio of the two volumes at both side of the discontinuous transition is also a function only of S.

The parameter S can be interpreted by the following argument. Consider an effective polymer chain consisting of n freely jointed segments of radius a and persistent length b. The volume of solvent molecules is assumed to be a^3. When the interaction is neglected among the segments constituting the polymer, the average end-to-end distance of the chain is $R \approx n^{1/2}b$. Since R is also the mean distance between neighboring cross-links, $\nu \approx ba^2/R^3$, and $\phi_0 \approx nba^2/R^3$. From these relations, the parameter S is expressed as

$$S \approx \left(\frac{b}{a}\right)^3 (2f+1)^4 \qquad (7)$$

This equation shows that the S value depends only on the number f of counterions per chain and the ratio of the persistent length b to the effective radius a of the polymer chain. The ratio b/a represents the stiffness of the chain. The S value varies rapidly with f and b/a, as shown by Eq.(7). In order for a gel to have a large S value and thus undergo a discontinuous phase transition, the constituent polymer chains must have sufficient stiffness or a sufficient amount of ionized groups. Both the counterions to the ionized groups and the stiffness of the polymer chains increase the osmotic pressure acting to expand the polymer network, resulting in a discontinuous volume change. This situation is similar to a gas liquid phase transition which can be either continuous or discontinuous depending on the external pressure exerted on the system. Thus, the equation of state (1) explain the existence of the discontinuous volume-phase transition, and the important effects of ionization on volume change and transition temperature or solvent composition.

Effect of pH and Salt Ions on Swelling

Reversible swelling and contraction of ionic gels respond to changes of pH and ion compositions [16,17]. The cycle of swelling-contraction phenomena was used to develop a mechano-chemical transducer [18]. Although large volume changes were observed in the gel upon varying pH or ionic composition, no discontinuous volume transition was observed.

Recently, discontinuous volume transitions have been observed in ionized acrylamide gels [11]. Considering the essential role of ionization in the phase transition, the effects of pH and ionic composition on the gel phase should be important. When the pH is varied or salt ions are added to the solvent, the effective number of counterions varies, as does the ionic osmotic pressure. The equilibrium swelling curves of an ionized acrylamide gel are shown in Figure 6 as a function of the pH in acetone-water mixtures of different composition. The gel undergoes a continuous or discontinuous volume transitions. This behavior is understood by examining the theoretical curves shown in Figure 5: the f-value, ie, the number of dissociated

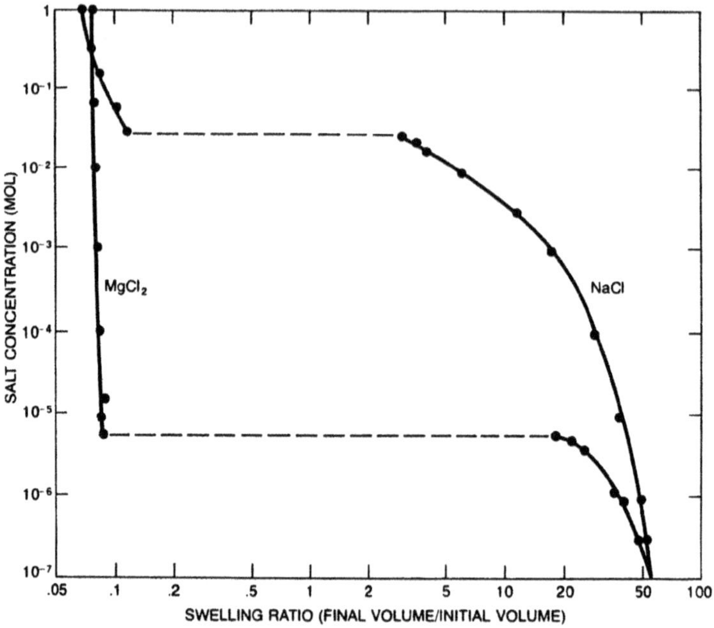

7 Dependence of gel volume on concentration of disolved salts, showing phase transition involving collapse of the gel. The salt concentrations needed to induce the phase transition is 4000 times smaller with divalent magnesium ions than it is with monovalent sodium ions.

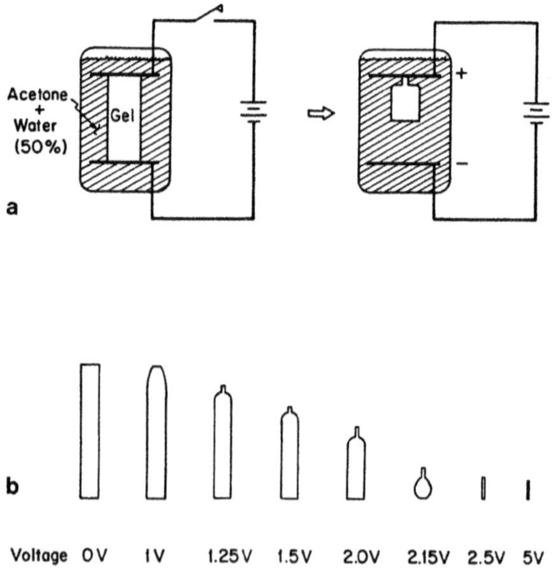

8 A hydrolyzed acrylamide gel undergoes a discontinuous volume transition in 50% acetone-water mixture when a small electric field is applied on the gel. A portion close to the positive electrode collapsed and the rest remains swollen. The relative portion of the collapsed gel increases with the amplitude of the electric field.

counterions per chain, changes with pH. In Figure 5, this change is represented by a horizontal line at a constant reduced temperature corresponding to a fixed temperature and solvent composition. Such horizontal line is continuous if the reduced temperature is higher than -0.36, but is discontinuous if lower than -0.36.

Changes in the ionic composition also induces the phase transition. The swelling curves of an acrylamide gel in 50% acetone-water mixture are shown in Figure7. The concentrations needed for the transition differ by four orders of magnitude between monovalent NaCl and divalent $MgCl_2$. This remarkable difference can be explained with a similar theoretical argument to that presented above [10].

Collapse of Gels under Electric Field

Phase transition can be induced by applying an electric field across the gel [18]. As shown in Figure 8, a gel is placed in a 50% acetone-water mixture and held in between two platinum electrodes. When an electric field of ~500mV/cm is applied, the gel undergoes a discontinuous volume change. At the critical position, the portion of the gel from that point to the positive electrode collapses and the remainder is swollen. The gel, therefore, has a bottle shape, as shown in Figure 8. The swollen portion becomes shorter as the electric field becomes stronger. The collapsed part swells again when the electric field is turned off.

The physical mechanism involved in this phenomenon is not yet understood. Among the possible explanations for the volume transition are pH gradient and electroosmotic- or electrostatic-stress gradient along the gel, possibly induced by the electric field.

Volume change has been electrically induced in collagen gels placed between two solvents with different ionic compositions [20]. The electric field causes different concentration gradients for different ions, depending on their mobilities and charges. The ion composition gradient alters the ionic pressure within the gel and induces the volume change. Such studies would have substantial impact on the technological and medical application of gels.

Universal Feature of Volume Phase Transition

According to the theory presented above, any gels may undergo the discontinuous volume transition if the polymer chain is sufficiently ionized or stiff. This universal feature has been confirmed with several different synthesized and natural gels. The theory predicts that the volume transition should not depend on the sign of the charge of counterions. It has indeed been demonstrated that the positively ionized gels also undergo the similar volume transition similar to that of the negatively charged gels (Figure 9) [21].

N-isopropylacrylamide polymer seems to have enough stiffness. Its gel in pure water has been found to undergo a small but discontinuous volume transition with a temperature change (Figure 10) [13].

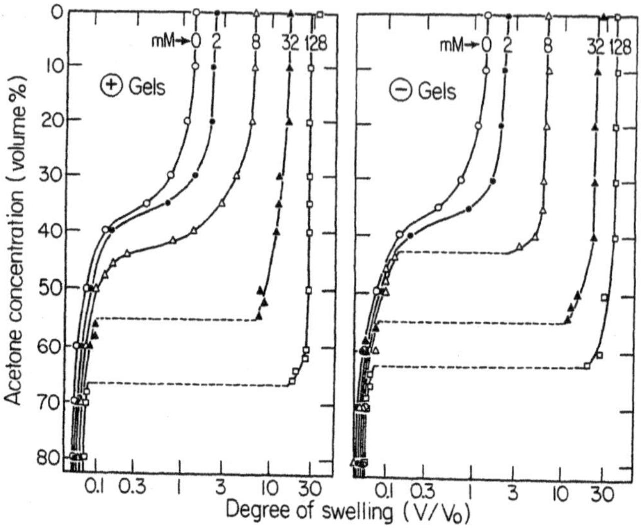

9 The volume transition is also observed in the positively charged acrylamide-MAPTAC copolymer gel immersed in acetone-water mixtures. The concentrations indicated in the Figure denote the amount of MAPTAC incorporated in the gels.

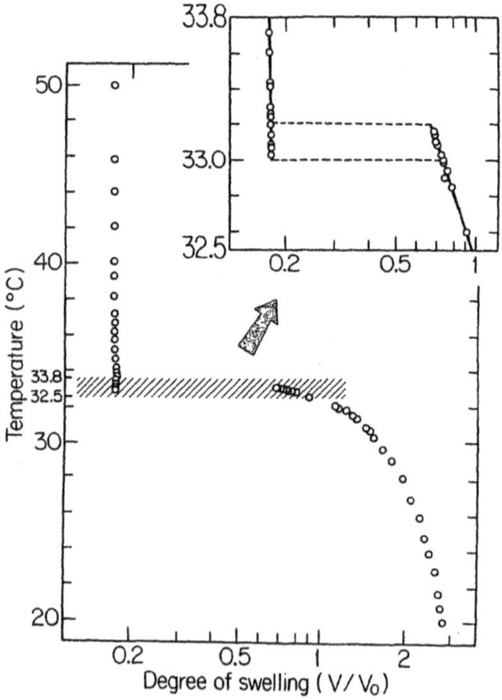

10 Non-ionic N-isopropylacrylamide gel shows a small, but discontinuous volume transition in pure water when the temperature is changed.

Conformation of Polymers Constituting a Gel

In order to evaluate the validity of equation of state independently from the swelling equilibrium, the conformation of polymer chains incorporated within a polymer network must be known. Conformation is measured by observing a small number of deutrium-labeled chains incorporated within a polymer network by neutron scattering [22-24]. For a wide range of gel swelling and shrinking, the conformation of incorporated chains is described by a Gaussian chain conformation. This observation enables improvement of the theoretical formulation of the equation of state.

Inhomogeneities in Polymer Network

Gels have some inhomogeneities in its densities of the cross-linkings and polymers. Density fluctuations may affect mechanical properties and the spatial resolution in electrophoresis and chromatography. If gelation takes place in a good solvent for the polymer, a transparent gel forms, whereas the gel formed in a poorer solvent is opaque. These phenomena were interpreted in terms of microsyneresis upon gelation, ie, the formation of liquid droplets within the gel [25]. The inhomogeneities of a gel, however, may be understood in a more general way using the concept of phase transition whereby the gel has two phases with different polymer concentrations which creates inhomogeneities in the gel.

Mechanical Properties

Mechanical properties of a gel are typically represented by the stress-strain relationship [1]. Both swelling and mechanical properties of gels have equilibrium, dynamic and kinetic aspects. Each can be determined by the frequency dependent stress-strain relationship.

The independence of the mechanical properties from phase properties have been clearly demonstrated in studies on the shear modulus of a gel undergoing phase transition [26]. The shear modulus changed discontinuously at the volume collapse of the gel, but if it was normalized by the polymer-network concentration, it is continuous and did not show any anomaly at the transition.

The mechanical properties of gels, being unaffected by the phase properties, depend primarily on the elasticity of the polymer network [1]. Such elasticity, known as rubber elasticity, is due to the capacity for polymer chains to adopt diverse configurations. The restoring force of the chain is owing to their tendency to assume the density at which the number of possible configurations becomes maximum. The early experiments on elastomers clearly demonstrated that rubber elasticity originates mainly within the chains of the network and not from the interactions among polymers or solvents. For this reason, the mechanical properties of gels can be well understood by knowing the physical basis underlying in the rubber elasticity. Detailed information on the present theoretical understanding of rubber elasticity is given in reference [22]. The understanding of rubber elasticity has been substantially improven with the help of the reevaluation of phantom network theory [27], of new concepts of entanglements [26] and of scaling concept [2].

Rubber elasticity also influences the swelling equilibrium of gels.

Mechanical responses are usually examined in linear regime where the deformation is proportional to the external mechanical stimulation. However, as the mechanical force applied to a gel exceeds a certain limit, the linear proportionality relationship does not hold (non-linear regime) and eventually the gel is broken. Such limit of gel breakage is also a characteristics of the gel and of particular importance from a technological point of view [29].

IV. DYNAMIC PROPERTIES

Collective Diffusion of Polymer Network

The polymer network of a gel is constantly in random thermal motions, giving rise to the space and time fluctuations of the polymer density. Such density fluctuations reveal fundamental properties of the gel, and can be observed by light scattering [30-33] or neutron scattering techniques [22-24].

An equation was derived from the continuum theory of gels describing the density fluctuations $\rho(\mathbf{r},t)$ within the gels.

$$\partial\rho/\partial t = (K/f)\Delta\rho \tag{8}$$

where K denotes the osmotic bulk modulus of the polymer network, f_r denotes the frictional coefficient between network and fluid, and Δ represents the Laplacian operator. This is a diffusion equation with a diffusion coefficient which is defined as a ratio of K to f_r. It describes the microscopic thermal-density fluctuations and the kinetics of macroscopic swelling and shrinking.

Light and Neutron Scattering from Collective Modes

The collective diffusion modes, as described by Eq.(11), give rise to light or neutron scattering. The space-time correlation function of the scattered light electric field is proportional to the spatial Fourier transform of the time-correlation function of the density fluctuations. It is given by an exponential function whose amplitude is inversely proportional to the bulk modulus K. Its decay rate is proportional to the collective diffusion coefficient K/f_r:

$$\langle E(\mathbf{q},t)E(\mathbf{q},0)\rangle = \frac{I_0}{c}\left(\frac{\omega_0}{c}\right)^4\frac{\sin^2\theta}{4\pi R^2}\left(\frac{\partial\varepsilon}{\partial\rho}\right)_T^2 \rho^2 \,\frac{LkT}{K}\, \exp\left(-\frac{K}{f}\, q^2 t\right) \tag{9}$$

Here I_0 is the incident intensity of light, L the illuminated length, θ the angle between the polarization of the incident light and scattering direction, ε the dielectric constant of the gel. Thus, from the time-correlation function of the scattered light, it is possible to determine both the osmotic bulk modulus K and the friction coefficient f_r independently.

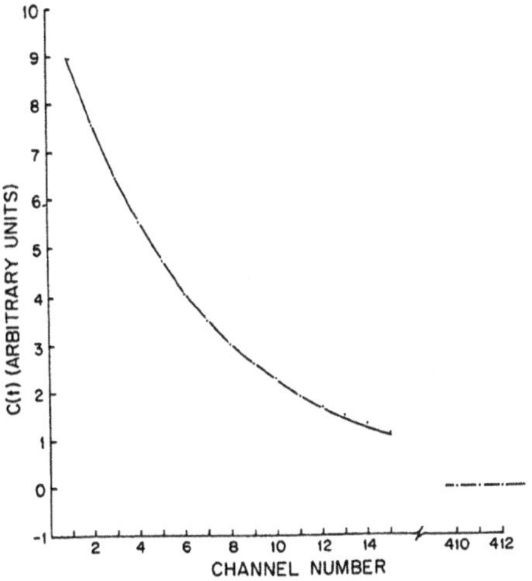

11 The measured correlation function of a 5% polyacrylamide gel at 25°C at a 90° scattering angle. The solid line is the best single expotential fit to the data. One channel corresponds to 10μsec.

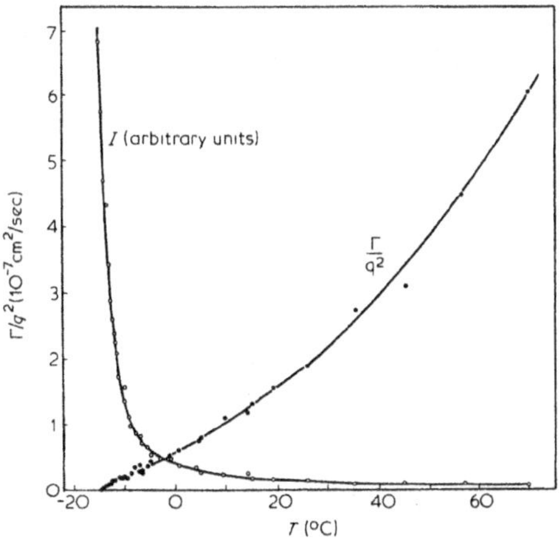

12 Intensity, I and the reciprocal correlation time (decay rate), Γ, of laser light scattered by a 2.5% polyacrylamide gel. Γ is divided by the square of the scattering vector, $|\vec{q}|^2$

The correlation function of light scattered from acrylamide gel in water is shown in Figure 11; it is indeed exponential. Using light and neutron scattering, the viscoelastic quantities, K and f_r have been investigated as a function of polymer-network concentration and temperature. Scaling arguments [2,34] predict the dependence of these quantities on the network concentration in good solvents, which agree well with the experiments [35,36]. The values of K and f_r obtained by macroscopic deswelling and microscopic light scattering experiments show excellent agreement.

Critical Behavior

The temperature dependence of the light scattering reveals the critical behavior of the gels. As the temperature approaches the critical point, the intensity of scattered light increases to infinity and the relaxation time slows down to zero (Figure 12) [37]. The divergence of the intensity can be quantitatively analyzed using the equation of state, since $K = (\partial\phi/\partial\pi)/\phi$. Simple calculation shows that K is written in the form $K = a(T - T_0)$, and thus can become zero at $T = T_0$. The critical slowing-down, describing the increase of relaxation time to infinity, has been explained using the mode-mode coupling theory [38].

Ultrasonic experiments also revealed the critical behavior of the acrylamide gels in water [39]. The absorption by the gel of the shear ultrasonic waves with different frequencies appears to diverge at the critical point. The result is consistent with the light scattering data.

Light scattering spectroscopy is also used to determine the shear modulus of a gel [40]. Agitating the gel generates standing shear waves with nodes at the gel wall. Light is incoherently scattered from inhomogeneities embedded in the gel which move together with the shear waves. By spectrum analysis, the shear modulus of the gel is determined.

NMR Measurements of Local Motions

In order to understand the molecular state and motions of polymer segments and solvent molecules in gels, the local conformational fluctuations of polymer network in a gel have been studied by NMR [41,42]. It has been found, for example, that the rate of conformational change of polystyrene gels in organic solvents depends on solvent composition, temperature and cross-linkings, as expected; network concentration is less important.

The motion of water molecules within a polymer network was also studied with the gelp of the ^{1}H-NMR relaxation time T_1, which increases steadily with the network concentration. Such experiments are important in elucidating the solvent-polymer interaction and evaluating the theory of rubber elasticity.

V. KINETICS OF GEL SWELLING AND SHRINKING

In spite of its importance both in science and technology, the kinetics of the volume change of gels has not been fully understood. In particular, the theory of the kinetics of gel swelling was incorrect until quite recently. Previously it was thought that gel swelling was determined by the diffusion of water molecules into a polymer network [44]. The swelling

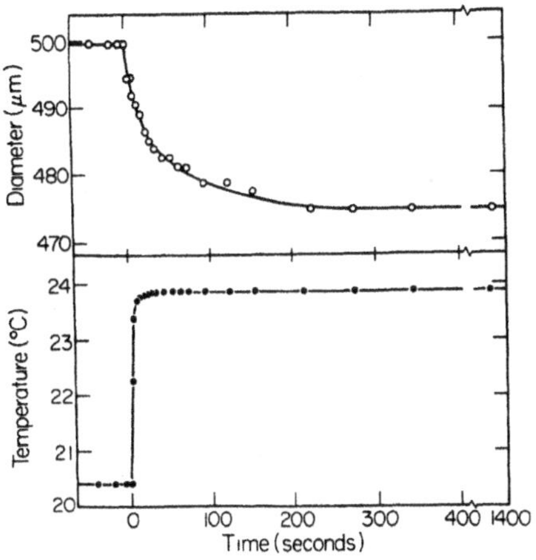

13 Time dependence of the radius of an N-isopropylacrylamide gel swelling in water in response to a sharp temperature jump as shown in the lower half of the figure. Except for the very beginning of the swelling process the radius increases exponentially.

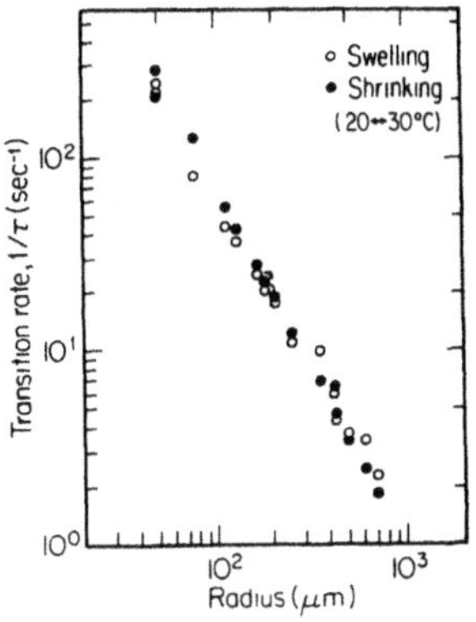

14 Characteristic time of swelling of the N-isopropylacrylamide gel in water is proportional to the square of the final radius of the gels.

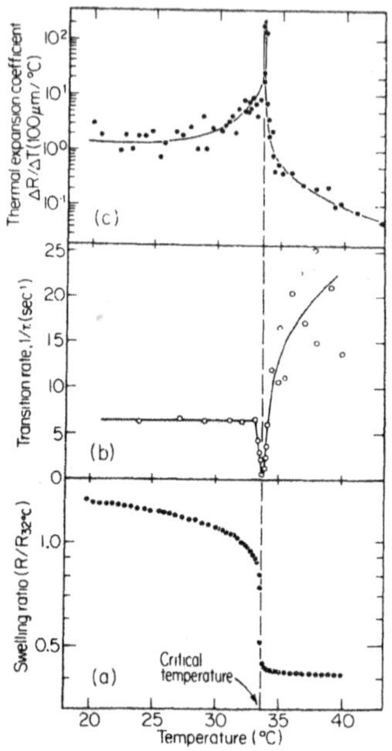

15 (a) Equilibrium radius of an isopropylacrylamide gel sphere as a function of temperature. At lower temperatures the gel is swollen and at higher temperatures it is shrunken. At about 34 °C the swelling curve becomes infinitely sharp, which corresponds to the critical point. (b) Relaxation time of gel volume change in response to a temperature jump, as a function of temperature. The relaxation becomes infinitely slow at the critical point. (c) Thermal expansion coefficient, the relative radius change per temperature increment, also diverges at the critical point.

was believed to be fast if the network was loose enough for water molecules to diffuse into it, and slow if the network was dense. It was demonstrated experimentally and theoretically, however, that the swelling process of a gel is determined by the diffusion of the polymer network [44].

As discussed above, the local motion of a polymer network obeys the diffusion equation with a diffusion coefficient defined by the ratio of the elastic modulus of the network K to the frictional coefficient f_r between the network and fluid [31]. Because it is a diffusion motion, the characteristic time taken for a gel to swell or shrink should be proportional to the square of characteristic linear dimension of the gel and inversely proportional to the collective diffusion coefficient: $\tau = a^2/D = a^2/(K/f)$.

A spherical acrylamide gel swells as a function of time when it is exposed to a sudden negative osmotic pressure (see Figure 13). The square relation between the relaxation time and the gel radius a is shown in Figure 14. The collective diffusion coefficient is obtained from the slope of the line and coincides well with the value obtained by dynamic light scattering experiments.

The collective diffusion coefficient determined by light scattering experiment becomes zero at the critical point. This indicates that the kinetics of gel volume change should also slow-down infinitely at the point. Such critical kinetic behavior has indeed been discovered recently.

CONCLUSION

Extended applications of gel phase transition is anticipated. Artificial muscles, robot actuaters, display units, switches, chemical memories, solvent specific absorbent, matrices for bioreactors and drug delivery systems are some of the examples of applications. Gels made of photosensitive, cunductive, or magnetic polymers may show exciting subjucts to explore. Gels consisting of rodlike polymers show interesting coupling phenomena between electrical and mechanical properties [45].

The understanding of this important form of material requires further efforts to combine the expertise from various interdisciplinary areas in chemistry, physics, medicine, and engineering.

REFERENCES

[1] P.J.Flory, **Principles of Polymer Chemistry**, Cornell University Press, Ithaca and London, (1953)

[2] P.G.deGennes, **Scaling Concepts of Polymer Physics**, Cornell University Press, Ithaca and London (1979)

[3] P.J.Flory, J. Am. Chem. Soc., **63** 3083,3091,3096 (1941)

[4] W.Stockmayer, J. Chem. Phys., **11** 45 (1943), **12** 125 (1944)

[5] D.Stauffer, J.C.S.,Faraday Trans. II, **72** 1354 (1976)

[6] P.G.deGennes, J.Phys. (Paris) Lett., **37L** 61 (1976)

[7] T.Lubernsky and D.Isaacson, Phys. Rev. **A20** 2130 (1979)

[8] A.Cognilio, H.E.Stanley and W.Klein. Phys. Rev. Letters, **42** 518 (1979), Phys. Rev. **B25 6805 (1982)**

[9] R.J.Cohen and G.B.Benedek, J. Phys. Chem., **86** 3696 (1982)

[10] T. Tanaka, Scientific American, **244** 110 (1981), Phys. Rev. Letters, **40** 820 (1978)

[11] T. Tanaka, D.J. Fillmore, S.-T. Sun, I. Nishio, G. Swislow and A. Shah, Phys. Rev. Letters, **45** 1636 (1980)

[12] M.Ilavsky, J.Hrouz, J.Stejskal and K.Bouchal, Macromol. **17** 2868 (1984)

[13] Y.Hirokawa and T.Tanaka, J. Chem. Phys., **81** 6379 (1985)

[14] K.Dusek and D.Patterson, J. Polym. Sci., A-2 **6** 1209 (1968)

[15] M.Huggins, J. Chem. Phys., **46** 151 (1942)

[16] A.Katchalsky, O.Kunzle, and W.Kuhn, J. Polym. Sci., **5** 283 (1950)

[17] A.Silberberg and W.Kuhn, Nature **170** 450 (1952)

[18] W.Kuhn, B.Hargitay, A.Katchalsky, and H.Eisenberg, Nature, **165** 514 (1950)

[19] T. Tanaka, I. Nishio, S.-T. Sun, and S. Ueno-Nishio, Science, **218** 467 (1982)

[20] A.Grodzinsky, Proc. 37th Conf. on Eng. Med. Biol., **37** 342 (1984)

[21] S.Katayama, Modern Chemistry (Japanese), **169** 42 (1985)

[22] R.Ullman, in **Elastomers and Rubber Elasticity** Edits J.E.Mark and J.Lal, Am. Chem. Soc., p.257 (1982)

[23] E.Geissler, R.Duplessix, A.M.Hecht, Macromol., **16** 712 (1983)

[24] P.Lutz, C.Picot, G.Hild and P Rempp, Br. Polym. J., **9** 151 (1977)

[25] K.Dusek, in **Polymer Network** A.Chompff and S.Newman Eds., Plenum Press, New York (1971)

[26] M.Ilavsky and J.Hrouz, Polym. Bull. (Berlin), **9** 159 (1983)

[27] P.J.Flory, Proc. Royal Soc., **A351** 351 (1976)

[28] M.Doi and S.F.J.Edwards, J. Chem. Soc. Faraday Trans. 2, **13** 1789, 1808, 1818 (1978)

[29] A.S.Argon, R.E.Cohen, B.Z.Jank, and J.B.Vandersande, J. Polym. Sci., **19** 253 (1981)

[30] T.Tanaka, L.O.Hocker and G.B.Benedek, J. Chem. Phys., **59** 5151 (1973)

[31] J.P.Munch, S.Candau, R.Duplessix, C.Picot, J.Herz and H.Benoit, J. Polym. Sci., Polym. Phys. Ed., **14** 1097 (1976)

[32] G.Patterson, Abstract No.27, **International Symposium on Macromolecules**, Strasbourg (1981)

[33] M.Adam and M.Delsanti, J. Phys. (Paris), **37** 1045 (1976)

[34] H.Benoit, et al, J. Polym. Sci., A2, **14** 2119 (1976)

[35] J.Bastide, S.Candau and L.Leibler, Macromol. **14** 719 (1980)

[36] A.M.Hecht and E.Gaessler, J. Chem. Phys., **73** 4077 (1980)

[37] T.Tanaka, S.Ishiwata and C.Ishimoto, Phys. Rev. Lett., **38** 771 (1977)

[38] T.Tanaka, Phys. Rev., **A17** 763 (1978)

[39] J.Bacri and R.Rajaonarison, J. Phys., Lett (Orsay, France) **40** L5 (1979)

[40] R.Nossal, J. Appl. Phys. **50** 3105 (1979)

[41] B.Schneider, D.Doskocilova and J.Dybal, Polymer, **26** 253 (1985)

[42] D.Live and S.B.H.Kent in **Elastomers and Rubber Elasticity** , Eds. J.Mark and J.Lal, Am. Chem. Soc., p.501 (1982)

[43] S.Katayama, Y.Arata and S.Fujiwara, Bull. Chem. Soc. Japan, **51** 1545 (1978)

[44] T.Tanaka and D.J.Fillmore, J. Chem. Phys., **70** 1214 (1979)

[45] K.C.Lim, F.R.Fincher and A.J.Heeger, Phys. Rev. Letters, **50** 1934 (1983)

LIST OF PARTICIPANTS TO THE NATO ADVANCED STUDY INSTITUTE ON "BIOPOLYMERS STRUCTURE AND DYNAMICS" - INTERNATIONAL SCHOOL OF PURE AND APPLIED BIOSTRUCURE
22 JUNE - 4 JULY 1986

F. ABOUL-ELA
315 Hildebrand Hall
Dept. Chemistry
University of California
BERKELEY, CA 94704 USA

P. ANDER
Seton Hall University
Department of Chemistry
SOUTH ORANGE, New Jersey 07079 USA

H. BAGCI
Dept. Medical Biology and Genetics
Medical Faculty
University of Akdeniz
ANTALYA, Turkey

R.L. BALDWIN
Department of Biochemistry
Stanford University Medical Center
STANFORD, CA 94305 USA

J. BENBASAT
Univ. British Columbia
Dept. Microbiology
Vancouver B.L.
CANADA V6TIW5

C. BENHAM
University of Kentucky
Mathematics Dept.
College of Arts and Sciences
Patterson Office Tower
LEXINGTON, Kentucky 40506-0027, USA

M. BERTOLOTTO
Cattedra di Biofisica
Universita' di Genova
Viale Benedetto XV, 2
16132 GENOVA , Italy

D. ERIE
The State University of New Jersey
RUTGERS
Department of Chemistry
New Brunswick
P.O. Box 939
PISCATAWAY, New Jersey 08854 USA

W. BURKART
Biologie & Umwelt, Abt. 81/SU
Swiss Federal Inst Reactor Res
EIR
CH-5303 WURENLINGEN, Switzerland

F. CARATOZZOLO
DIBE
Viale Francesco Causa, 13
16145 GENOVA, Italy

G. CHIRICO
Dipartimento di Fisica
Sezione Stati Aggregati
Via Celoria, 16
20133 MILANO Italy

M. DAUNE
IBMC - CNRS
15, rue Descartes
07004 STRASBOURG, France

G. DEMIREL
Dept. Chemistry
Middle East Technical University
ANKARA, Turkey

A. DIASPRO
Chair of Biophysics
University of Genova
Viale Benedetto XV, 2
16132 GENOVA, Italy

M. DONADONI OMODEO
Via Degan, 98
Roma, Italy

M. GRATTAROLA
DIBE
Viale Causa, 13
16145 GENOVA, Italy

A. ELGSAETER
Division of Biophysics
Univ. of Trondheim
N-7034 TRONDHEIM-NTN, Norway

A. FEIS
Laboratorio di Spettroscopia Molecolare
Dipartimento di Chimica
Via Gino Capponi, 9
50121 FIRENZE, Italy

C. FRONTALI
Istituto Superiore di Sanita´
Viale Regina Margherita 299
00198 GENOVA, Italy

GUZSK CALIK
Chemical Eng. dept.
Ankara Univ.
ANKARA, Turkey

P. GERMANO
Cattedra di Biofisica
Universita´ di Genova
16100 GENOVA, Italy

G. GIANNETTI
Dipartimento di Ingegneria,
Biofisica ed Elettronica
Via F. Causa 13
16145 GENOVA, Italy

G. LULECI
Department of Medical Biol. and Genetics
Medical Faculty
University of Akdeniz
ANTALYA, Turkey

M. MACCARONE
Via San Sisto, 87
67100 L´AQUILA, Italy

G.S. MANNING
Department of Chemistry
Rutgers University
P.O. Box 939
PISCATAWAY, NJ 08854, USA

A.M. MARTINS
Laboratorio Nacional de Engenharia
e Tecnologia Industrial
Estrada das Palmeiras
Queluz de Baixo
2745 QUELUZ, Portugal

F. HAUGEN
Institute of Biotechnology
Unit/NTH
N 7034 TRONDHEIM-NTH, Norway

C.W. HILBERS
Laboratory of Biophysical Chemistry
Katholieke Universiteit
Toernooiveld
6225 ED NIJMEGEN, The Netherlan

R. JOHNSTON
Dept. Experimental Pathology
Life Sciences Division
Los Alamos National Laboratory
P.O. Box 1663 Mail Stop
LOS ALAMOS, New Mexico, USA

K. KNUDSEN
Division of Biophysics
Univ. of Trondheim
N-7034 TRONDHEIM-NTH, Norway

M. LE BRET
Institute Gustave-Roussy
Physiochimie Macromomoleculaire
rue Camille Desmoulins
9480 VILLEJUIF, France

D.M.J. LILLEY
Department of Biochemistry
University of Dundee
DUNDEE DD1 4HN, UK

C. NICOLINII
Cattedra di Biofisica
Facolta´ di Medicina
Universita´ di Genova
16100 GENOVA, Italy

B. OZDEN
Dept. Biology
Middle East Technical University
ANKARA, Turkey

S. PAOLETTI
Dipartimento Biochimica e Biof.
Univ. di Trieste
34127 TREISTE, Italy

I. PEPE
Universita´ di Genova
Dipartimento di Fisica
Via Dodecaneso 33
16146 GENOVA, Italy

K. MARX
Department of Chemistry
University of Lowell
LOWELL, MA 01854, USA

L. Mc INTOSH
University of Oregon
Inst. Molecular Biology
EUGENE, OR 97403-1229
USA

N. MERT
Vice Dean
School of Veterinary
100. Yil University
VAN, Turkey

J.D. PUGLISI
Department of Chemistry
321 Hildebrand Hall
University of California
BERKELEY, California 94720 USA

M. ROCCO
Jersey
Laboratorio di Mutagenesi
Istituto Nazionale Ricerca Cancro
Viale Benedetto XV, 19
16132 GENOVA, Italy

M. SANDVOLD
Division of Biophysics
Univ. of Trondheim
N-7034 TRONDHEIM-NTH, Norway

W. SCHREIL
Institute of Zooology
Univ. Heidelberg
HEIDELBERG, FRG

D. SCHUMMER
Laboratoire de Biochimie
Universite' Catholique de Louvain
1, Place Louis Pasteur
1348 LOUVAIN-LA NEUVE, Belgium

S. TAKACI
Ankara University
Fen F. Dept. of Chemical Eng.
06100 ANKARA, Turkey

T.TANAKA
Department of Physics -MIT
CAMBRIDGE MA 02139, USA

L. PIANTONI
Via Madre Cabrini 9
20122 MILANO, Italy

E. PISKIN
Hacettepe University
Chemical Eng. Dept.
P.K. 716 Kizilay
ANKARA, Turkey

D. PORSCHKE
Max-Planck-Institute
fur Biophysikalische Chemie
34, GOTTIGEN, FRG

I. TINOCO
Department of Chemistry
University of California
BERKELEY, CA 94720 USA

I. TOBIAS
The State University of New Jersey
Department of Chemistry
P.O. Box 939
PISCATAWAY New Jersey 08854, USA

M. TORTONESE
P.zza G. Verdi, 1/1
16121 GENOVA
Italy

G. VARANI
Universita' degli Studi di Milano
Dipartimento di Fisica
Via Celoria 16
20133 MILANO, Italy

P. PEDERIVA
Universita' di Trento
Facolta' di Matematica
Via Dosso Dossi
38110 TRENTO, Italy

Index

GPSR Compliance

The European Union's (EU) General Product Safety Regulation (GPSR) is a set of rules that requires consumer products to be safe and our obligations to ensure this.

If you have any concerns about our products, you can contact us on ProductSafety@springernature.com

In case Publisher is established outside the EU, the EU authorized representative is:

Springer Nature Customer Service Center GmbH
Europaplatz 3
69115 Heidelberg, Germany

Batch number: 10761599

Printed by Printforce, the Netherlands